Current Topics in Behavioral Neurosciences

Volume 70

Current Topics in Behavioral Neurosciences provides critical and comprehensive discussions of the most significant areas of behavioral neuroscience research, written by leading international authorities. Each volume in the series represents the most informative and contemporary account of its subject available, making it an unrivalled reference source. Each volume will be made available in both print and electronic form.

With the development of new methodologies for brain imaging, genetic and genomic analyses, molecular engineering of mutant animals, novel routes for drug delivery, and sophisticated cross-species behavioral assessments, it is now possible to study behavior relevant to psychiatric and neurological diseases and disorders on the physiological level. The Behavioral Neurosciences series focuses on translational medicine and cutting-edge technologies. Preclinical and clinical trials for the development of new diagnostics and therapeutics as well as prevention efforts are covered whenever possible. Special attention is also drawn on epigenetical aspects, especially in psychiatric disorders.

CTBN series is indexed in PubMed and Scopus.

Marc Wittmann • Irina A. Strigo
Alan N. Simmons
Editors

The Bodily Self, Emotion, and Subjective Time

Exploring Interoception through the Contributions of A.D. (Bud) Craig

Springer

Editors
Marc Wittmann
Institute for Frontier Areas of Psychology and Mental Health
Freiburg im Breisgau, Baden-Württemberg, Germany

Irina A. Strigo
UCSF Weill Institute for Neurosciences, School of Medicine
University of California, San Francisco
San Francisco, CA, USA

Alan N. Simmons
San Diego Veterans Healthcare System
University of California, San Diego
San Diego, CA, USA

ISSN 1866-3370 ISSN 1866-3389 (electronic)
Current Topics in Behavioral Neurosciences
ISBN 978-3-031-94034-7 ISBN 978-3-031-94035-4 (eBook)
https://doi.org/10.1007/978-3-031-94035-4

This Springer imprint is published by the registered company Springer Nature Switzerland AG
The registered company address is: Gewerbestrasse 11, 6330 Cham, Switzerland

Preface

The goal of this volume on *The Bodily Self, Emotion, and Subjective Time: Exploring Interoception Through the Contributions of A.D. (Bud) Craig* is to provide a comprehensive overview of the neural mechanisms underlying emotion, pain, interoception, and subjective time, inspired by the pioneering work of Bud Craig. Renowned for his groundbreaking research, Craig significantly advanced our understanding of the neuroanatomical foundations of the bodily self, emphasizing the insula's central role in integrating signals from the entire body. His work elucidated the intricate pathways of interoception, demonstrating how bodily sensations—such as pain and temperature—are deeply interconnected with emotional experiences and homeostatic regulation. By mapping thalamocortical pathways and highlighting the insula's predictive functions, he provided a transformative framework linking physiological states to subjective emotional and temporal experiences. This volume brings together contributions from researchers influenced by Craig's insights, continuing the exploration of these fundamental processes.

In the opening chapter, we the editors reflect on Craig's profound impact on our own research and set the stage for the collective contributions that follow that focus on three key areas of exploration: the bodily self, emotion, and subjective time. To lay the foundation of study into the bodily self, the chapter by Anders Blomqvist and Jonathan Dostrovsky delves into Craig's life and pioneering work in understanding interoceptive processes. Henry Evrard examines the insula and its associated interoceptive pathways, while Hugo Critchley and Joel Patchitt focus on the role of interoception and the insula in autonomic integration, particularly concerning psychiatric symptoms. Rebecca Böhme explores the connection between interoception and touch and her chapter is augmented by Emily Choquette and Sahib Khalsa who focus on the insula and the body image. To continue this work further into emotion and psychiatry, the relationship between interoception and pain anticipation is discussed by Irina Strigo and Alan Simmons, followed by Wolf E. Mehling' chapter, which emphasizes the importance of interoceptive awareness in chronic pain and mind-body interventions. Martin Paulus investigates interoceptive processing in substance-use disorders through neuroanatomical insights and predictive coding frameworks. April May and Susan Tapert look at the development of interoception during adolescence and its implications for addiction. The book draws to a close with a focus on subjective time, with Marc Wittmann and Sylvie Droit-Volet in one chapter and Alice Teghil in a following chapter addressing the role of interoceptive feelings in subjective time perception, both during ordinary and altered states of consciousness, as well as in prospective and retrospective timing. Nina Sooter, Federico Seragnoli, and Fabienne Picard examine how the insula contributes to altered experiences such as subjective time and its disturbances in epileptic auras. Clemens Wöllner provides an interoceptive perspective on how musical emotions and awareness relate to time perception. Julian Kiverstein offers a philosophical view of embodied and enactive time. Finally, the editors of this volume synthesize the insights from these diverse areas and chart new frontiers for future research in interoception.

Freiburg im Breisgau, Baden-Württemberg, Germany — Marc Wittmann
San Francisco, CA, USA — Irina A. Strigo
San Diego, CA, USA — Alan N. Simmons

Contents

A Personal and Scientific Introduction to the Work of A.D. (Bud) Craig on Interoception and the Insular Cortex

Irina A. Strigo, Alan N. Simmons, and Marc Wittmann

Contents

Abstract A.D. (Bud) Craig was best known to us colloquially as Bud Craig; his groundbreaking work has advanced our understanding of the neuroanatomical basis of the bodily self, emotion, and subjective time. His research elucidated the intricate pathways of interoception – the brain's processing of internal body signals – and highlighted the insula's pivotal role in integrating these signals. Craig's pioneering insights demonstrated that bodily sensations, such as pain and temperature, are deeply intertwined with emotional experiences and homeostatic needs. By mapping the thalamocortical pathways and emphasizing the insula's function in predicting future bodily states, he provided a comprehensive framework that connects physiological states to emotional and temporal experiences. This fundamental work has profoundly influenced a new generation of scientists and inspired the current volume. In this book, we have invited researchers influenced by Craig's theories to explore the neural correlates of emotion, pain, interoception, and time. In this chapter, we present how the editors of this book were directly influenced by Craig's

I. A. Strigo (✉)
Veterans Affairs & University of California, San Francisco, San Francisco, CA, USA
e-mail: irina.strigo@ucsf.edu

A. N. Simmons (✉)
Veterans Affairs & University of California, San Diego, San Diego, CA, USA
e-mail: ansimmons@health.ucsd.edu

M. Wittmann (✉)
Institute for Frontier Areas of Psychology and Mental Health, Freiburg, Germany
e-mail: wittmann@igpp.de

Curr Topics Behav Neurosci (2025) 70: 1–13
https://doi.org/10.1007/7854_2025_583
Published Online: 09 May 2025

findings and lay the groundwork for the large collective contribution of the authors of the subsequent chapters that continue to explore this ever-evolving scientific landscape.

Keywords Emotions · Insula · Interoception · Pain · Time perception

1 Introduction

In this book "The Bodily Self, Emotion, and Subjective Time: Exploring Interoception Through the Contributions of A.D. (Bud) Craig," we seek to provide a comprehensive exploration of the interplay between the bodily self, emotion, and subjective time perception, consciousness, as well as clinical implications with a primary focus on Bud's groundbreaking work. By delving into Bud's contributions to interoception and his insights into the sense of a time within the context of the bodily self, this volume offers an in-depth understanding of the neural and psychological mechanisms underlying these phenomena. While he passed away in 2023, his broad impact in a vast array of domains ensures that his intellectual legacy will continue. This book brings together leading experts from various disciplines such as neurophysiology, medicine, cognitive neuroscience, psychology, and philosophy, who will explore in their respective chapters specific aspects of (1) interoception, (2) emotions, (3) time perception, (4) consciousness, (5) mental health, and their intersections.

There is no doubt that Bud was a visionary seeker of truth, driven by a passion that was impossible to miss if you were fortunate enough to work alongside him. His mind wasn't just brilliant – it was alive with an electric curiosity that never dimmed. Even in his teenage letters from Michigan State University, you can see the spark that would later illuminate our understanding of interoception and pain. At just 18, he wrote to his parents with a wisdom beyond his years: 'I want to make damned sure where I want to go, what I want to do, first.' That same deep intentionality would later guide his groundbreaking scientific work.

His brain worked like a finely tuned instrument, but what made Bud truly extraordinary was how he wove together the artistic and scientific threads of his life. Whether he was losing himself in jazz at his college radio show, performing with his band "Tides of the Solway Firth," speaking German, or exploring Europe on his Norton motorcycle at 18, he brought the same intensity and joy to understanding the world around him. He was that rare combination – a free spirit with the rigorous mind of a scientist, someone who could both dream big and drill down to the finest details of neural networks.

To the next generation of scientists, Bud's journey shows us that groundbreaking science isn't just about intelligence – it's about having the courage to question, the resilience to persevere, and the wisdom to stay true to your inner compass. His legacy isn't just in his discoveries, but in showing us how a truly great scientific

Fig. 1 1976: Bud on his Norton motorcycle Cornell Ithaca NY

mind develops – through unwavering curiosity, deep reflection, and the determination to seek truth wherever it leads (Figs. 1 and 2).

2 Temperature, Pain, and Emotion: Irina Strigo

I probably should not describe my friendship with Bud as *a solar power fueling my extraordinary journey to self-discovery,* but it was pretty damn close! Bud was the sole reason I stayed in academia, a responsibility he never took lightly, especially when my grants were 'Not Discussed,' or papers were editorially rejected. Yet here we are. Bud will always be my mentor!

My first experience testing human pain occurred during my undergraduate thesis. I aimed to replicate the work of Cabanac et al. (1972) and Mower (1976), using slightly more advanced methods than those applied by James D. Hardy to test the effects of ambient temperature on human temperature sensitivity. I focused on both pain intensity (discriminative sensory capacity) and unpleasantness (affective impact).

The experiments took place during winter at the Royal Victoria Hospital in Montreal, Canada – an outdated structure built in 1893. Briefly opening the windows

Fig. 2 1983: Bud with his koala buddy in Australia

during Montreal's winter was enough to cool the small room to 15° ± 0.5°C, but turning on the heaters in such an old building kept blowing fuses, including in my advisor's gynecology office. Eventually, I was banned from conducting studies in the building!

My work demonstrated that thermal pain is not just a sensation, but also an emotion, and these two components are not always closely linked. As Bud Craig eloquently stated in 2002 (Craig 2002), my first study – unbeknownst to me at the time – was actually connected to his life's work: the close relationship between bodily sensations of temperature and pain and emotional feelings. I could not articulate this connection as clearly until Bud published his seminal paper in *Nature Reviews Neuroscience* in 2002 (Craig 2002), as discussed below.

Thermosensory activity provides a clear example of the connection between pain and emotion. We usually think of temperature perception as a sensory sensation on the skin. However, with each temperature sensation, we also experience an emotion of pleasantness or unpleasantness. N.B. "Hedonic" refers to pleasantness alone. This emotion provokes us to act in a way that helps the body maintain equilibrium when it cannot appropriately adjust the temperature. This underscores the fundamental significance of temperature sensation for homeostasis because its valence depends directly on the body's thermoregulatory needs (Cabanac et al. 1972; Mower 1976; Strigo et al. 2000). For example, a cool glass of water that is highly pleasant if you are overheated feels highly unpleasant if you are cold. These affective feelings accompany behavioral motivations driven by the homeostatic needs of the body.

During my doctoral training, I continued to explore the distinction between emotion and sensation by comparing behavior, neural substrates, and the pharmacology of deep (visceral) and superficial (cutaneous) pain in healthy humans. I discovered that, for the same perceived intensity, deep (visceral) non-painful and painful sensations elicited much stronger unpleasant emotions than superficial (cutaneous) sensations (Strigo et al. 2002). These emotional differences were reflected in the functional magnetic resonance imaging (fMRI) activation patterns (Strigo et al. 2003) and in the effects of intravenous ketamine infusions (Strigo et al. 2005).

It was around this time in 2002 that Bud published his seminal article "*How do you feel? Interoception: the sense of the physiological condition of the body, Nature Reviews Neuroscience (2002),*" wherein he revised and deepened the definition of interoception, in a way that would strongly influence later developments and understanding of the sense of interoception and provide an overarching view of implications of practical applications in multiple, converging fields of study. In the conventional view, the different sensations of temperature, itch, and pain were associated with an *exteroceptive* somatosensory system, whereas the less distinct visceral feelings of vasomotor activity, hunger, thirst, and internal sensations were associated with a separate *interoceptive* system. Bud Craig redefined interoception by providing neuroanatomical evidence that interception is the sense of physiological condition of the entire body, not just the viscera, and categorized pain and temperature as part of the redefined interoceptive system. He introduced a new thalamocortical afferent pathway that transports interoceptive information, including temperature and pain messages, to the brain, ultimately reaching the limbic sensory (insula) and limbic motivational (anterior cingulate) cortices. The results from all my work started to make sense within his framework.

After joining Dr. Bud Craig's laboratory at the Barrow Neurological Institute (BNI) as a post-doctoral fellow, I began exploring the possibilities of extending the study on interoceptive challenges to psychiatric populations. I specifically aimed to merge the fields of pain and mood to understand the significant connection between pain and psychiatric disorders. I focused my studies of pain and temperature in individuals suffering from major depression as a joint post-doc in Martin Paulus' laboratory in the Department of Psychiatry at the University of California in San Diego (UCSD). At that time, most published research indicated that individuals with depression were insensitive to experimental pain, which contradicted our clinical observations.

My research again demonstrated that thermal sensations and associated emotions are not always closely linked. A key finding was that depressed individuals report feelings of unpleasantness in response to thermal stimuli that healthy individuals do not perceive as unpleasant. In the first study, we administered brief stimuli that ranged from warm to painfully hot to the left forearm of patients diagnosed with major depressive disorder (MDD) who at the time were not taking prescribed medications and matched them with healthy comparison (HC) subjects. We asked the participants to rate the quality of each stimulus using three separate, validated visual analog scales, one for intensity ("How strong was the stimulus?"), one for pain

("How painful was the stimulus?"), and one for unpleasantness ("How bothersome was the stimulus?"). We found that the relationship between the discriminative sensory scales and unpleasantness was strikingly different among MDD patients. They rated warm stimuli as quite unpleasant, even at temperatures they did not report as painful. The same warm stimuli were neither unpleasant nor painful for the HC participants. The leftward shift in the unpleasantness ratings of MDD patients was surprising. Their unpleasantness rating of a normally non-painful stimulus was as high as the unpleasantness rating of a clearly painful heat stimulus for HCs. Statistical analyses showed that the temperatures at which MDD participants began to report unpleasantness were (1) significantly lower than those of HCs and (2) significantly lower than temperatures at which MDDs began to report pain (Strigo et al. 2008b).

We call this phenomenon *emotional allodynia* because stimuli that are normally not unpleasant produce an unpleasant feeling in MDD subjects. (Clinical *allodynia* means unpleasant light touch.) Their affective feeling of unpleasantness was uncoupled from their sensory discrimination of intensity or pain, in stark contrast to HCs, who began to report unpleasantness at the pain threshold (Strigo et al. 2013; Strigo et al. 2008a, b).This distinctly differs from hyperalgesia or sensitization, which means enhanced pain sensation, because even though MDD subjects can have lower pain thresholds for heat than HCs, they report normal intensity and unpleasantness at normally painful higher temperatures. We further found that neither attention nor mood altered emotional allodynia. We found the same pattern of significantly lower temperatures that produced unpleasantness in individuals with MDD despite the fact that they reported similar levels of sad or happy moods as the HCs (Ushinsky et al. 2013). In other words, the temperatures at which the MDD subjects reported unpleasantness were not changed by their prevailing moods; they were abnormally reduced in both sad and happy moods. In contrast, we observed that the temperatures that were deemed unpleasant were lower in a sad than in a happy mood in the HCs (Ushinsky et al. 2013). These findings support the notion that emotional allodynia in MDD is not simply a reflection of the lower baseline affect or increased self-focus found in this disorder. They imply that emotional allodynia is a stable characteristic of major depression, i.e., a measurable manifestation of emotional dysfunction that is inherent to this disorder.

Several lines of evidence suggest that emotional allodynia is homeostatically driven. Considering the above discussion that a thermal stimulus can be perceived as pleasant or unpleasant, depending on one's homeostatic needs, and that core temperatures of depressed individuals tend toward hyperthermia, warm stimuli become homeostatically aversive. Warm stimuli applied to a body that is already too warm potentially threaten to increase the deviation from thermoregulatory neutrality. The result is the observed phenomenon of emotional allodynia. Homeostatic sensory integration is pivotal for thermoregulation (Craig 2003) and for emotional dysfunction in MDD.

3 Interoception, Emotion, and Anticipation: Alan Simmons

My first exposure to the work of Bud Craig came early in my research career. When I had the great fortune of getting a post-doctoral position with Martin Paulus and Murray Stein in 2002, it gave me the opportunity to study anxiety and trauma using fMRI. My initial understanding of the insula was a very rudimentary concept of a mysterious island that took care of the gut. My perspective quickly changed when Martin recognized the gravity of Bud's seminal work in interoception (Craig 2002). His exploration into the uncharted regions of the insula provided a road map that I continued to follow over the years. Working with one of his wonderful proteges, Irina Strigo, provided the chance to see the parallels of pain and anxiety. Martin's guidance in viewing anxiety and interoception through a lens of anticipation propelled me further into the pages of Bud's unfolding and expanding interoceptive model. This path of discovery also gave me the opportunity to work with Marc Wittmann on his projects regarding the overlap of anticipation and interoception and time and eventually to this volume.

The importance of comprehending how the body processes the signals regarding our body state cannot be overemphasized. Understanding of the outside world is built on our interaction with the points and places we encounter. Awareness of internal bodily sensations is influenced by the processing of bodily stimuli and the feelings generated (i.e., interoception), in other words, interoceptive feelings and conjoint motivations as homeostatic emotion.

There is a substantial school of thought that posits that emotions do not exist without the physiological signals that undergird this experience. This is where the natural progression of the afferent path of interoceptive signal and the natural progression of the work product of Bud Craig's career projected into the realm of understanding emotions. It is particularly through the lens of this interoceptive pathway and the specific role of the insula that a new and exciting construct of emotion can be viewed.

There has been vastly greater recognition of the role of the insula in the processing of emotion over the last two decades, due to the work of a growing number of researchers. The understanding of how and why the insula plays such a critical role can be linked to its unique structure and role in various brain networks. Bud Craig's fundamental work in mapping out the unique pathways of the interoceptive system into and through the insula and outlining the implications of these pathways are crucial to understanding the insula.

Interoceptive information primarily transported to the brain through afferent Aδ- and C-type fibers that are relayed by lamina I contains homeostatic information. Small-diameter afferents that innervate tissues in parallel with sympathetic and parasympathetic paths provide input to lamina I to be further transported to the nucleus of the solitary tract (NTS). These are integrated in the parabrachial nucleus, which projects to the insular cortex by way of the ventromedial thalamic and ventromedial nuclei. This contralateral homeostatic afferent input then projects topographically to the mid/posterior dorsal insula. Further integration and representation of this information occur in the mid/anterior insula (Craig 2002). For a more detailed and robust discussion of this path, please see the chapters by Anders Blomqvist and by Henry Evrard in this volume. Subsequently, the insula projects this information to the prefrontal and orbital frontal regions for the evaluative aspects that are critical to the determination of emotional state (Strigo and Craig 2016). The construction of the insula is crucial for reintegration of these signals. The posterior aspect of the insula (i.e., granular cortex) has a more robust IV layer in its cortical columns containing an abundance of granule cells which help distribute the projections from the thalamus. Conversely, the anterior aspect of the insula (i.e., agranular) does not have this clear layer IV and lacks the density of thalamic projections of the granular insula (Barrett and Simmons 2015). Essentially the interoceptive signal in the anterior insula is neuroanatomically separated from the direct signal. The gradient structure is common across cortical regions as the cortical columns move from sensory inputs to representative systems (Huntenburg et al. 2018). Once information reaches the anterior aspects of the insula, it uses the information that has been delayed either anatomically through signaling across unmyelinated fibers or though sensory limitations of the signal frequency (e.g., breathing rate or heart rate). For these reasons, the insula takes the role of a prediction engine (Craig 2009; Singer et al. 2009). The graduated structure of the insula leads directly to it needing to predict the future signals to provide a valid sense of the body's current interoceptive state. This is where the vital role of the insula and the associated networks plays in presenting a predictive affective state (Menon and Uddin 2010).

The significant role of the anterior insula in the prediction of interoceptive states and the consequent affective states have elucidated our understanding of the experiences of pain, fear, and anxiety (Paulus and Stein 2006, 2010). While all emotional states are highly dependent on the insula due to its role in interoceptive processing, this is even more true in states that intrinsically have a predictive element. In anxious states, the role of uneasiness and apprehension that are cardinal aspects of this emotion is strongly linked to the prediction, or fear, that the future experience will produce a negative interoceptive experience. We have repeatedly shown that the insula is active in preparing for a negative experience, such as pain (Strigo et al. 2013, 2024) and negative images (Simmons et al. 2006, 2011). This insula response is greater in those with post-traumatic stress disorder (PTSD) (Simmons et al. 2008, 2013; Strigo et al. 2010). In PTSD, an abrupt and disruptive interoceptive signal induces profound and enduring changes to an individual's emotional state. PTSD presents with a wide array of comorbid neurocognitive symptoms including disruption to working memory, attention, fear conditioning, and executive function (Hayes et al. 2012). These combine with a disruption of interoceptive signaling (Joshi et al. 2023) that is emblematic of how disruptions to interoceptive systems can provide a foundation for negative emotional states.

4 Embodied Time Consciousness: Marc Wittmann

Between 2004 and 2009, I worked as a post-doc in the laboratory of Martin Paulus in the Department of Psychiatry at the University of California at San Diego (UCSD), where research on decision-making and emotion regulation was conducted in patients with anxiety disorders, depression, and drug addiction. Studies of those patient groups directed by Martin Paulus and his coworkers Alan Simmons and Irina Strigo repeatedly showed under- or overactivation of the insular cortex, which provided a neurobiological marker for an imbalanced cognitive-emotional valuation system. During my time at the UCSD, where I conducted fMRI studies on time perception, Bud Craig was formulating a new theory of subjective time based on his insights into the functional anatomy of the interoceptive system of the body and brain. I learned of this during Bud's frequent visits to the UCSD, where he collaborated with Martin and Irina. According to Bud, the judgment of duration and passage of time evolves through the experienced progression of successive moments of bodily self-realization, a series of emotional moments which constitutes the "material me." Bud formulated his ideas on the bodily origins of subjective time in detail for the first time (Craig 2009) in a thematic issue on the experience of time in the Philosophical Transactions of the Royal Society B, which Virginie van Wassenhove and I edited (Wittmann and Van Wassenhove 2009).

While working in Martin's lab, I often conversed with Bud. My time perception studies were the first to demonstrate empirically and explicitly describe how the insular cortex is the main brain region responsible for sensing time passage. The interpretation came naturally because I worked in a lab studying interoception in relation to the insular cortex. I had applied the duration-reproduction paradigm with multiple-second tone intervals (3, 9, 18 s) while taking fMRI measurements (Wittmann et al. 2010a, 2011). Participants were first presented with an auditory tone (encoding phase) which they had to reproduce by pressing a button when they thought that a second comparison tone was as long as the first one had been (reproduction phase). A ramp-like increase in fMRI activation in the dorsal posterior insula was detected across intervals which terminated with the end of the tone during the presentation of the encoding phase. The second interval, which had to be reproduced in the duration of the first tone, exhibited climbing activity in the anterior insula.

I remember well when Bud came up with an idea about how to further investigate insular-cortex activation in an event-related fMRI study on subjective time utilizing a "time-dilation effect." Virginie van Wassenhove from Caltech in Pasadena had behaviorally designed a task where a disc moving toward an observer is subjectively perceived as longer in duration than the same object that is static or moving away (van Wassenhove et al. 2008). Together with Martin and Bud, Virginie and I conducted the time-dilation task while measuring brain activity with fMRI (Wittmann et al. 2010b). Participants were presented with a series of five visual discs (all stimuli lasting around 500 ms), four of which were static and of identical duration except the fifth one, which was a deviant target, either a virtually looming (becoming bigger) or a receding disc (becoming smaller) and varying in duration. The time-dilation effect was found in the contrast between looming and static and between looming and receding stimuli. We discovered that, in comparison to the static discs, both the looming and the receding targets induced increased activation in the anterior insula and anterior cingulate cortices. The contrast between looming and receding targets showed activation of cortical midline structures (the "default network").

In 2012, I had an email exchange with Bud concerning a study with the patient Roger, who had suffered from *herpes* simplex encephalitis (HSE) and whose bilateral insular cortices were destroyed (Damasio et al. 2013). The authors claimed that Roger "had an average range of emotions and feelings, before and after the illness," which contradicted the idea that the insula's prime function was for self-awareness and emotions. At that time, Bud argued that portions of the anterior insula had survived, which would explain why Roger seemed emotional and self-aware (Craig 2015), see p. 208, footnote 1. In an email to Bud on June 25, 2012, I argued that the way Roger responded to questions was rather repetitive and evasive, at least in the quotes recorded and transcribed for the article. When asked about his introspective feelings, he repeatedly answered, "I am happy […], "I have a strong feeling of happiness, that we are here together working together [...] and feeling happy together […] I am very happy and delighted [...] and to be working together […] I think I am happy to be here [...] I do not like to think about that. I am happy to be here with you today." Other responses were evasive. Investigator, "Am I aware of myself?" Answer Patient: "You look very handsome. I think you know what to do here." Roger did not seem to respond normally.

Bud's insula theory of time perception was confirmed in 2023. Two independent meta-analyses including more than 100 available neuroimaging experiments concentrated on two sites, the (pre-) supplementary motor area (SMA) and bilateral insula (Mondok and Wiener 2023; Naghibi et al. 2024). Many brain regions had been postulated to govern the judgment of duration. The insula had rarely been mentioned in studies investigating the neurophysiological basis of time perception. The meta-analytic results concurred with the idea that subjective time is driven by sensorimotor or enactive (SMA) as well as interoceptive or embodied experience (insula). These meta-analyses present the strongest accumulated evidence so far for a dominant involvement of the insular cortex across all time ranges, from the milliseconds to the multiple-seconds range. This bodily basis of time perception had already been predicted by Bud Craig 15 years earlier.

I am glad that Bud learned about these meta-analyses as well as two other studies which proved he was right. Based on phenomenal awareness, I believed that the embodied basis of time perception mainly referred to intervals of at least a few seconds' duration when an observer consciously senses herself over time. Two studies were published in 2023 that showed how the heart rate was decisive in the perception of very short time intervals of a few hundred milliseconds (Arslanova et al. 2023; Sadeghi et al. 2023). The subjective duration was relatively expanded when the interval between heartbeats before the presented stimulus was longer as compared to when it was shorter (Sadeghi et al. 2023). A relative overestimation of duration happened when a short stimulus was presented during the diastole, namely the period of heart relaxation (Arslanova et al. 2023). The state of the heartbeat thus determines our perception of short time intervals.

5 Conclusion

Bud Craig's work not only advanced our understanding of pain, emotion, anticipation, and time but also highlighted the commonalities among these diverse study areas. In the following chapters, we gathered contributions from leading experts in various fields, showcasing the unified understanding that emerges from the interoceptive constructs of Bud's intellectual legacy. This includes neuroanatomy, pain, emotion, time, anticipation, substance use, development, mindfulness, music, and even the concept of self. It was a pleasure to work with Bud and to take part in the exploration of interoception! (Fig. 3).

Fig. 3 2001. Bud Craig. MD, Honoris Causa conferred at Linköping University (LIU), Linköping, Sweden

References

Arslanova I, Kotsaris V, Tsakiris M (2023) Perceived time expands and contracts within each heartbeat. Curr Biol 33:1389–1395. e1384

Barrett LF, Simmons WK (2015) Interoceptive predictions in the brain. Nat Rev Neurosci 16:419–429

Cabanac M, Massonnet B, Belaiche R (1972) Preferred skin temperature as a function of internal and mean skin temperature. J Appl Physiol 33:699–703

Craig AD (2002) How do you feel? Interoception: the sense of the physiological condition of the body. Nat Rev Neurosci 3:655–666

Craig AD (2003) A new view of pain as a homeostatic emotion. Trends Neurosci 26:303–307

Craig AD (2009) How do you feel [mdash] now? The anterior insula and human awareness. Nat Rev Neurosci 10:59–70

Craig AD (2015) How do you feel?: an interoceptive moment with your neurobiological self. Princeton University Press

Damasio A, Damasio H, Tranel D (2013) Persistence of feelings and sentience after bilateral damage of the insula. Cereb Cortex 23:833–846

Hayes JP, VanElzakker MB, Shin LM (2012) Emotion and cognition interactions in PTSD: a review of neurocognitive and neuroimaging studies. Front Integr Neurosci 6:89

Huntenburg JM, Bazin P-L, Margulies DS (2018) Large-scale gradients in human cortical organization. Trends Cogn Sci 22:21–31

Joshi SA, Aupperle RL, Khalsa SS (2023) Interoception in fear learning and posttraumatic stress disorder. Focus 21:266–277

Menon V, Uddin LQ (2010) Saliency, switching, attention and control: a network model of insula function. Brain Struct Funct 214:655–667

Mondok C, Wiener M (2023) Selectivity of timing: a meta-analysis of temporal processing in neuroimaging studies using activation likelihood estimation and reverse inference. Front Hum Neurosci 16:1000995

Mower GD (1976) Perceived intensity of peripheral thermal stimuli is independent of internal body temperature. J Comp Physiol Psychol 90:1152

Naghibi N, Jahangiri N, Khosrowabadi R, Eickhoff CR, Eickhoff SB, Coull JT, Tahmasian M (2024) Embodying time in the brain: a multi-dimensional neuroimaging meta-analysis of 95 duration processing studies. Neuropsychol Rev 34:277–298

Paulus MP, Stein MB (2006) An insular view of anxiety. Biol Psychiatry 60:383–387

Paulus M, Stein M (2010) Interoception in anxiety and depression. Brain Struct Funct 214:451–463

Sadeghi S, Wittmann M, De Rosa E, Anderson AK (2023) Wrinkles in subsecond time perception are synchronized to the heart. Psychophysiology 60:e14270

Simmons A, Strigo I, Matthews SC, Paulus MP, Stein MB (2006) Anticipation of aversive visual stimuli is associated with increased insula activation in anxiety-prone subjects. Biol Psychiatry 60:402–409

Simmons AN, Paulus MP, Thorp SR, Matthews SC, Norman SB, Stein MB (2008) Functional activation and neural networks in women with posttraumatic stress disorder related to intimate partner violence. Biol Psychiatry 64:681–690

Simmons AN, Stein MB, Strigo IA, Arce E, Hitchcock C, Paulus MP (2011) Anxiety positive subjects show altered processing in the anterior insula during anticipation of negative stimuli. Hum Brain Mapp 32:1836–1846

Simmons AN, Flagan TM, Wittmann M, Strigo IA, Matthews SC, Donovan H, Lohr JB, Paulus MP (2013) The effects of temporal unpredictability in anticipation of negative events in combat veterans with PTSD. J Affect Disord 146:426–432

Singer T, Critchley HD, Preuschoff K (2009) A common role of insula in feelings, empathy and uncertainty. Trends Cogn Sci 13:334–340

Strigo IA, Craig ADB (2016) Interoception, homeostatic emotions and sympathovagal balance. Philos Trans R Soc Lond Ser B Biol Sci:371

Strigo IA, Carli F, Bushnell MC (2000) Effect of ambient temperature on human pain and temperature perception. Anesthesiology 92:699–707

Strigo IA, Bushnell MC, Boivin M, Duncan GH (2002) Psychophysical analysis of visceral and cutaneous pain in human subjects. Pain 97:235–246

Strigo IA, Duncan GH, Boivin M, Bushnell MC (2003) Differentiation of visceral and cutaneous pain in the human brain. J Neurophysiol 89:3294–3303

Strigo IA, Duncan GH, Bushnell MC, Boivin M, Wainer I, Rodriguez Rosas ME, Persson J (2005) The effects of racemic ketamine on painful stimulation of skin and viscera in human subjects. Pain 113:255–264

Strigo IA, Simmons AN, Matthews SC, Craig AD, Paulus MP (2008a) Association of major depressive disorder with altered functional brain response during anticipation and processing of heat pain. Arch Gen Psychiatry 65:1275–1284

Strigo IA, Simmons AN, Matthews SC, Craig AD, Paulus MP (2008b) Increased affective bias revealed using experimental graded heat stimuli in young depressed adults: evidence of "emotional allodynia". Psychosom Med 70:338–344

Strigo IA, Simmons AN, Matthews SC, Grimes EM, Allard CB, Reinhardt LE, Paulus MP, Stein MB (2010) Neural correlates of altered pain response in women with posttraumatic stress disorder from intimate partner violence. Biol Psychiatry 68:442–450

Strigo IA, Matthews SC, Simmons AN (2013) Decreased frontal regulation during pain anticipation in unmedicated subjects with major depressive disorder. Transl Psychiatry 3:e239–e239

Strigo IA, Kadlec M, Mitchell JM, Simmons AN (2024) Identification of group differences in predictive anticipatory biasing of pain during uncertainty: preparing for the worst but hoping for the best. Pain 165(8):1735–1747

Ushinsky A, Reinhardt LE, Simmons AN, Strigo IA (2013) Further evidence of emotional allodynia in unmedicated young adults with major depressive disorder. PLoS One 8:e80507

van Wassenhove V, Buonomano DV, Shimojo S, Shams L (2008) Distortions of subjective time perception within and across senses. PLoS One 3:e1437

Wittmann M, Van Wassenhove V (2009) The experience of time: neural mechanisms and the interplay of emotion, cognition and embodiment. R Soc Lond:1809–1813

Wittmann M, Simmons AN, Aron JL, Paulus MP (2010a) Accumulation of neural activity in the posterior insula encodes the passage of time. Neuropsychologia 48:3110–3120

Wittmann M, van Wassenhove V, Craig AD, Paulus MP (2010b) The neural substrates of subjective time dilation. Front Hum Neurosci 4:2

Wittmann M, Simmons AN, Flagan T, Lane SD, Wackermann J, Paulus MP (2011) Neural substrates of time perception and impulsivity. Brain Res 1406:43–58

The Life and Legacy of Arthur D. Craig, Jr.: A Pioneer in Interoception

Anders Blomqvist and Jonathan O. Dostrovsky

Contents

Abstract A.D. (Bud) Craig (1951–2023) redefined the concept of interoception and provided a novel, revolutionary understanding of the neural basis for human awareness. In unsurpassed anatomical-physiological studies in monkeys, Craig showed that the insular cortex is the primary sensory cortex for interoception, or the image of the "material me" that provides a homeostatic representation of the physiological condition of the body. He showed that the insula contains a postero-anteriorly organized somatotopic map of the interoceptive sensations, and that it encodes both the localization and the intensity discrimination of interoceptive sensations. In seminal work in humans, he demonstrated that the interoceptive feelings are re-represented, and multimodally integrated, in anterior portions of the insula in sequence of increasingly homeostatically efficient representations that integrate all salient neural activity. He further showed that subjective awareness is associated with activation of the anterior insular cortex and suggested that this brain region also is critical for fluid intelligence and the perception of time. His work has led to a paradigm shift in our understanding of interoception and how interoceptive sensations underlie consciousness, a topic that long has been considered elusive, or even beyond our comprehension.

A. Blomqvist (✉)
Division of Neurobiology, Department of Biomedical and Clinical Sciences, Linköping University, Linköping, Sweden
e-mail: anders.blomqvist@liu.se

J. O. Dostrovsky (✉)
Department of Physiology, University of Toronto, Toronto, ON, Canada
e-mail: j.dostrovsky@utoronto.ca

Curr Topics Behav Neurosci (2025) 70: 15–31
https://doi.org/10.1007/7854_2024_574
Published Online: 19 March 2025

Keywords Homeostatic afferent pathway · Insular cortex · Interoception · Lamina I neurons · Spinothalamic tract · VMpo

1 Introduction

Arthur D. Craig, Jr., generally known as Bud Craig, died on 15 July 2023 at age 71 (Blomqvist et al. 2023), after a decade long struggle with multiple sclerosis and a brief subsequent illness. He did groundbreaking studies on the organization of the primate pain and temperature pathways, identifying these as an important component of a larger interoceptive system that by means of inputs from thin sensory fibers constitutes a homeostatic afferent pathway that provides information about the status of the body. He showed that this system, via two adjacent relay nuclei in the ventromedial thalamus, the basal and posterior portions, respectively, of the ventromedial nucleus (VMb and VMpo), projects to the posterior part of the insular cortex. He further demonstrated that the information represented in this primary interoceptive cortex was re-represented and integrated with salient multimodal activities in more anterior portions of the insula, substantiating subjective feelings, hence providing a neurobiological substrate for the James-Lange theory of emotion and the "somatic marker" hypothesis.

Both of us knew Bud for over 40 years. We each collaborated with him, but he was also a personal friend. It is a privilege to have known him. He was a very warm, intense person, and while being very demanding in his studies, always striving for utmost rigor, he was also very generous, and very loyal to his friends. We both have many dear memories of mutual visits to our laboratories (Figs. 1, 2, and 3). Bud's scientific career should serve as an example for the aspiring young scientist: To formulate an important question, stick to it, and, from various angles and with different techniques, try to solve it. We strongly recommend reading his monograph (Craig 2014a): It beautifully presents his scientific journey, and how exploring an unexplained paradox regarding how the somatosensory pathways were organized (why fibers serving pain and temperature sensations took a different route in the spinal cord from those serving low threshold tactile information if they were to converge on the same site in the thalamus) grew into a completely new view on how the nervous system was organized.

2 Early Life and Scientific Foundations

Bud, who was the oldest of four brothers (Fig. 4), was born on August 31, 1951, in Lansing, Michigan, on his father's birthday, and was named after his father, Arthur Dewitt. Both Bud's parents (Fig. 5) were veterans of World War II, his father subsequently working as an electrical engineer for the Chrysler Cooperation,

Fig. 1 Bud (left) and Anders, in Linköping ca. 1995

whereas his mother, Nelda, was a registered nurse. He adopted his nickname "Bud" at the age of three after having seen a sitcom television show, called Father Knows Best, which featured a teenage son, James "Bud" Anderson, Jr.

After having graduated from high school at age 16 (Fig. 6), Bud enrolled at Michigan State University, earning a bachelor's degree in mathematics. However, not being thrilled by the prospects of working as an actuary for an insurance company, or becoming a math teacher, he entered a graduate level course in functional neuroanatomy. Taking part in an electrophysiological demonstration of the somatotopic representation of light touch in the sensory thalamus, he was hooked on neuroscience and began graduate studies in neurophysiology, neuroanatomy, and electrical engineering at Cornell University under the supervision of Daniel Tapper. After his doctoral degree in 1978 (Craig 1978) (Fig. 7), he pursued postdoctoral

Fig. 2 Bud, Lecture in Linköping 2001

studies with Harold Burton in the Dept of Anatomy at Washington University, St Louis, Missouri, which was headed by the preeminent neuroanatomist Maxwell Cowan. There, Bud started exploring the anatomy and physiology of the spinothalamic tract (Craig and Burton 1981, 1985), well-established to be critical for pain and temperature sensations. The functional organization of this pathway would be the focus of much of his career, and the findings from his studies of its distinct thalamocortical connections in primates would lay the foundations for his subsequent ideas on interoception. In 1981, Bud moved to Germany, to join the Institute of Physiology, directed by Robert Schmidt, a student of John Eccles, first in Kiel and from 1983 in Würzburg.

The interest of the Schmidt research group was the neurophysiological basis of nociception in the musculoskeletal system (Schmidt 1996). Along this line Bud showed that thin afferent fibers from muscles and joints terminated in the spinal cord in laminae I and V (Craig et al. 1988; Mense and Craig 1988), i.e. in a pattern clearly distinct from that of innocuous cutaneous afferents, but similar to that observed for cutaneous nociceptors. These observations led to the conclusion that the neurons in the superficial lamina were critical for pain transmission, which in turn prompted examination of the response properties of those neurons, i.e., the spinothalamic tract

Fig. 3 Dinner in Phoenix in 2014. From left: Bud, his wife Lora, Jonathan's wife Esther, and Jonathan

cells that forwarded the nociceptive signals to higher brain centers. Together with Klaus-Dietrich Kniffki (Fig. 8), Bud performed the first comprehensive analysis of these cells (Craig and Kniffki 1985). Craig and Kniffki showed that the lamina I spinothalamic tract neurons consisted of select populations of distinct physiological classes that predominantly reflected the functional specificity of peripheral fibers in that they were either nociceptive or thermoreceptive specific; however, polymodal cells that responded to both cold and noxious stimulation were also present. Later on, Bud would also identify lamina I spinothalamic cells that were selectively responsive to warming of the skin (Andrew and Craig 2001a) and to the injection of histamine (Andrew and Craig 2001b) (hence demonstrating a central pathway for itch), as well as lamina I spinobulbar neurons selectively activated by static muscle contraction (Wilson et al. 2002). He also demonstrated that the functionally different lamina I neurons were morphologically distinct (Craig et al. 1999; Han et al. 1998).

3 Scientific Challenges and Development of the Homeostatic Theory

Bud's demonstration of the presence of "labeled lines" for pain, temperature, and other sensations from the body challenged the view held by many in the field that pain in particular was the result of patterned activity of the so-called wide dynamic range (WDR) neurons that responded in a graded fashion to first innocuous and then

Fig. 4 Bud (4th from front to back) with his brothers, father and grandfather, 1959

noxious stimuli (Wall 1989). Bud's findings were therefore not taken well by many who had devoted their efforts to studying WDR neurons and the responses they evoked in the somatosensory cortex, and during a large part of his career Bud had to face fierce opposition to the work he presented.

The stay in Germany was formative for Bud also because it exposed him to the work on somatosympathetic reflexes that were done by Robert Schmidt's disciple Wilfrid Jänig (Jänig 1996). While most investigators in the pain field stuck to Sherrington's categorization of pain and temperature as exteroceptive sensations (Sherrington 1900), Bud came to realize that there were considerable interactions between these two modalities and the autonomic nervous system. After his return to the USA in 1986 to take up a position at the Barrow Neurological Institute in Phoenix, AZ (Fig. 9), Bud began examining the projections of lamina I neurons from this perspective, using anterograde tracing with localized phaseolus vulgaris leukoagglutinin (PHA-L) iontophoretic injections, guided by electrophysiological

Fig. 5 Bud's father and mother, 1940s

Fig. 6 Bud 1968

recordings, into the superficial lamina of the spinal dorsal horn. This approach, which he took to a still unsurpassed level of precision, made it possible for him to selectively target lamina I cells without involving cells in other spinal cord layers. He

Fig. 7 Bud, PhD 1978

Fig. 8 Bud and Klaus-Dietrich Kniffki, Germany 1982

Fig. 9 Bud at the Barrow Neurological Institute, Phoenix, AZ 1987

showed that lamina I neurons form segmentally organized somatovisceral reflex pathways in the spinal cord (Craig 1993) thus providing a direct pathway for the somatosympathetic reflex effects of thermal and noxious stimuli, and that they also projected onto all the autonomic relay nuclei in the brainstem (Craig 1995). Based on these observations, Bud suggested that the lamina I projection neurons made up a general homeostatic afferent pathway that distributed activity originating from small-diameter fibers to regions involved in the physiological maintenance of the integrity of the body (Craig 1996).

While the brain stem projections from lamina I were found to be similar in cats and monkeys (Craig 1995), Bud demonstrated a unique termination pattern in the primate thalamus of lamina I neurons that corroborated the idea of the lamina I projection as a homeostatic afferent system: The main termination site was located in an area adjacent to the basal part of the ventral medial nucleus (VMb), which in primates, but not in other orders, receives direct input from the nucleus of the solitary tract, which transmit information from taste buds and visceral afferents. In analogy with the term VMb, Bud named the spinothalamic termination site the posterior part of the ventromedial nucleus, or VMpo (Craig et al. 1994). He showed that VMpo contains neurons that maintain the response properties of the lamina I cells, and that it, similar to the VMb, projects to the posterior portion of the insular cortex (Craig 2014b; Craig et al. 1994). He subsequently showed with functional imaging that painful heat stimuli produced a somatotopically organized activation pattern in the insular cortex (Brooks et al. 2005) and that pain-related laser-evoked potentials in the human insular cortex could encode both the somatic location and the intensity of pain sensations (Baumgartner et al. 2006). Similarly, he also showed a somatotopic representation of innocuous cooling activation in the human dorsal posterior insular cortex (Hua et al. 2005). He conceptualized the posterior insular cortex as being primary interoceptive cortex, containing a representation of all afferent signals from small-diameter fibers, including, in addition to pain and temperature, those mediating the sensations of itch, hunger, thirst, etc., creating what German nineteenth century authors called *Gemeingefühl* (common sensation), or the feeling of the "material me" (Craig 2002, 2003a).

The interaction of homeostatic signals and in particular the antagonistic functions of pain and temperature also attracted Bud's interest. Utilizing the thermal grill illusion, first described by Thunberg in 1896 (Thunberg 1896), Bud made seminal contributions to our understanding of the interactions that occur in the CNS between the innocuous thermal inputs and the pain pathway (Craig and Bushnell 1994; Craig et al. 1996; Morin et al. 2002; for further details, see Anders Blomqvist 2023). In recent years, many other groups have utilized the thermal grill as a useful tool for investigating pain perception (Forstenpointner et al. 2020).

4 Advancing Understanding of Pain and Interoception

Bud's idea of pain as an interoceptive sensation that signals information about the integrity of the body (Craig 2003b) helps explain why the same tissue damaging stimulus can evoke perceptions ranging from excruciating pain to no pain at all, such as what is reported by soldiers who experience no pain from their wounds until brought to safety: This is because the perception is related to the homeostatic needs. In a fight/flight situation attention to the tissue damage could jeopardize the chances of survival, whereas in safety it is important to attend to the wound. Similarly, pain varies with hormonal status, with higher testosterone levels in males, who fight for dominance, suppressing pain perception. In the same vein, our temperature perception, also represented in the insular cortex, is dependent on the homeostatic needs: When you are overheated, you seek cold, which then is perceived as pleasant, whereas if you are chilled, cold will be felt as unpleasant.

On the basis of reports from imaging studies on how various subjective feelings evoked activation in the anterior insula, Bud suggested that the primary sensations were re-represented and integrated with other activities at successively more rostral levels in the insula to create subjective feelings that guided behavior (Craig 2009a, 2010). He supported this idea by a seminal imaging study on thermosensory activation of the insular cortex by showing that the graded cooling stimuli correlated with activation in the middle/posterior insula, whereas perceived thermal intensity correlated with activation in the anterior insula (Craig et al. 2000).

Bud's finding that the insula harbors the sensory cortex for pain is corroborated by virtually every imaging study, which clearly shows that pain stimuli result in activation of the posterior insular cortex (Segerdahl et al. 2015). It was also corroborated by a contemporary brain stimulation study (Mazzola et al. 2009), expanding on Penfield's seminal studies of the 1930s. Penfield rarely could evoke anything close to pain sensations in his subjects by stimulation of the cortical surface (Penfield and Boldrey 1937). The same observations were obtained when Penfield's experiment was repeated; however, when the insula was stimulated, pain sensations were evoked from its posterior region and were somatotopically organized with the legs represented most posterior and the head more anteriorly (Laure Mazzola et al. 2011), as postulated from Bud's studies. Notably, pain was not evoked from any other brain region, including the somatosensory cortex.

Despite all these data, many researchers in the field refused to accept Bud's findings. It was erroneously suggested that the VMpo did not exist (Jones 2002), and that the activation of the insular cortex was relayed via the S1 and S2 cortices (Willis et al. 2002; see also Craig and Blomqvist 2002). What many had difficulty accepting was that the insular cortex could account for the haptic component of pain, and it was maintained that only the S1 cortex could substantiate a fine topographic representation of pain. Bud addressed these objections in several ways. In a detailed anatomical tract tracing study he demonstrated the topographically organized projection from the VMpo nucleus to the posterior insular cortex (Craig 2014b). With retrograde tracing techniques, including also antidromic activation of spinal lamina I nociceptive and thermoreceptive neurons by stimulation in VMpo, he showed that whereas the spinothalamic projection to the VMpo nucleus almost exclusively originated from lamina I neurons (Craig and Zhang 2006; Dostrovsky and Craig 1996), the spinothalamic projection to the ventral posterior nuclei, the main relays for innocuous cutaneous and proprioceptive information from the body that is transmitted via the medial lemniscus, originated from spinal lamina V neurons (Craig 2006). Since the lamina V neurons largely are WDR neurons (Price et al. 2003), the spinothalamic tract (STT) input to the ventral posterior nuclei is modality ambiguous. While some have argued that the response properties of the WDR neurons were necessary to encode the discriminative aspects of pain (Coghill 2020), Bud showed that the lamina I STT nociceptive specific cells, with input primarily from nociceptive Aδ-fibers, displayed small receptive field and responses in line with the psychophysics of mechanical pain and sharpness, and he further demonstrated that lamina I STT multimodal heat, pinch, cold (HPC) cells, with input primarily from C-fibers, encoded second pain (Andrew and Craig 2002; Craig et al. 2001). Furthermore, in a direct comparison of lamina I STT HPC cells and lamina V STT WDR cells in the monkey, using a repeated brief-contact heat paradigm, which in humans elicits a strongly augmenting sensation of burning pain and a unique "reset" phenomenon, in which the omission of one or two stimuli in a train of brief contacts causes the pain to suddenly disappear and the temporal augmentation to begin again from near baseline (Vierck et al. 1997), he showed that the responses of the HPC cells but not those of the WDR cells matched the human psychophysics (Craig 2004).

Bud pointed out that the WDR neurons in addition to responding to exteroceptive stimuli also displayed proprioceptive sensitivity to limb position, and that they were organized in the spinal cord in a musculotopic rather than somatotopic pattern, suggesting a role in somatomotor integration (Craig 2006). Alternatively, Bud suggested that the WDR neurons, while not being able to convey modality specific information, could play some role in intensity coding (Craig 2004). Notably, the spinal projection to the ventroposterior nucleus, which mainly originates in lamina V (Craig 2006), terminates in peripheral regions of that nucleus as well as along the fiber laminae that separates the different body representations and not in the core of the nucleus as do the medial lemniscal fibers (Rausell et al. 1992). Furthermore, whereas the medial lemniscal terminals form triadic relationships or complex glomerular arrangements with GABAergic profiles and relay cell dendrites (Ralston and Ralston 1994) (as do lamina I terminals in the VMpo nucleus (Beggs et al. 2003)), characteristic for thalamic sensory relays, most STT terminals make only simple synaptic contacts along projection neuron dendrites implying a more modulatory function (Ralston and Ralston 1994). This idea is also in line with the finding that the cells in the STT termination field in ventral posterior nucleus project onto the superficial layers of the S1 cortex and not to the middle and deep layers as do the cells driven by medial lemniscal input (Rausell et al. 1992). Interestingly, it was recently shown that postsynaptic dorsal column neurons, which take their origin in the deep dorsal horn, mediate wide dynamic range force intensity tuning in the dorsal column nuclei (Turecek et al. 2022), opening up the possibility of a similar role for the WDR neurons in the spinothalamic tract.

From the above account it should be obvious that Bud's contribution to the field of pain was a game changer and of fundamental importance and extended also to a better understanding of the processes involved in other feelings related to the body. As with all paradigm changing ideas, it is not surprising that his work evoked strong resistance from many that had devoted their careers to the conventional concept and, unfortunately, it has not yet been generally accepted. Almost all textbooks in neuroscience still deal with pain and temperature as haptic somatosensory sensations, and there are still many investigators who are either unaware of or simply neglect Bud's work. What also is not often understood is that the primate pain pathways differ fundamentally from those in non-primates such as cats and rodents (A. D. Craig 2009b). Similar to the encephalization of the gustatory pathways with direct projection from the solitary tract to the thalamus in primates (Beckstead et al. 1980), the pain pathways, which in rodents seem to be mainly relayed via the parabrachial nucleus (Polgar et al. 2010), have undergone a corresponding development including a vast expansion of the thalamocortical interoceptive connections.

Bud's discoveries of the organization of the primate pain and temperature pathways lay the foundation for his later work, much of it being of theoretical nature, on interoception and how interoceptive information was processed in the insular cortex. His ideas were fully developed in some highly influential review articles (Craig 2002, 2009a) and in his monograph (Craig 2014a). On the basis of a thorough compilation of the literature from various fields of neuroscience, Bud provided compelling evidence that the posterior insular cortex is the primary sensory cortex for interoception, i.e., the cortical region where the body's ability to sense and interpret its internal states is represented. He supported his suggestion that the interoceptive feelings are re-represented, and multimodally integrated, in anterior portions of the insula by meticulous neuroanatomical studies of the organization of the primate insular cortex, showing a modular organization consonant with his suggested model of homeostatic afferent processing (Evrard et al. 2014). He further proposed, with support of experimental studies of patients with epileptic auras (Landtblom et al. 2011; Picard and Craig 2009), as well as of human volunteers (Engström et al. 2014; van Wassenhove et al. 2011), that subjective awareness is associated with activation of the anterior insular cortex, and that this brain region also is critical for general intelligence and the perception of time. He also explored how lamina I sympathetic and vagal parasympathetic emotional processes were arranged asymmetrically in the bicameral brain, with the left forebrain subserving functions associated with positive affect, approach behavior, and energy nourishment, and the right forebrain involved in negative affect, avoidance behavior, and energy expenditure (Strigo and Craig 2016).

Bud's work on interoception has caused a paradigm shift in our understanding of the neurological basis of consciousness, a field that long has been considered elusive, or even beyond our comprehension. His discoveries have also helped explain pathological medical conditions, such as so-called central pain that occurs after stroke, and the loss of awareness in several neurodegenerative diseases, in which the insula is lesioned. His finding that the insula is the site of our emotions has inspired a wealth of new studies into its involvement in various conditions such as drug addiction, alcoholism, anxiety and eating disorders, to which he also contributed (Goldstein et al. 2009; Larsson et al. 2012; Strigo et al. 2008).

5 Conclusions

It is very rare today that a scientist can articulate a new concept with the breadth and depth that Bud did. It is something we generally associate with earlier times when a scientific endeavor was like entering terra incognita. These days, even the brightest minds are generally just adding a brick or two to an already existing building. Bud was indeed a rare exception. In a review of Bud's monograph for the British Journal of General Practice (https://bjgplife.com/review-an-interoceptive-moment-with-your-neurobiological-self/), the author, Alistair Dobbin, compared Bud's contributions to our understanding of the brain to those of William Harvey's on blood circulation. Perhaps an even better comparison would be to Charles Sherrington. The award ceremony speech for the Nobel Prize in Physiology or Medicine of 1932 credited Sherrington's discoveries to "have ushered in a new epoch in the physiology of the nervous system" (https://www.nobelprize.org/prizes/medicine/1932/

ceremony-speech/). Bud Craig's pioneering discoveries on interoception have done similarly: He is the Sherrington of our time.

Acknowledgment We thank Lora Sanders for providing photographs and information about Bud Craig's family history.

References

Andrew D, Craig AD (2001a) Spinothalamic lamina I neurones selectively responsive to cutaneous warming in cats. J Physiol 537(Pt 2):489–495. Retrieved from http://www.ncbi.nlm.nih.gov/htbin-post/Entrez/query?db=m&form=6&dopt=r&uid=11731580

Andrew D, Craig AD (2001b) Spinothalamic lamina I neurons selectively sensitive to histamine: a central neural pathway for itch. Nat Neurosci 4(1):72–77. Retrieved from http://www.ncbi.nlm.nih.gov/htbin-post/Entrez/query?db=m&form=6&dopt=r&uid=11135647

Andrew D, Craig AD (2002) Responses of spinothalamic lamina I neurons to maintained noxious mechanical stimulation in the cat. J Neurophysiol 87(4):1889–1901. Retrieved from http://www.ncbi.nlm.nih.gov/htbin-post/Entrez/query?db=m&form=6&dopt=r&uid=11929909

Baumgartner U, Tiede W, Treede RD, Craig AD (2006) Laser-evoked potentials are graded and somatotopically organized anteroposteriorly in the operculoinsular cortex of anesthetized monkeys. J Neurophysiol 96(5):2802–2808. Retrieved from http://www.ncbi.nlm.nih.gov/entrez/query.fcgi?cmd=Retrieve&db=PubMed&dopt=Citation&list_uids=16899640

Beckstead RM, Morse JR, Norgren R (1980) The nucleus of the solitary tract in the monkey: projections to the thalamus and brain stem nuclei. J Comp Neurol 190(2):259–282. https://doi.org/10.1002/cne.901900205

Beggs J, Jordan S, Ericson AC, Blomqvist A, Craig AD (2003) Synaptology of trigemino- and spinothalamic lamina I terminations in the posterior ventral medial nucleus of the macaque. J Comp Neurol 459(4):334–354. https://doi.org/10.1002/cne.10613

Blomqvist A (2023) Pain and temperature, and human awareness: the legacy of Bud Craig. Temperature 10(04):395–401. https://doi.org/10.1080/23328940.2023.2265946

Blomqvist A, Evrard HC, Dostrovsky J, Strigo I, Jänig W (2023) A.D. (Bud) Craig, Jr. (1951–2023). Nat Neurosci 26(11):1835–1836.

Brooks JC, Zambreanu L, Godinez A, Craig AD, Tracey I (2005) Somatotopic organisation of the human insula to painful heat studied with high resolution functional imaging. Neuroimage 27(1):201–209. Retrieved from http://www.ncbi.nlm.nih.gov/entrez/query.fcgi?cmd=Retrieve&db=PubMed&dopt=Citation&list_uids=15921935

Coghill RC (2020) The distributed nociceptive system: a framework for understanding pain. Trends Neurosci 43(10):780–794. https://doi.org/10.1016/j.tins.2020.07.004

Craig AD (1978) Anatomic and Electrophysiologic studies on the lateral cervical nucleus in cat and dog. Cornell University

Craig AD (1993) Propriospinal input to thoracolumbar sympathetic nuclei from cervical and lumbar lamina I neurons in the cat and the monkey. J Comp Neurol 331(4):517–530. Retrieved from http://www.ncbi.nlm.nih.gov/htbin-post/Entrez/query?db=m&form=6&dopt=r&uid=8509508

Craig AD (1995) Distribution of brainstem projections from spinal lamina I neurons in the cat and the monkey. J Comp Neurol 361:225–248

Craig AD (1996) An ascending general homeostatic afferent pathway originating in lamina I. Prog Brain Res 107:225–242

Craig AD (2002) How do you feel? Interoception: the sense of the physiological condition of the body. Nat Rev Neurosci 3(8):655–666. Retrieved from http://www.ncbi.nlm.nih.gov/htbin-post/Entrez/query?db=m&form=6&dopt=r&uid=12154366

Craig AD (2003a) Interoception: the sense of the physiological condition of the body. Curr Opin Neurobiol 13(4):500–505. https://doi.org/10.1016/s0959-4388(03)00090-4

Craig AD (2003b) A new view of pain as a homeostatic emotion. Trends Neurosci 26(6):303–307. https://doi.org/10.1016/S0166-2236(03)00123-1

Craig AD (2004) Lamina I, but not lamina V, spinothalamic neurons exhibit responses that correspond with burning pain. J Neurophysiol 92(4):2604–2609. https://doi.org/10.1152/jn.00385.2004

Craig AD (2006) Retrograde analyses of spinothalamic projections in the macaque monkey: input to ventral posterior nuclei. J Comp Neurol 499(6):965–978. Retrieved from http://www.ncbi.nlm.nih.gov/entrez/query.fcgi?cmd=Retrieve&db=PubMed&dopt=Citation&list_uids=17072832

Craig AD (2009a) How do you feel – now? The anterior insula and human awareness. Nat Rev Neurosci 10(1):59–70. https://doi.org/10.1038/nrn2555

Craig AD (2009b) A rat is not a monkey is not a human: comment on Mogil (Nature Rev. Neurosci. 10, 283-294 (2009)). Nat Rev Neurosci 10(6):466. https://doi.org/10.1038/nrn2606-c1

Craig AD (2010) The sentient self. Brain Struct Funct 214(5–6):563–577. https://doi.org/10.1007/s00429-010-0248-y

Craig AD (2014a) How do you feel?: An interoceptive moment with your neurobiological self. Princeton University Press

Craig AD (2014b) Topographically organized projection to posterior insular cortex from the posterior portion of the ventral medial nucleus in the long-tailed macaque monkey. J Comp Neurol 522(1):36–63. https://doi.org/10.1002/cne.23425

Craig AD, Blomqvist A (2002) Is there a specific lamina I spinothalamocortical pathway for pain and temperature sensations in primates? J Pain 3(2):95–101; discussion 113–104. https://doi.org/10.1054/jpai.2002.122953

Craig AD, Burton H (1985) The distribution and topographical organization in the thalamus of anterogradely-transported horseradish peroxidase after spinal injections in cat and raccoon. Exp Brain Res 58(2):227–254. https://doi.org/10.1007/bf00235306

Craig AD, Bushnell MC (1994) The thermal grill illusion: unmasking the burn of cold pain. Science 256:252–255

Craig AD Jr, Burton H (1981) Spinal and medullary lamina I projection to nucleus submedius in medial thalamus: a possible pain center. J Neurophysiol 45(3):443–466. https://doi.org/10.1152/jn.1981.45.3.443

Craig AD, Kniffki KD (1985) J Physiol (Lond) 365:197–221. https://doi.org/10.1113/jphysiol.1985.sp015767

Craig AD, Zhang ET (2006) Retrograde analyses of spinothalamic projections in the macaque monkey: input to posterolateral thalamus. J Comp Neurol 499(6):953–964. Retrieved from http://www.ncbi.nlm.nih.gov/entrez/query.fcgi?cmd=Retrieve&db=PubMed&dopt=Citation&list_uids=17072831

Craig AD, Heppelmann B, Schaible HG (1988) The projection of the medial and posterior articular nerves of the cat's knee to the spinal cord. J Comp Neurol 276(2):279–288. https://doi.org/10.1002/cne.902760210

Craig AD, Bushnell MC, Zhang E-T, Blomqvist A (1994) A thalamic nucleus specific for pain and temperature sensation. Nature 372:770–773

Craig AD, Reiman EM, Evans A, Bushnell MC (1996) Functional imaging of an illusion of pain. Nature 384:258–260

Craig AD, Zhang E-T, Blomqvist A (1999) A distinct thermoreceptive sub-region of lamina I in nucleus caudalis of the owl monkey. J Comp Neurol 404:221–234

Craig AD, Chen K, Bandy D, Reiman EM (2000) Thermosensory activation of insular cortex. Nat Neurosci 3(2):184–190. Retrieved from http://www.ncbi.nlm.nih.gov/cgi-bin/Entrez/referer? http://library.neurosci.nature.com/server-java/Propub/neuro/nn0200_184.fulltext

Craig AD, Krout K, Andrew D (2001) Quantitative response characteristics of thermoreceptive and nociceptive lamina I spinothalamic neurons in the cat. J Neurophysiol 86(3):1459–1480.

Retrieved from http://www.ncbi.nlm.nih.gov/htbin-post/Entrez/query?db=m&form=6&dopt=r&uid=11535691
Dostrovsky JO, Craig AD (1996) Cooling-specific spinothalamic neurons in the monkey. J Neurophysiol 76(6):3656–3665. https://doi.org/10.1152/jn.1996.76.6.3656
Engström M, Karlsson T, Landtblom AM, Craig AD (2014) Evidence of conjoint activation of the anterior insular and cingulate cortices during effortful tasks. Front Hum Neurosci 8:1071. https://doi.org/10.3389/fnhum.2014.01071
Evrard HC, Logothetis NK, Craig AD (2014) Modular architectonic organization of the insula in the macaque monkey. J Comp Neurol 522(1):64–97. https://doi.org/10.1002/cne.23436
Forstenpointner J, Berry D, Baron R, Borsook D (2020) The cornucopia of central disinhibition pain – an evaluation of past and novel concepts. Neurobiol Dis 145:105041. https://doi.org/10.1016/j.nbd.2020.105041
Goldstein RZ, Craig AD, Bechara A, Garavan H, Childress AR, Paulus MP, Volkow ND (2009) The neurocircuitry of impaired insight in drug addiction. Trends Cogn Sci 13(9):372–380. https://doi.org/10.1016/j.tics.2009.06.004
Han Z-S, Zhang E-T, Craig AD (1998) Nociceptive and thermoreceptive lamina I neurons are anatomically distinct. Nat Neurosci 1:218–225
Hua le H, Strigo IA, Baxter LC, Johnson SC, Craig AD (2005) Anteroposterior somatotopy of innocuous cooling activation focus in human dorsal posterior insular cortex. Am J Physiol Regul Integr Comp Physiol 289(2):R319–R325. Retrieved from http://www.ncbi.nlm.nih.gov/entrez/query.fcgi?cmd=Retrieve&db=PubMed&dopt=Citation&list_uids=15805097
Jänig W (1996) Spinal cord reflex organization of sympathetic systems. Prog Brain Res 107:43–77. https://doi.org/10.1016/s0079-6123(08)61858-0
Jones EG (2002) A pain in the thalamus. J Pain 3(2):102–104. https://doi.org/10.1054/jpai.2002.122952
Landtblom AM, Lindehammar H, Karlsson H, Craig AD (2011) Insular cortex activation in a patient with "sensed presence"/ecstatic seizures. Epilepsy Behav 20(4):714–718. https://doi.org/10.1016/j.yebeh.2011.01.031
Larsson MB, Tillisch K, Craig AD, Engström M, Labus J, Naliboff B et al (2012) Brain responses to visceral stimuli reflect visceral sensitivity thresholds in patients with irritable bowel syndrome. Gastroenterology 142(3):463–472.e3. https://doi.org/10.1053/j.gastro.2011.11.022
Mazzola L, Isnard J, Peyron R, Guénot M, Mauguière F (2009) Somatotopic organization of pain responses to direct electrical stimulation of the human insular cortex. Pain 146(1):99–104. https://doi.org/10.1016/j.pain.2009.07.014
Mazzola L, Isnard J, Peyron R, Mauguière F (2011) Stimulation of the human cortex and the experience of pain: Wilder Penfield's observations revisited. Brain 135(2):631–640. https://doi.org/10.1093/brain/awr265
Mense S, Craig AD Jr (1988) Spinal and supraspinal terminations of primary afferent fibers from the gastrocnemius-soleus muscle in the cat. Neuroscience 26(3):1023–1035. https://doi.org/10.1016/0306-4522(88)90117-0
Morin C, Bushnell MC, Luskin MB, Craig AD (2002) Disruption of thermal perception in a multiple sclerosis patient with central pain. Clin J Pain 18(3):191–195. https://doi.org/10.1097/00002508-200205000-00008
Penfield W, Boldrey E (1937) Somatic motor and sensory representation in the cerebral cortex of man studied by electrical stimulation. Brain 60(4):389–443. https://doi.org/10.1093/brain/60.4.389
Picard F, Craig AD (2009) Ecstatic epileptic seizures: a potential window on the neural basis for human self-awareness. Epilepsy Behav 16(3):539–546. https://doi.org/10.1016/j.yebeh.2009.09.013
Polgar E, Wright LL, Todd AJ (2010) A quantitative study of brainstem projections from lamina I neurons in the cervical and lumbar enlargement of the rat. Brain Res 1308(5):58–67. https://doi.org/10.1016/j.brainres.2009.10.041

Price DD, Greenspan JD, Dubner R (2003) Neurons involved in the exteroceptive function of pain. Pain 106(3):215–219. Retrieved from https://journals.lww.com/pain/fulltext/2003/12000/neurons_involved_in_the_exteroceptive_function_of.2.aspx

Ralston HJ 3rd, Ralston DD (1994) Medial lemniscal and spinal projections to the macaque thalamus: an electron microscopic study of differing GABAergic circuitry serving thalamic somatosensory mechanisms. J Neurosci 14(5 Pt 1):2485–2502. https://doi.org/10.1523/jneurosci.14-05-02485.1994

Rausell E, Bae CS, Vinuela A, Huntley GW, Jones EG (1992) Calbindin and parvalbumin cells in monkey VPL thalamic nucleus: distribution, laminar cortical projections, and relations to spinothalamic terminations. J Neurosci 12(10):4088–4111. https://doi.org/10.1523/JNEUROSCI.12-10-04088.1992

Schmidt RF (1996) The articular polymodal nociceptor in health and disease. Prog Brain Res 113: 53–81. https://doi.org/10.1016/s0079-6123(08)61081-x

Segerdahl AR, Mezue M, Okell TW, Farrar JT, Tracey I (2015) The dorsal posterior insula subserves a fundamental role in human pain. Nat Neurosci 18(4):499–500. https://doi.org/10.1038/nn.3969

Sherrington CS (1900) Cutaneous sensations. In: Shäfer EA (ed) Text-book of physiology. Pentland, Edinburgh, pp 920–1001

Strigo IA, Craig AD (2016) Interoception, homeostatic emotions and sympathovagal balance. Philos Trans R Soc Lond B Biol Sci 371(1708):20160010. https://doi.org/10.1098/rstb.2016.0010

Strigo IA, Simmons AN, Matthews SC, Craig AD, Paulus MP (2008) Association of major depressive disorder with altered functional brain response during anticipation and processing of heat pain. Arch Gen Psychiatry 65(11):1275–1284. https://doi.org/10.1001/archpsyc.65.11.1275

Thunberg T (1896) Förnimmelserna vid till samma ställe lokaliserad, samtidigt pågående köld- och värmeretning. Upsala Lakareforen Forh 1:488–555

Turecek J, Lehnert BP, Ginty DD (2022) The encoding of touch by somatotopically aligned dorsal column subdivisions. Nature 612(7939):310–315. https://doi.org/10.1038/s41586-022-05470-x

van Wassenhove V, Wittmann M, Craig AD, Paulus MP (2011) Psychological and neural mechanisms of subjective time dilation. Front Neurosci 5:56. https://doi.org/10.3389/fnins.2011.00056

Vierck CJ Jr, Cannon RL, Fry G, Maixner W, Whitsel BL (1997) Characteristics of temporal summation of second pain sensations elicited by brief contact of glabrous skin by a preheated thermode. J Neurophysiol 78(2):992–1002. https://doi.org/10.1152/jn.1997.78.2.992

Wall PD (1989) Introduction. In: Wall PD, Melzack R (eds) Textbook of pain. Churchill Livingstone, Oxford, pp 1–18

Willis WD Jr, Zhang X, Honda CN, Giesler GJ Jr (2002) A critical review of the role of the proposed VMpo nucleus in pain. J Pain 3(2):79–94. https://doi.org/10.1054/jpai.2002.122949

Wilson LB, Andrew D, Craig AD (2002) Activation of spinobulbar lamina I neurons by static muscle contraction. J Neurophysiol 87(3):1641–1645. Retrieved from http://www.ncbi.nlm.nih.gov/htbin-post/Entrez/query?db=m&form=6&dopt=r&uid=11877534

The Isle of Craig: Neuroanatomical and Functional Evidence for a Role of the Insular Cortex in Subjective Feelings

Henry C. Evrard

Contents

Abstract At the start of the twenty-first century, Arthur D. (Bud) Craig brought back to the fore the Island of Reil (insular cortex or insula). He did so by following, step by step, with rigor and tenacity, the afferent sensory pathway that informs the

---

H. C. Evrard (✉)
Functional & Comparative Neuroanatomy of the Dynamic Embodied Brain Lab, International Center for Primate Brain Research, Center for Excellence in Brain Science and Intelligence Technologies, Institute of Neuroscience, Chinese Academy of Sciences, Shanghai, China

Werner Reichardt Center for Integrative Neuroscience, Karl Eberhard University of Tübingen, Tübingen, Germany

Max Planck Institute for Biological Cybernetics, Tübingen, Germany

Curr Topics Behav Neurosci (2025) 70: 33–61
https://doi.org/10.1007/7854_2025_590
Published Online: 28 May 2025

forebrain about the ongoing physiological status of the organs and tissues of the body. Along with his demonstration of the existence of a primate-specific ascending interoceptive pathway and his subsequent re-interpretation of Sherrington's concept of interoception, Bud Craig's seminal experiments and profound interpretations led him to make the groundbreaking proposals that the dorsal posterior insular cortex provides an ideal substrate for James's concept of emotional embodiment, that the insular cortex contextualizes interoception across a posterior-to-mid-to-anterior integration with multimodal activities, and that the anterior insular cortex has a crucial role in the evolutionary emergence of the awareness of subjective feelings in humans, for the purpose of optimizing metabolic energy usage. Bud Craig's unique work paves the path for further elucidation of the role of the insula and other brain regions in subjective feelings. His discoveries and proposals rest on implacable attention to neuroanatomical and neurophysiological details and a serendipitous quest for the fundamental evolutionary Logic of Life. This chapter provides a detailed description of the ascending interoceptive pathway and the functional and comparative neuroanatomy of the insular cortex in primates. Building on Bud Craig's work, our recent findings suggest that the primary interoceptive cortex serves as a representation of the spino-solitary-parabrachial neuraxis, merging with posterior-to-mid-anterior and dorsal-to-ventral processing streams that form a latticework integration pattern. At the ventral anterior tip of this integration, the von Economo neuron area closes the corticofugal interoceptive-autonomic loop of the sensory-motor homeostatic system through projections to all brainstem nuclei integrating interoceptive afferences.

Keywords Awareness · Embodiment · Homeostasis · Interoception · Von Economo neuron

1 Introduction

The scientific career of Arthur D. (Bud) Craig (1951–2023) is distinguished by a remarkable series of groundbreaking discoveries, culminating in a major paradigm shift in how we understand pain, subjective feelings, and their evolutionary underpinnings (Blomqvist et al. 2023; Strigo et al. 2024). In this transformative framework, the insular cortex emerges as a crucial cortical hub continuously mapping the ongoing physiological condition of the body – interoception – and integrating it with salient multimodal information to construct a cohesive, embodied sense of self. This integration aligns cognitive appraisal, autonomic regulation, and behavioral outputs with the overarching goal of optimizing metabolic, energetic efficiency (Craig 2002). This dynamic integration underscores how interoception serves not merely as a measure of bodily needs but as the cornerstone of the subjective awareness of any and all feelings across time, making it a prerequisite for embodied consciousness (Craig 2010, 2015).

While most readers of Bud's work find inspiration in his revolutionary theories on interoception and the insula's role in subjective experience, it is essential to realize that these conclusions stem first and foremost from his unrelenting examination of structural and functional details defining the ascending interoceptive pathway and laying the foundation of his model for the insula. Qualified by Bud as a "work in progress," the map of the ascending interoceptive pathway represents over 25 years of mostly single-handed, tenacious, and rigorous neuroanatomical tract-tracing and neurophysiological recording in the macaque monkey (Craig 2015; Blomqvist 2023; Blomqvist and Dostrovsky 2025). The progressive construction of this map is marked by key breakthroughs: *(1)* Lamina 1 (SpL1) spinothalamic tract (STT) neurons – but not spinal lamina 5 wide dynamic range (WDR) neurons – encode the intensity of thermal and nociceptive stimuli along discrete modality-specific "labeled lines," aligning closely with human psychophysical ratings (Craig 2003a, 2004a). *(2)* Small-diameter afferents to SpL1 encode, beyond cutaneous thermoreceptors and nociceptors, a broad diversity of homeostatically relevant activities that continuously inform the brain about the health of bodily tissues (Craig and Kniffki 1985; Craig et al. 1988; Han et al. 1998; Andrew and Craig 2001; Wilson et al. 2002; Craig 2002). *(3)* SpL1 projects to spinal and brainstem autonomic centers, further relating small-diameter afferences to homeostasis (Craig 1992, 1993, 1995a, b; Westlund and Craig 1996). *(4)* In primates, but not other mammals, SpL1 projects directly to a phylogenetically novel nucleus of the thalamus (which Bud named the *posterior part of the ventral medial nucleus*; VMpo) located adjacent to the basal part of the ventral medial nucleus (VMb; also known as VPMpc) that receives visceral inputs from the vagus nerve via the solitary tract nucleus (nTS) and parabrachial nucleus (PB) (Craig et al. 1994; Craig 1995a, 2004a, b, 2006; Blomqvist et al. 2000; Craig and Blomqvist 2002; Craig and Zhang 2006). *(5)* VMpo projects primarily to the posterior dorsal fundus of the insula (IDFP, formerly abbreviated dpIns) (Craig et al. 1995; Craig 2014), next to the projection of VMb to a more anterior portion of the dorsal fundus (Pritchard et al. 1986; Ito and Craig 2008; Strigo and Craig 2016). *(6)* VMpo and IDFP have a posterior-to-anterior bodily map (foot-to-face) that is orthogonal to the classical homunculus of VP and S1 (Craig 2004b, 2014; Hua et al. 2005; Baumgartner et al. 2006).

This body of evidence refutes the long-standing incongruity of the *Gate Control Theory of Pain*, which misclassifies pain and temperature as exteroceptive modalities represented in VP and S1 alongside touch (Melzack and Wall 1965; Craig 2003a). More critically, it led Bud to redefine Sherrington's concept of interoception as the physiological condition of all body tissues – not just the viscera – represented along a phylogenetically novel ascending sensory pathway in primates (Sherrington 1948; Craig 2002, 2003b), serving homeostasis as the ascending counterpart to Langley's autonomic pathway (Langley 1921; Craig 1996) and providing a neurobiological substrate for Weber's "common sensation" (*Gemeingefühl*), Sherrington's "material me," and the "emotional embodiment" underlying subjective feelings as envisioned in the *James-Lange Theory of Emotion* (Weber 1846; Lange 1885; James 1894; Craig 2002) – an undeniable paradigm shift.

Bud's seminal PET study revealed that objective cooling of the right hand correlated primarily with activity in the contralateral IDFP, while subjective ratings of cooling intensity engaged the mid-insula bilaterally and, more prominently, the right anterior insular cortex (AIC) (Craig et al. 2000). This posterior-to-anterior pattern has since been replicated across various interoceptive modalities (Drzezga et al. 2001; Olausson et al. 2002; Wattendorf et al. 2016; Hassanpour et al. 2018), reinforcing the AIC's consistent association with subjective feelings (Craig et al. 2000; Craig 2002, 2009a). From this foundation, Bud formulated his influential model of posterior-to-mid-to-anterior integration: interoceptive signals are first mapped in the dorsal posterior insula (the primary interoceptive cortex) and then progressively enriched in the mid-insula through integration with multisensory, hedonic, and homeostatic information, generating what he termed "homeostatic sentience." This evolving representation culminates in the AIC, where it merges with other salient brain activities into a unified "global emotional moment," encompassing all bodily sensations, social emotions, and cognitive states linked to attention, thought, and introspection (Craig 2009a). Through a "cinemascopic" sequencing of these moments across time, the AIC supports the continuous and embodied sense of self, manifest in our awareness of bodily states, emotional shifts, fluid thoughts, and the passing of time (Craig 2009a, b, 2011, 2015; Engström et al. 2014).

Following Bud's footsteps, this chapter examines the functional neuroanatomy of the macaque insula while identifying possible homologies with the human insula. After reviewing the macroscopic organization of the insula and the refined subdivision of each of its three vast architectonic sectors – granular, dysgranular, agranular – into smaller areas and "stripes," this chapter focuses on the convergence of functional activations of the dorsal fundus with topographic features of the ascending interoceptive pathway to tentatively propose that the primary interoceptive cortex in macaque represents the progressive integration of interoception along the spino-solitary-parabrachial hierarchy. Building on this and the partitioning of the macaque dysgranular insula into integration stripes, this chapter proposes a novel latticework model of insular integration, complementing Bud's framework by adding a dorso-ventral axis and linking the posterior-to-mid-to-anterior progression to a representation of the spino-solitary-parabrachial neuraxis – a new idea that Bud and I discussed in recent years. Finally, new evidence on the von Economo neuron (VEN) is presented, indicating that VEN belongs to a unique architectonic area that projects to brainstem homeostatic centers, hereby closing one of the corticofugal interocepto-autonomic loops sitting atop the interoceptive neuraxis.

2 Folds of the Insular Cortex: Evolutionary and Comparative Landmarks

2.1 The Dorsal Fundus as an Evolutionary and Development Cortical Anchor

The insular cortex, named for its location deep within the lateral (Sylvian) fissure, is hidden beneath the orbital prefrontal cortex anteriorly, the frontoparietal operculum dorsally, and the temporal operculum ventrally (Reil 1809; Naidich et al. 2004; Afif and Mertens 2010). Once these opercula are removed, the insula appears, outlined by the anterior, superior, and inferior circular sulci (Fig. 1a, b). Bud identified the dorsal anterior and posterior fundal regions, straddling the superior circular sulcus, as the primary targets of the interoceptive pathway via VMpo and VMb (Craig 2002). Drawing on the tension-based morphogenesis theory (Van Essen 1997, 2020), and noting that other VMpo projections terminate in deep fundi (areas 3a and 24d), Bud proposed that these interoceptive cortices act as tensile anchors, pulling themselves into close proximity to optimize energy-efficient cortico-cortical communication (Craig 2011, 2015), noting also that the dorsal fundus, the last insular feature to emerge evolutionarily, forms first during development (Afif et al. 2007; Butti and Hof 2010; Mallela et al. 2023), suggesting an evolutionary late orchestration of insular functions by the ascending interoceptive pathway (Craig 2015). While most tension-based models emphasize corticocortical tensions (Van Essen 2020), the anchoring might also result from thalamocortical axon tension (Striedter et al. 2015), compensating for the slow conduction of unmyelinated C-fibers during early development when interoceptive inputs already engage the cortex (Evrard 2019).

2.2 *Insular Sulci and Dimples: A Quest for Structural Homologies*

Beyond the dorsal fundus, the human insula is divided into anterior and posterior lobules by the central insular sulcus (Fig. 1a). The posterior lobule contains two long gyri, while the anterior typically has three short gyri, flanked anteriorly by the accessory and transverse gyri (Ture et al. 1999; Afif and Mertens 2010), separated from the short gyri by an accessory anterior circular sulcus (Evrard 2019).[1] The accessory and transverse gyri, collectively known as frontoinsula, are distinctively enriched in VENs and their companion bifid fork neurons (FNs) (Von Economo and Koskinas 1925; Ngowyang 1932; Allman et al. 2010). In light of Bud's view of the dorsal fundus as major interoceptive anchor, which tensile anchors create these additional sulci and whether these anchors have precursors in non-hominin monkeys remain to be explored.

[1] Bud taught that the primate brain is like a face, varying in shape but with essential elements always found where expected (Craig 2015, p. 111). At the 2009 SfN meeting, Jack Johnson and colleagues reported that the number and orientation of insular gyri vary markedly between individuals and hemispheres (Johnson et al. 2009; Craig 2010; Wysiadecki et al. 2018). Group fMRI analyses should carefully consider this variability in their registration and normalization procedures. John "Jack" Irwin Johnson was an inspirational functional neuroanatomist at Michigan State University (Prof. Johnson Memorial Source: https://www.youtube.com/watch?v=2CQUjbK7kZc). Bud always credited Jack for opening him the doors to the fields of neuroscience by authorizing him to attend his course in functional neuroanatomy (Craig 2015). They remained in close contact ever since. Jack passed away in 2017.

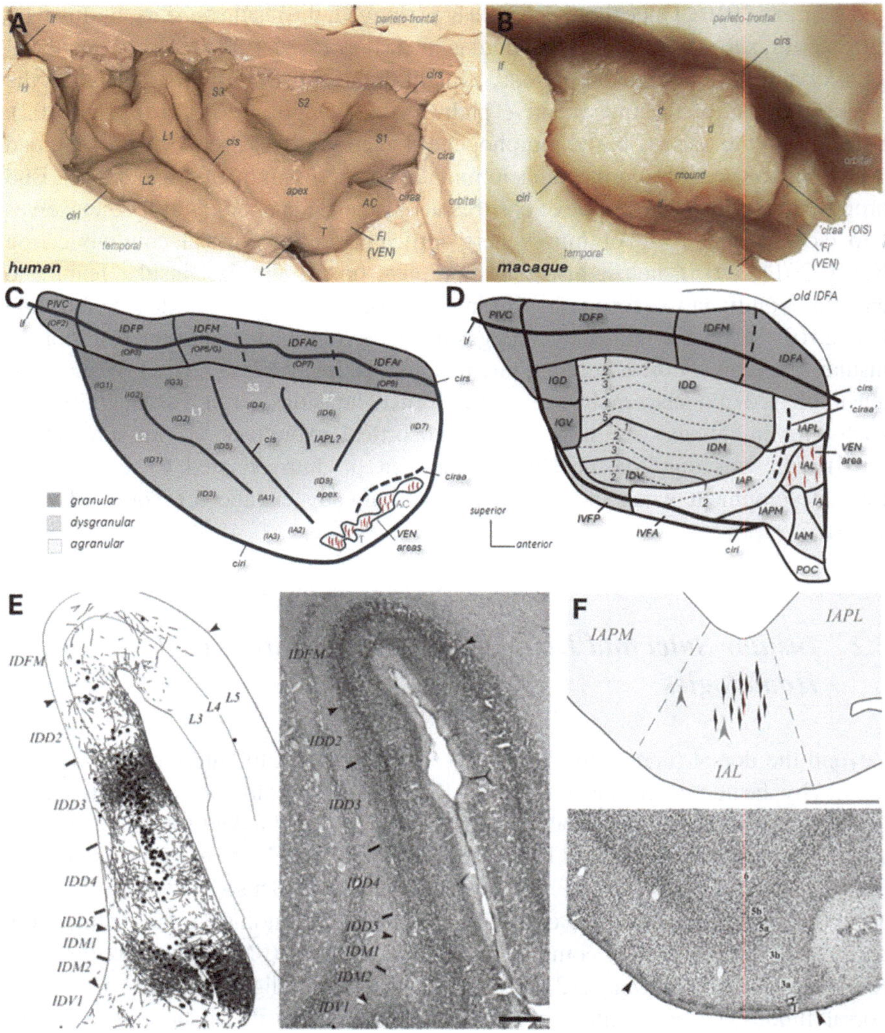

Fig. 1 (**a**, **b**): Lateral views of the human (**a**) and cynomolgus macaque monkey (**b**) insula exposed after excision of the frontoparietal and temporal opercula. (**c**, **d**): Flat map views of the architectonic parcellation of the human (**c**) and cynomolgus macaque monkey (**d**) insula. (**e**): Comparison of the distribution of anterograde (lines) and retrograde (dots) labeling (left panel) with the Nissl stain-based cytoarchitectonic parcellation (right panel) in the insula in case OM85 with an injection of lucifer yellow in area 24 of the cingulate cortex. Adapted from Krockenberger et al. (2023) with permission. (**f**): Comparison of the distribution of spindle-shaped von Economo neurons (elongated rhombi) and bifid fork (Y-shaped symbols) neurons (top panel) with the Nissl stain-based cytoarchitectonic parcellation (bottom panel) in the ventral anterior insula in one macaque monkey. Adapted from Horn (2020) with permission. *Abbreviations*: *aals* accessory anterior limiting sulcus (human), *AC* accessory gyrus, *cis* central insular sulcus, *cira* circular sulcus, anterior branch, *ciraa* circular sulcus, accessory anterior branch, *ciri* circular sulcus, inferior branch, *cirs* circular sulcus, superior branch, *d* dimple, *IAL* lateral agranular area of the insula, *IAPL* posterior lateral agranular area of the insula, *IAPM* posterior median agranular area of the insula, *IDD1–5* dorsal dysgranular area stripes 1–5 of the insula, *IDFM* middle dorsal fundus area of the insula, *IDM1–3* mound

The macaque insula is largely lissencephalic, complicating the direct application of Bud's posterior-to-mid-to-anterior model to macaque data. It features a rudimentary ventral horizontal gyrus – the "mound" (Fig. 1b) – which is dysgranular, receives VMpo projections (Craig 2014), and lacks a clear human counterpart. Additionally, a distinct anterior vertical groove, previously termed the "orbitoinsular sulcus" (Mesulam and Mufson 1985; Mufson et al. 1997) borders an elevated region rich in VENs and FNs (Evrard et al. 2012), suggesting homology with the human frontoinsula and accessory anterior circular sulcus (Evrard 2019). The macaque insula also shows subtle dimples that may represent precursors of human insular sulci, as suggested by recent findings on prefrontal dimples (Amiez et al. 2023). In fact, species close to macaques, such as baboons, display an incipient central insular sulcus (Mufson et al. 1997).

3 Architectonic Parcellation: The Highly Organized Brain

3.1 *IDFP: Another Region Discovered by Bud, as Important as VMpo*

Since the dorsal posterior fundus is the main target of VMpo projections, Bud sought to examine its cellular organization using classical cyto- and myelo-architectonics. At the time, the insula was broadly divided into posterior granular, middle dysgranular, and anterior agranular sectors, with the border to secondary somatosensory cortex (S2) variably placed around the superior circular sulcus (Roberts and Akert 1963; Sanides 1968; Burton and Jones 1976; Mesulam and Mufson 1982). Bud's anterograde labeling from VMpo injections consistently straddled the fundus (Craig 1995a, 2014; Craig et al. 1995), similar to VMb projections to the dorsal anterior fundus (Pritchard et al. 1986; Strigo and Craig 2016). In my very first hour in Bud's lab in Phoenix, AZ, Bud placed under the stereomicroscope a thin section of the macaque insula histologically stained to reveal myelinated fibers. Pointing to the dorsal posterior fundus, he asked with excitement: "Isn't that obvious?" Yes, it was obvious: the dorsal posterior fundus forms a single area – which we named the *dorsal fundal posterior area of the insula* (IDFP; Fig. 1d) – with sharp medial and lateral boundaries, coinciding precisely with the projection field (Craig 2014; Evrard et al. 2014). At the fundus midpoint, a boundary marked the transition from VMpo to

Fig. 1 (continued) dysgranular area stripes 1–3 of the insula, *IDV1–2* ventral dysgranular area stripes 1–2 of the insula, *FI* frontoinsula, *H* Heschl gyrus, *L* limen insula, *L1–2* posterior long gyri 1–2, *lf* fundus of the lateral fissure (denotes the fundus posterior to the insula and circular sulcus), *PIVC* posterior insular vestibular cortex, *POC* piriform cortex, *S1–3* anterior short gyri 1–3, *T* transverse gyrus, *VEN* von Economo neuron, *(IA1–3)* Jülich agranular areas 1–3, *(ID1–9)* Jülich dysgranular areas 1–9, *(IG1–3)* Jülich granular areas 1–3. *Scale bars*: 2 mm in **a**–**D**, 500 μm in **e**–**f**

VMb projections. We initially designated the VMb-recipient region as the *dorsal fundal anterior area* (IDFA) and considered it the macaque's gustatory cortex (Kadohisa et al. 2005; Evrard et al. 2014). Later, this "old IDFA" was subdivided into IDFA and IDFM (Fig. 1d) to reflect fMRI evidence that the primary gustatory cortex lies in the fundus' mid-portion (IDFM), possibly like in humans (Hartig 2019; Evrard 2021; Hartig et al. 2023) (see Sect. 4).

3.2 The Macaque Insula Contains Numerous Sharply Delimited and Highly Reproducible Areas

How is the rest of the insula organized? Inspired by Joel Price's re-examination of the prefrontal cortex (Carmichael and Price 1994; Ongur et al. 2003), Bud and I went through several iterations of the macaque insula's architectonic map. My first map, drawn in Phoenix, identified 21 areas – a number we worked hard to reduce a more "realistic" value. Yet, when revisiting the project together at the *Max Planck Institute for Biological Cybernetics* in Tübingen, where I had taken a position, we concluded that the insula is indeed highly organized, revealing up to 24 sharply defined areas (Fig. 1d). To satisfy reviewers, we consolidated this into 15 areas: 2 fundal granular areas (IDFP and IDFA, later joined by IDFM), 2 posterior granular areas (IGD and IGV), 4 dysgranular areas – dorsal (IDD), mound (IDM), ventral (IDV), and posterior fundal (IVFP), and 7 agranular areas, including 5 pre-limen (IAI, IAM, IAL, IAPL, IAPM) and 2 post-limen areas (IAP, IVFA), as initially recognized by Carmichael and Price (1994). We also suggested that 3 dysgranular areas (IDD, IDM, IDV) may each contain 2 to 5 subtle subareas, which we termed "stripes" (Evrard et al. 2014).

3.3 Dysgranular Stripes: Overlap Between Architectonics and Tract-Tracing

While training my eye for neuroanatomical details, Bud taught me to trust the subjective impression of consistent structural patterns. Although we initially presented the dysgranular stripes as hypothetical (Evrard et al. 2014), subsequent tract-tracing studies confirmed their reality. Labeling from tracer injections in other cortical areas consistently aligned with the architectonic stripes, showing distinct patterns between adjacent stripes (Krockenberger et al. 2023). For instance, a dual anterograde and retrograde tracer injection in area 24b' of the middle cingulate cortex produced dense labeling precisely matching the independently defined third IDD and second IDM stripes (Fig. 1e). This striped subdivision of the dysgranular insula, parallel to the dorsal fundus, suggests a re-representation of the primary interoceptive map into multiple secondary dysgranular maps, each integrating distinct high-order brain activities (Evrard et al. 2014; Krockenberger et al. 2023) (see Sect. 5).

3.4 Von Economo Neuron Area: Novel Elemental Localization in the Primate Cortex

VENs (and co-mingled FNs[2]) are found predominantly in the anterior insular cortex (AIC) and anterior cingulate cortex (ACC) of humans and great apes (Nimchinsky et al. 1999; Allman et al. 2010) as well as in non-hominin primates like macaques (Evrard et al. 2012; Steiner et al. 2024) and a few large-brained non-primates (Hakeem et al. 2009; Butti et al. 2011; Raghanti et al. 2015). We first identified the macaque VEN and FN prior to developing the insular architectonic map (Evrard et al. 2012) consistently observing a small cluster within the agranular insula. This cluster precisely matches the lateral agranular area (IAL), now also called the VEN area (Horn et al. 2017; Horn 2020) (Fig. 1f). Its strict confinement, what Brodmann would have termed an "elemental" localization (Brodmann 1909), provides a valuable entry point to explore the developmental, evolutionary, and functional roles of the VEN and its host area (Evrard 2018). Alongside the dysgranular stripes, this elemental VEN localization underscores the macaque insula's fine-grained organization, suggesting an equal or even higher differentiation in humans.

3.5 Architectonic Parcellation of the Human Insular Cortex

Figure 1c shows my interpretative transposition of the Jülich-Brain Project parcellation of the human posterior lobule and part of the anterior insula (Kurth et al. 2010; Amunts et al. 2020; Duong et al. 2023). This parcellation identifies 5 granular opercular[3] areas (OP2, 3, 5, 7, and 9), 3 posterior granular areas (IG1–3), 10 dysgranular areas, and 3 small agranular areas (IA1–3) at the limen base, resembling Brockhaus' early map (Brockhaus 1940) but differing from maps recognizing 5 agranular areas as in macaques (Carmichael and Price 1994; Ongur et al. 2003). We previously suggested that OP2, OP3, and OP5 correspond to the macaque PIVC, IDFP, and IDFM, respectively (Eickhoff et al. 2006a, b; Evrard et al. 2014;

[2]The literature emphasizes VENs over FNs, due to the lower number of FNs compared to VENs. The two neurons are co-mingled. They express the same immunohistochemical and gene markers (e.g., theta subunit of the GAB receptor) and have the same projections. While their morphology is distinct, whether they have distinct functions is not known (Dijkstra et al. 2016).

[3]The Jülich Brain Project uses an opercular (OP) nomenclature for their dorsal fundus areas. Recognizing that the dorsal fundus belongs to the insular cortex and is distinct from the secondary somatosensory cortex (S2), Bud and I prefer using an insular nomenclature (e.g., IDFP instead of OP3) (Evrard et al. 2014).

Evrard 2019), while OP7 and OP9 may represent an expanded and subdivided IDFA, with OP9 possibly being hominin-specific. The comparison between the other human areas (IGs, IDs, and IAs) with the macaque areas remains to be determined. The human insulas Bud and I collected are now under comparative analysis. Preliminary observations confirm, as in macaques and the Jülich map, multiple sharply delimited areas (Fig. 2a). We expect a more differentiated anterior insula in humans, reflecting its hyperallometric expansion (Bauernfeind et al. 2013). Our ongoing examination suggests that the VEN and FN in humans occur in several distinct frontoinsular areas, unlike the single VEN area in macaques (Fig. 2b) (Horn et al. 2017; Horn 2020).

4 Primary Interoceptive Cortex: A Representation of the Interoceptive Neuraxis

In our earlier model, the macaque dorsal fundus was divided into two domains: IDFP and the "old" IDFA, interpreted primarily as gustatory based on earlier macaque recordings (Ogawa et al. 1989; Yaxley et al. 1990; Rolls 2005; Evrard 2015). In contrast, the human dorsal fundus is partitioned into at least three domains – IDFP, IDFM, and IDFA – with human IDFP (OP3) and IDFM (OP5) homologous to macaque IDFP and old IDFA, respectively, and with human IDFA, absent in macaques, being associated with higher-order integrations supporting "homeostatic sentience" and "global emotional moments."

However, in a recent fMRI study using comparable paradigms in anesthetized macaques and humans, my colleague Renée Hartig and I found robust taste-evoked BOLD responses in a middle fundus region (IDFM) and, more variably, at the anterior tip (Fig. 3a) (Hartig et al. 2017, 2023; Hartig 2019). Similar clusters were found in awake macaques (Kaskan et al. 2019); where the anterior cluster, modulated by thirst, likely encodes taste value or expectation as well (Veldhuizen et al. 2011). These findings led us to revise the macaque model, splitting old IDFA into a middle gustatory area (IDFM), similar to human IDFM, and a smaller anterior IDFA, possibly a primordial form of the expanded human IDFA representing homeostatic value (Fig. 1c) (Evrard 2021).

This updated IDFP–IDFM–IDFA sequence in macaques may reflect a cortical map of hierarchical interoceptive integration following the SpL1–nTS–PB sequence. Whether the "primary interoceptive cortex" should be limited to IDFP and IDFM (posterior fundus) or extended to the full posteroanterior fundus in both species remains an open question, though a conservative approach favors restricting it to regions with di-synaptic ascending projections. Importantly, the SpL1–nTS–PB representation could mirror Bud's posterior-to-mid-to-anterior insular integration model (Craig 2009a, b), where each insular stage reflects a corresponding step in the subcortical interoceptive hierarchy. Supporting evidence lies in the topography of SpL1, nTS, and PB projections to VMpo and VMb, and homotopic projections to

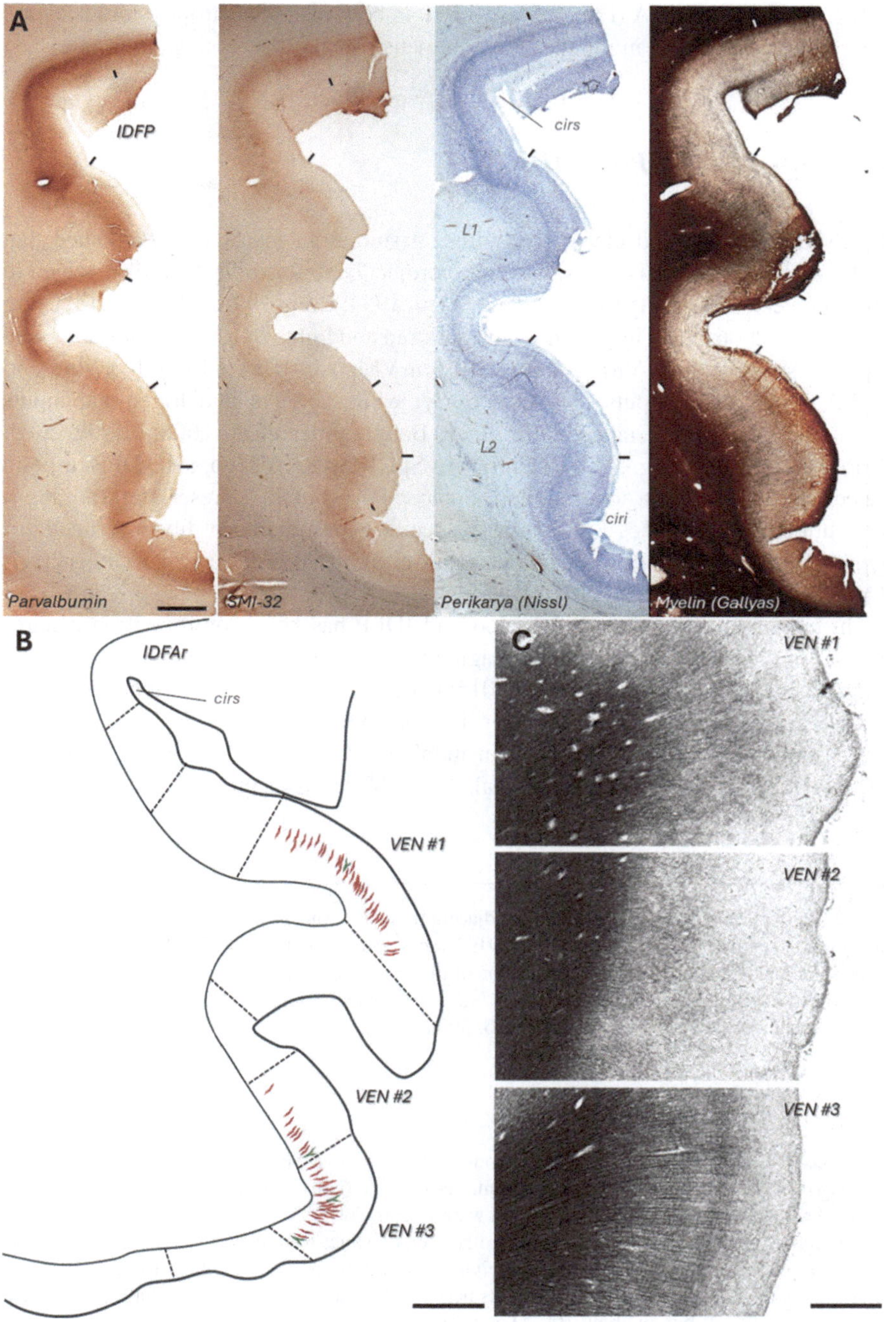

Fig. 2 (**a**) Photomicrographs of a consecutive series of coronal sections of a human posterior insula stained with four distinct methods. The tick marks indicate architectonic boundaries. In the preliminary view, only IDFP is named. The photomicrographs and the figure panel were originally prepared by A.D. (Bud) Craig and later completed by HCE. (**b**) Chart of the distribution of von Economo neurons (red) and fork neurons (green) and of the position of architectonic boundaries in a coronal section of the anterior insula in one human subject. (**c**) Photomicrographs of subregions of a

IDFP, IDFM, and IDFA (Fig. 3b), as well as in convergent electrophysiological and neuroimaging patterns in monkeys and humans.

4.1 SpL1 → VMPo → IDFP

Bud's tract-tracing and electrophysiological studies in macaques established that SpL1 and its trigeminal extension (Sp5C) project contralaterally to VMpo, which in turn projects ipsilaterally to IDFP (Craig et al. 1994; Craig 1996, 2004b, 2014; Craig and Zhang 2006). (In Fig. 3, follow the dark red and light red arrows from SpL1 and Sp5c, respectively, to VMpo and then from VMpo to IDFP.) In both VMpo and IDFP, lumbosacral inputs map posteriorly, while cervical and trigeminal inputs terminate anteriorly (Craig 2004b, 2014; Baumgartner et al. 2006; Hartig 2019; Hartig et al. 2019) (Fig. 3b, c). Neurons in SpL1, Sp5C, VMpo, and IDFP respond selectively to distinct thermal and nociceptive stimuli,[4] preserving the high-resolution "labeled lines" of modality-specific small-diameter fibers, crucial for generating discrete bodily sensations (Han et al. 1998; Andrew and Craig 2001; Craig 2002, 2004a; Hartig 2019; Hartig et al. 2019).

In humans, the existence of VMpo and IDFP has been confirmed via electrophysiology, microstimulation, and imaging (Craig et al. 2000; Blomqvist et al. 2000; Iannetti et al. 2005; Segerdahl et al. 2015; Chien et al. 2017).[5] IDFP, localized in the dorsal fundus capping the posterior insular gyri (Fig. 3d), displays the same posterior-to-anterior somatotopy as in monkeys (Olausson et al. 2002; Hua et al. 2005; Brooks et al. 2005; Mazzola et al. 2009). VMpo is disproportionately larger in

Fig. 2 (continued) myelin-stained section adjacent to the section shown in panel B, illustrating the myelination pattern in each of the three VEN areas illustrated in panel B. *Abbreviations*: *cirs* superior circular sulcus, *IDFAr* rostral portion of the anterior dorsal fundus area of the insula, *IDFP* posterior dorsal fundus area of the insula, *L1–2* posterior long gyri 1–2, *VEN #1–3* von Economo neuron areas #1–3. *Scale bars*: 1 mm in **a**–**b**, 300 μm in **c**

[4]COOL, HPC, and NS SpL1 STT cells account for ~90% of all SpL1 STT cells in macaque, with the remaining 10% including WARM, histamine-selective (itch), mustard oil-selective cells and a small fraction of "unidentified" cells (Andrew and Craig 2001). Most other projecting SpL1 cells are spinobulbar, including COOL, HPC, and NS cells distinct from the STT cells, cells receiving muscle afferents, and most likely cells receiving visceral afferents that remain to be explored in monkeys (Craig 2015). In rodents, the vagus nerve is the main source of visceral afferences, with the spinal cord contributing about 10% (Janig 2022).

[5]In a fraction of the human literature, VMpo remains either ignored (Blomqvist and Evrard 2024; Mandonnet et al. 2024), or misinterpreted as being indistinct from VP (Lenz et al. 2024; Evrard and Blomqvist 2025) or the anterior pulvinar nucleus (Vartiainen et al. 2016), which does not receive SpL1 projections (Craig 2004b; Dum et al. 2009).

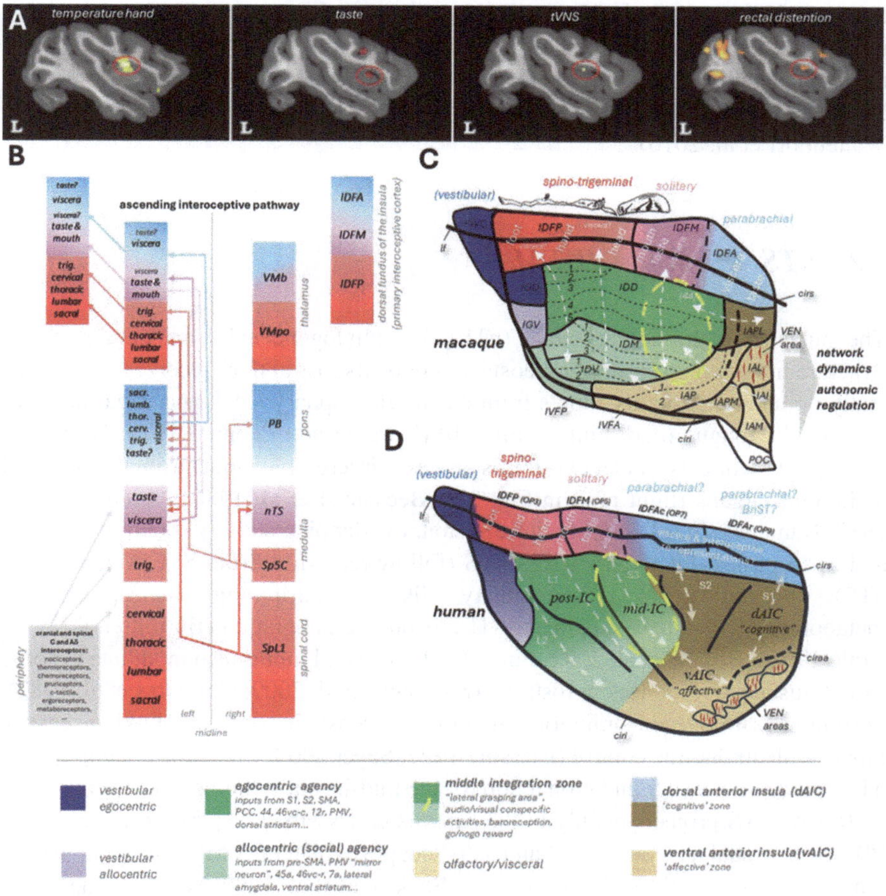

Fig. 3 (**a**) Distribution of BOLD signal responses to distinct interoceptive stimuli in one representative macaque brain ($p < 0.05$ uncorrected). (**b**) Schematic of the topography of projections in the macaque ascending interoceptive pathway. Each box represents one level of the neuraxis, from the spinal cord to the cerebral cortex. The left and right sides are shown to illustrate also projections crossing the midline. The name of the structures is shown in the boxes on the right side (e.g., SpL1) and the topographic organization within each structure is shown in boxes on the left side (e.g., sacral to cervical, from posterior to anterior SpL1). Lines with arrows indicate projections ascending from a caudal level (e.g., SpL1) to a more rostral level (e.g., nTS). To avoid cluttering, projections are shown mainly from/to one side, chosen arbitrarily. There is no known asymmetry in the overall organization of the ascending pathway. The color code is kept consistent to represent the overall representation of neuraxis level in the thalamus and in cortex. The figure highlights three main projection streams: In dark red, SpL1 projects bilaterally throughout nTS, contralaterally to PB (with an inverted topography, foot/lumbar anterior to hand/cervical posterior) and contralaterally to VMpo (foot/lumbar posterior to hand/cervical anterior). In lighter red, Sp5C projects bilaterally to nTS and PB (posterior to SpL1's cervical representation), contralaterally to VMpo (anterior to SpL1's cervical representation). The SpL1 and Sp5C representations in VMpo project homotopically to the SpL1 and Sp5C representations of ipsilateral IDFP. In light pink, the rostral gustatory portion of nTS projects ipsilaterally mainly to caudolateral VMb and sparsely to PB. The caudolateral region of VMb receiving direct inputs from nTS projects to IDFM. In dark pink, the caudal visceral portion of nTS projects ipsilaterally to PB; and, in blue, PB mainly projects to the rostromedial portion of VMb, which projects to IDFA. **C-D:** Flat map models of the functional organization of the macaque (**c**) and human (**d**) insula. The color code in the dorsal fundus areas

humans (Strigo and Craig 2016), likely reflecting a broader range of interoceptive modalities activating IDFP, thereby enabling a more refined sense of the "material me" and contributing to richer subjective experience (Drzezga et al. 2001; Henderson et al. 2007; Bjornsdotter et al. 2009; Spetter et al. 2014; Craig 2015; Wattendorf et al. 2016).

4.2 nTS → VMb → IDFM

The nucleus of the solitary tract (nTS) is a phylogenetically ancient brainstem structure integrating sensory homeostatic inputs. Its rostral third receives topographically organized gustatory inputs from facial, glossopharyngeal, and vagal innervations of the tongue (dark pink in Fig. 3b) (Norgren and Leonard 1973; Beckstead et al. 1980), while its caudal two-thirds process viscerotopic inputs from thoracic and abdominal organs (light pink in Fig. 3b) (Beckstead et al. 1980; Altschuler et al. 1989; Ran et al. 2022). Despite this division, overlapping activity is common; SpL1 and Sp5c project bilaterally across nTS (follow red arrow from SpL1 and Sp5c to nTS), without a clear topography, likely contributing nociceptive and metaboreceptive signals supporting viscerosomatic integration (Burton et al. 1979; Craig 1995a; Boscan et al. 2002; Janig 2022). Visceral inputs also modulate to some extent gustatory responses in rostral nTS (Huang et al. 2009; Contreras et al. 1982). Forming a primitive substrate for computing homeostatic needs and linking sensory inputs with autonomic control (Blessing 1997; Saper 2002; Craig 2002; Janig 2022), nTS could serve as a foundational hub for the mid-insula's "homeostatic sentience."

Rostral nTS projects mainly to the caudolateral VMb and sparsely to the caudal PB, while caudal nTS targets PB and a distinct posterolateral VMb zone (follow dark pink arrow from nTS to VMb) (Fig. 3b) (Beckstead et al. 1980; Ito and Craig 2008; Strigo and Craig 2016). In primates, taste signals reach VMb mostly monosynaptically, while visceral inputs are relayed via PB, though partial overlap exists. Taste

Fig. 3 (continued) (IDFP, IDFM, and IDFA) uses the same color code as in Fig. 3a. In monkeys, IDFP receiving inputs from SpL1 and Sp5c via VMpo, with a posterior (foot/lumbar) to anterior (trigeminal/face) somatotopy. Sparse visceral inputs from the spinal cord may reach IDFP, although most SpL1 visceral afferences in monkeys stop in PB. IDFM receives inputs from nTS, mainly from the rostral gustatory nTS, with possibly some visceral inputs as well; whether these inputs are co-mingled or separated along a gradient remains to be explored. In monkeys, IDFA receives inputs from PB via VMb. Whether IDFA in humans also receives inputs from PB via VMb as well as possibly other inputs remains hypothetical. For the rest of the insula, see the color code at the bottom of the figure and see the text. *Abbreviations*: *IDFA, IDFM, IDFP* anterior, middle, and posterior dorsal fundus area of the insula, respectively, *nTS* solitary tract nucleus, *PB* parabrachial nucleus, *Sp5c* trigeminal lamina 1, *SpL1* spinal lamina 1, *VMb* ventral medial nucleus of the thalamus, basal part, *VMpo* ventral medial nucleus of the thalamus, posterior part. For other abbreviations, see the legend of Fig. 1. (Panels **b–d** were adapted from similar figures in Evrard (2021), with permission)

activity is well documented in VMb of both monkeys (VPMpc) and humans (ventral caudal parvicellular interne; Vcpci) (Pritchard et al. 1989; Lenz et al. 1997). Posterolateral VMb neurons selectively encode gustatory, thermal, and mechanical oral stimuli, with sparse anticipatory modulation (Yaxley et al. 1990; Verhagen et al. 2004; Rolls 2005), and project to IDFM, putatively forming a gusto-visceral gradient (Fig. 3b, c). IDFM itself responds robustly to oral and taste stimuli in macaques and humans (Ogawa et al. 1989; Jantsch et al. 2005; Ifuku et al. 2006; Small 2010; Trulsson et al. 2010; Nakamura et al. 2011; Hartig 2019; Avery et al. 2019; Hartig et al. 2023), while also displaying overlapping or nearby responses to vagal and rectal stimuli in macaques (Fig. 3a) (Hartig et al. 2017; Hartig 2019) and attention to heartbeat in humans (Avery et al. 2017). Gustatory and visceral activations consistently localize to the middle of the dorsal fundus, aligning with Bud's mid-insular integration stage (Fig. 3d) (Craig 2009a, 2015). This region is further implicated in processes such as craving (DelParigi et al. 2005), thirst (Becker et al. 2017), pain unpleasantness (Schreckenberger et al. 2005), uncertainty (Rubio et al. 2015), vividness (Todd et al. 2012), emotional processing (Duerden et al. 2013), repetition suppression (Zweynert et al. 2011), and encoding of time intervals (Wittmann et al. 2010). While the boundary between gustatory and visceral representations remains to be fully resolved, the evidence suggests that IDFM, in addition to purely gustatory signals, could embody homeostatic needs pre-processed in nTS and supports the construction of homeostatic sentience. Bilateral SpL1 projections to nTS (Craig 1995a) and transcallosal communication (Craig 2009a, b) could contribute to the bilateral mid-insular activation commonly observed.

4.3 PB → VMb → IDFA

The parabrachial nucleus (PB) integrates homeostatic, interoceptive, and affective information (Saper 2002), receiving dense inputs from SpL1 (and Sp5c) and nTS, including nociceptive, visceral, and chemosensory signals (Fig. 3b) (Saper 2002; Craig 1995b, 2015; Beckstead et al. 1980; Kitamura et al. 2001). SpL1 projects contralaterally along the rostrocaudal extent of PB with an inverted anteroposterior somatotopy relative to VMpo (Craig 1995b, 2015). Rostral nTS projects sparsely to caudal PB, while caudal nTS densely innervates the entire ipsilateral PB, mainly in lateral PB, with possible separation from SpL1 afferences that also terminate in lateral PB (Beckstead et al. 1980; Herbert et al. 1990; Craig 1995a; Feil and Herbert 1995; Hermanson and Blomqvist 1996, 1997). PB, in turn, projects to the rostromedial VMb, which encodes vagal and cardiorespiratory signals and projects predominantly to IDFA, with moderate input to IDFM (Pritchard et al. 2000; Ito and Craig 2008; Strigo and Craig 2016). PB's inverted somatotopy may account for secondary gustatory activation at the rostral tip of VMb and IDFA, distinct from the main taste zone in VMb and IDFM (Fig. 3b, c) (Craig 2015, 2018; Strigo and Craig 2016; A.D. Craig personal communication).

PB serves as a dynamic integrator, amplifying salient interoceptive signals to prioritize bodily urgency and link physiological states to motivational and affective processes (Saper 2002; Craig 2003b; Palmiter 2018). It likely achieves this through integrating ascending afferents with descending inputs from AIC, ACC, amygdala, BnST, and hypothalamus (Moga et al. 1990; Chiba et al. 2001; Evrard 2018; Grady et al. 2020). This has been proposed to provide PB with real-time monitoring of global brain states and to enable it to influence network dynamics (e.g., sleep-wake transitions) underlying cortical arousal and internally driven cognitive processes (Munk et al. 1996; Parvizi and Damasio 2003; Fuller et al. 2011; Ramirez-Villegas et al. 2020).

In humans, IDFA – often considered part of the dAIC – is consistently activated during bodily, emotional, and cognitive feelings, giving rise to "global emotional moments" through the sequential "cinemascopic" registration of such moments across time (Craig 2009a, 2015; Harrison et al. 2010; Zaki et al. 2012; Terasawa et al. 2013). The AIC contributes to global brain dynamics by switching functional networks to optimize cognitive control and metabolic efficiency (Menon and Uddin 2010; Ham et al. 2013). More recently, dAIC has been proposed to gate the access of sensory information to consciousness by weighing on subsequent prefrontal activation bound to perception (Warnaby et al. 2016; Wu et al. 2019; Huang et al. 2021; Molnar-Szakacs and Uddin 2022), a critical step anticipated by Bud's model.

The role of PB as a salience amplifier aligns with the salience detection and gating function of human IDFA (and dAIC). Although the presence of a dorsal salience network in macaques is debated (Touroutoglou et al. 2016), monkey IDFA is recruited during both internal state processing and cognitive appraisal (Wang et al. 2015; Kaskan et al. 2019; Yang et al. 2022), suggesting it may represent a precursor of the hyperallometrically expanded human IDFA (Bauernfeind et al. 2013). In our model, the human IDFA caps the dorsal fundus, encompassing the anterior and middle short gyri, the frontoinsula, and possibly part of the posterior short gyrus (Fig. 3d). The presence of VENs in the ventral AIC in both macaques and humans (Fig. 3c, d) suggests that the anterior insula expanded rostrally in a fan-like manner with the VEN region as a pivot, which itself may have expanded (Figs. 1e and 2b). A systematic analysis of the dAIC is needed to identify potential functional and connectional loci unique to humans, possibly involving additional subcortical representations anterior to the putative PB-recipient zone of monkey IDFA – one of which, as Bud highlighted, could be the BnST (Craig 2018).

5 Latticework Model of Insular Integration

The "latticework" model of insular integration extends Bud's posterior-to-mid-to-anterior model by incorporating two key hypotheses: first, that the dorsal fundus, or primary interoceptive cortex, maps the hierarchical interoceptive neuraxis; and second, that a serial re-representation occurs across the dorsoventrally stacked dysgranular stripes (Evrard 2021). Figure 3c, d illustrates this model on flat maps of the macaque and human insula, respectively, with dorsal fundus subdivisions color-coded as in Fig. 3b. The ascending interoceptive pathway and the dorsal fundus, as detailed earlier, provide the organizational foundation for the entire insular cortex.

5.1 Posterior Insula

In monkeys, the dysgranular stripes form two integration domains (Fig. 3c). The dorsal "egocentric" domain (dark green) at the level of IDFP receives somatotopically aligned projections from somatosensory and higher-order proprioceptive cortices, supporting interoceptive-propriomotor integration (Evrard 2019). The ventral "allocentric" domain (light green) connects with audio- and visuo-motor regions, likely processing environmental and social cues (Evrard 2019). This dorsal-ventral dichotomy parallels the body- vs. world-centered integration seen in the IGD and IGV granular areas. In humans, microstimulation and recordings reveal a similar posteroanterior gradient of vestibular and somatic activities across the posterior long gyri, roughly preserved along the dorsoventral axis (Stephani et al. 2011; Mazzola et al. 2019; Duong et al. 2023) (Fig. 3d).

5.2 Middle Insula

Ventral to IDFM, the dorsal "egocentric" region hosts a "lateral grasping network" that integrates homeostatic states and motor execution into vitality forms proposed to reflect internal feelings (Borra et al. 2017; Di Cesare et al. 2020, 2021). A similar representation is found at the junction of the third short gyrus in humans, ventral to IDFM (Di Cesare et al. 2021) (Fig. 3d). In monkeys, the ventral middle dysgranular region supports affective audio-visual communication, such as lip-smacking, and its gray matter volume correlates with social network size (Remedios et al. 2009; Caruana et al. 2011; Ku et al. 2011; Testard et al. 2022; Simone et al. 2025). This region also responds to convergent baroreceptive and somatosensory inputs and during reward-based go/no-go tasks (Zhang et al. 1998; Asahi et al. 2006). Collectively, these findings suggest that the middle insula integrates self- and world-centered actions with homeostatic needs, processed via nTS and relayed through VMb to IDFM. In humans, taste and vitality-related activities overlap along the central insular sulcus, with the taste axis likely positioned slightly posterior to the vitality axis (Porcu et al. 2020; Avery et al. 2021; Di Cesare et al. 2021).

5.3 Anterior Insula

Dorsal Anterior Insula Human dAIC activity is associated with attention to emotion and interoception, integration of bodily feelings, emotions, and thoughts, self-reflection, feeling-of-knowing, error detection, word processing, visual awareness, and decision-making (Craig 2015). It also contributes to managing cognitive and multiple-demand networks (Sridharan et al. 2008; Duncan 2010), regulating information capacity and timing (Wilk et al. 2012; Wu et al. 2019), with possible implications in salience gating (Smith et al. 2015; Warnaby et al. 2016; Huang et al. 2021), aligning with the role of PB in gating interoceptive inputs to the forebrain. In macaques, neuroimaging and electrophysiology implicate dAIC in cognition (Wang et al. 2015; Yang et al. 2022) and in a multiple-demand network (Mitchell et al. 2016); its involvement in cognitive appraisal predates humans, consistent with the notion that cognition evolved around the representation of homeostatic and metabolic needs (Evrard 2018).

Ventral Anterior Insula Human vAIC is primarily linked to emotional intensity and perception, social empathy, self-other distinction, and the prediction and adaptation of affective bodily states (Harrison et al. 2010; Critchley and Seth 2012; Zaki et al. 2012; Craig 2015). The presence of VENs and a ventral salience network in both human and macaque vAIC, combined with the induction of autonomic and emotional motor responses by electrical stimulation (Kaada 1951; Caruana et al. 2011; Evrard et al. 2012; Touroutoglou et al. 2016), suggests evolutionary continuity, despite the conspicuous expansion of the human AIC and its likely greater number of VEN areas compared to the smaller macaque IDFA and AIC.

In macaques, VENs and FNs in vAIC and ACC project massively to PAG, PB, and nTS (Evrard 2018; Chavez et al. 2023), hereby partly closing the highest hierarchical loop of the sensorimotor homeostatic system formed by ascending interoceptive and descending autonomic pathways. Other insular or cortical projections likely regulate spinal sympathetic efferences as well, in line with the idea that the insula is divided along a spino-medullar axis in which dorsal "sensory" fundi feed integrated output to the ventral "motor" or regulatory mound and VEN area protuberance (Evrard 2019). From a predictive coding perspective, vAIC projections may provide PB with real-time cortical state monitoring for top-down autonomic regulation, attenuating interoceptive prediction errors and reducing free-energy (Parvizi and Damasio 2003; Seth and Friston 2016; Ramirez-Villegas et al. 2020). Notably, dAIC is the last cortical region active before anesthesia-induced unresponsiveness (Warnaby et al. 2016; Huang et al. 2021), while vAIC and PB are functionally connected in healthy humans but disconnected in coma (Fischer et al. 2016). This pattern suggests that AIC operates much like a homeostatically-governed waterwheel system: driven by ascending interoceptive salience while simultaneously gating the propagation of interoceptive afferences along the neuraxis – including, crucially, PB – and forward perceptual representations in prefrontal cortex. This dual role would ensure the continuous and adaptive construction of subjective awareness.

6 Conclusion

This chapter proposes a refinement of our model of the primate insula, grounded in neuroanatomical and neurophysiological details that reveal both shared features between the human and macaque insula and the remarkable expansion of the human insula – likely supporting uniquely human cognitive and perceptual capacities. The prospect of a small, primal IDFA in macaques – if confirmed – may provide a precious opportunity to probe the roots of interoceptive integration, which in humans becomes the very substrate of feelings.

Bud's work was never about accumulating facts for their own sake or merely extending existing models. Instead, it charted a bold trajectory, providing deeply insightful evidence that systematically challenged dogma and led to a paradigm shift with profound implications for understanding the human condition. His contributions extend and refine the foundational works of Weber, Langley, James, and Sherrington, and I fully concur with my colleague that his legacy rivals, if not surpasses, that of the seminal discoveries on the molecular basis of interoception (Blomqvist 2023). More than two decades after his landmark *Nature Reviews Neuroscience* paper, Bud's scientific and intellectual legacy continues to shape the field, guiding the community's deepening engagement with interoception. His work will inspire for decades, with certainly still much to "digest" and appreciate in his writings, beyond the principal tenets of his theories.

While some of Bud's proposals may have appeared to be rather non-compromising, he was well aware that much remains to be done to further elucidate interoception, the insular cortex, and the nature of the self. While strongly advocating – after having provided implacable evidence – for the inclusion of pain and temperature in an expanded definition of interoception and for the role of interoception and the insula in the emergence of subjective feelings, Bud was not dogmatic; he would readily accept rigorous scientific evidence from others and adapt his thinking and interpretations. While recognizing his models as "work in progress," Bud had the merit to stick his neck out with challenging but lucid proposals that can be tested in the laboratory. His models continue to be supported by an increasing amount of new data, which hopefully will also lead to refined concepts that Bud would have gladly embraced if supported by strong and unflawed evidence. In his book, he listed many questions that remain to be addressed, for example, regarding the fine topographic organization the primary interoceptive cortex beyond what is currently known from tract-tracing and functional data; or regarding the overlap or separation of functional loci in dAIC and vAIC, which are known to contain multiple architectonic areas; or regarding how the two different sides of the insula interacts to contribute to a unified sense of self; and whether every feeling and thought must have an autonomic correlate and homeostatic value.

In recent years, Bud and I continued our discussions, too rarely, with the shared goal of publishing a remarkable collection of unpublished neuroanatomical and neurophysiological data. This effort is ongoing, with the hope that further characterization of these pathways will continue to illuminate "what goes on in these amazing bodies and in our remarkable brains" (Craig 2015).

Acknowledgments The author thanks the editors and Prof. Alan Blomqvist for their comments on an earlier version of this text.

Funding The author's research is funded by the Shanghai Municipal Science & Technology Major Project (2019SHZDZX02), the National Natural Science Foundation of China Mianshang Project (32271049), and the Belgian Fonds Wetenschappelijk Onderzoek-Vlaanderen (G0E0520N).

References

Afif A, Mertens P (2010) Description of sulcal organization of the insular cortex. Surg Radiologic Anatomy 32:491–498. https://doi.org/10.1007/s00276-009-0598-4

Afif A, Bouvier R, Buenerd A et al (2007) Development of the human fetal insular cortex: study of the gyration from 13 to 28 gestational weeks. Brain Struct Funct 212:335–346. https://doi.org/10.1007/s00429-007-0161-1

Allman JM, Tetreault NA, Hakeem AY et al (2010) The von Economo neurons in frontoinsular and anterior cingulate cortex in great apes and humans. Brain Struct Funct 214:495–517. https://doi.org/10.1007/s00429-010-0254-0

Altschuler SM, Bao XM, Bieger D et al (1989) Viscerotopic representation of the upper alimentary tract in the rat: sensory ganglia and nuclei of the solitary and spinal trigeminal tracts. J Comp Neurol 283:248–268. https://doi.org/10.1002/cne.902830207

Amiez C, Sallet J, Giacometti C et al (2023) A revised perspective on the evolution of the lateral frontal cortex in primates. Sci Adv 9:1–14. https://doi.org/10.1126/sciadv.adf9445

Amunts K, Mohlberg H, Bludau S, Zilles K (2020) Julich-brain: a 3D probabilistic atlas of the human brain's cytoarchitecture. Science 369:988–992. https://doi.org/10.1126/science.abb4588

Andrew D, Craig AD (2001) Spinothalamic lamina I neurons selectively sensitive to histamine: a central neural pathway for itch. Nat Neurosci 4:72–77. https://doi.org/10.1038/82924

Asahi T, Uwano T, Eifuku S et al (2006) Neuronal responses to a delayed-response delayed-reward go/nogo task in the monkey posterior insular cortex. Neuroscience 143:627–639. https://doi.org/10.1016/j.neuroscience.2006.08.008

Avery JA, Gotts SJ, Kerr KL et al (2017) Convergent gustatory and viscerosensory processing in the human dorsal mid-insula. Hum Brain Mapp 38:2150–2164. https://doi.org/10.1002/hbm.23510

Avery JA, Liu AG, Ingeholm JE et al (2019) Taste quality representation in the human brain. J Neurosci. https://doi.org/10.1523/JNEUROSCI.1751-19.2019

Avery JA, Liu AG, Ingeholm JE et al (2021) Viewing images of foods evokes taste quality-specific activity in gustatory insular cortex. Proc Natl Acad Sci USA 118:e2010932118. https://doi.org/10.1073/pnas.2010932118

Bauernfeind AL, de Sousa AA, Avasthi T et al (2013) A volumetric comparison of the insular cortex and its subregions in primates. J Hum Evol 64:263–279. https://doi.org/10.1016/j.jhevol.2012.12.003

Baumgartner U, Tiede W, Treede RD, Craig AD (2006) Laser-evoked potentials are graded and somatotopically organized anteroposteriorly in the operculoinsular cortex of anesthetized monkeys. J Neurophysiol 96:2802–2808. https://doi.org/10.1152/jn.00512.2006

Becker CA, Flaisch T, Renner B, Schupp HT (2017) From thirst to satiety: the anterior mid-cingulate cortex and right posterior insula indicate dynamic changes in incentive value. Front Hum Neurosci 11:234. https://doi.org/10.3389/fnhum.2017.00234

Beckstead RM, Morse JR, Norgren R (1980) The nucleus of the solitary tract in the monkey: projections to the thalamus and brain stem nuclei. J Comp Neurol 190:259–282. https://doi.org/10.1002/cne.901900205

Bjornsdotter M, Loken L, Olausson H et al (2009) Somatotopic organization of gentle touch processing in the posterior insular cortex. J Neurosci 29:9314–9320. https://doi.org/10.1523/JNEUROSCI.0400-09.2009

Blessing W (1997) The lower brainstem and bodily homeostasis. Oxford University Press, Oxford

Blomqvist A (2023) Pain and temperature, and human awareness: the legacy of Bud Craig. Temperature (Austin) 10:395–401. https://doi.org/10.1080/23328940.2023.2265946

Blomqvist A, Dostrovsky JO (2025) The life and legacy of Arthur D. Craig, Jr: a pioneer in interoception. In: Wittmann M, Strigo IA, Simmons AN (eds) The bodily self, emotion, and subjective time: exploring interoception through the contributions of A.D. (Bud) Craig. Springer

Blomqvist A, Evrard HC (2024) The thalamic projection of pain sensations to the posterior dorsal fundus in the insula: comment on Mandonnet et al. Pain 165:e15–e16. https://doi.org/10.1097/j.pain.0000000000003164

Blomqvist A, Zhang ET, Craig AD (2000) Cytoarchitectonic and immunohistochemical characterization of a specific pain and temperature relay, the posterior portion of the ventral medial nucleus, in the human thalamus. Brain 123(Pt 3):601–619. https://doi.org/10.1093/brain/123.3.601

Blomqvist A, Evrard HC, Dostrovsky JO et al (2023) A. D. (Bud) Craig, Jr. (1951–2023). Nat Neurosci 26:1835–1836. https://doi.org/10.1038/s41593-023-01463-9

Borra E, Gerbella M, Rozzi S, Luppino G (2017) The macaque lateral grasping network: a neural substrate for generating purposeful hand actions. Neurosci Biobehav Rev 75:65–90. https://doi.org/10.1016/j.neubiorev.2017.01.017

Boscan P, Pickering AE, Paton JF (2002) The nucleus of the solitary tract: an integrating station for nociceptive and cardiorespiratory afferents. Exp Physiol 87:259–266. https://doi.org/10.1113/eph8702353

Brockhaus H (1940) Die Cyto- und Myeloarchitektonik des Cortex claustralis und des Claustrum beim Menschen. J Psychol Neurol 49:249–348

Brodmann K (1909) Vergleichende Lokalizationlehre der Grosshirnrinde. Johann Ambrosius Barth, Leipzig

Brooks JC, Zambreanu L, Godinez A et al (2005) Somatotopic organisation of the human insula to painful heat studied with high resolution functional imaging. NeuroImage 27:201–209. https://doi.org/10.1016/j.neuroimage.2005.03.041

Burton H, Jones EG (1976) The posterior thalamic region and its cortical projection in New World and Old World monkeys. J Comp Neurol 168:249–301. https://doi.org/10.1002/cne.901680204

Burton H, Craig AD, Poulos DA, Molt JT (1979) Efferent projections from temperature sensitive recording loci within the marginal zone of the nucleus caudalis of the spinal trigeminal complex in the cat. J Comp Neurol 183:753–777. https://doi.org/10.1002/cne.901830406

Butti C, Hof PR (2010) The insular cortex: a comparative perspective. Brain Struct Funct 214:477–493. https://doi.org/10.1007/s00429-010-0264-y

Butti C, Raghanti MA, Sherwood CC, Hof PR (2011) The neocortex of cetaceans: cytoarchitecture and comparison with other aquatic and terrestrial species. Ann N Y Acad Sci 1225:47–58. https://doi.org/10.1111/j.1749-6632.2011.05980.x

Carmichael ST, Price JL (1994) Architectonic subdivision of the orbital and medial prefrontal cortex in the macaque monkey. J Comp Neurol 346:366–402. https://doi.org/10.1002/cne.903460305

Caruana F, Jezzini A, Sbriscia-Fioretti B et al (2011) Emotional and social behaviors elicited by electrical stimulation of the insula in the macaque monkey. Curr Biol 21:195–199. https://doi.org/10.1016/j.cub.2010.12.042

Chavez G, Horn F, Saleh TO, Evrard HC (2023) Von Economo neuron projections to brainstem autonomic and interoceptive nuclei in the macaque monkey

Chiba T, Kayahara T, Nakano K (2001) Efferent projections of infralimbic and prelimbic areas of the medial prefrontal cortex in the Japanese monkey, Macaca fuscata. Brain Res 888:83–101

Chien JH, Korzeniewska A, Colloca L et al (2017) Human thalamic somatosensory nucleus (Ventral Caudal, Vc) as a locus for stimulation by INPUTS from tactile, noxious and thermal sensors on an active prosthesis. Sensors (Basel) 17. https://doi.org/10.3390/s17061197

Contreras RJ, Beckstead RM, Norgren R (1982) The central projections of the facial, glossopharyngeal, and vagus nerves: an autoradiographic study in the rat. Journal of Comparative Neurology 207:465–485. https://doi.org/10.1002/cne.902070406

Craig AD (1992) Spinal and trigeminal lamina I input to the locus coeruleus anterogradely labeled with Phaseolus vulgaris leucoagglutinin (PHA-L) in the cat and the monkey. Brain Res 584: 325–328. https://doi.org/10.1016/0006-8993(92)90915-v

Craig AD (1993) Propriospinal input to thoracolumbar sympathetic nuclei from cervical and lumbar lamina I neurons in the cat and the monkey. J Comp Neurol 331:517–530. https://doi.org/10.1002/cne.903310407

Craig AD (1995a) Supraspinal projections of lamina I neurons. In: Besson JM, Guilbaud G, Ollat H (eds) Forebrain areas involved in pain processing. John Libbey Eurotext, Paris, pp 13–25

Craig AD (1995b) Distribution of brainstem projections from spinal lamina I neurons in the cat and the monkey. J Comp Neurol 361:225–248. https://doi.org/10.1002/cne.903610204

Craig AD (1996) An ascending general homeostatic afferent pathway originating in lamina I. Prog Brain Res 107:225–242. https://doi.org/10.1016/s0079-6123(08)61867-1

Craig AD (2002) How do you feel? Interoception: the sense of the physiological condition of the body. Nat Rev Neurosci 3:655–666. https://doi.org/10.1038/nrn894

Craig AD (2003a) Pain mechanisms: labeled lines versus convergence in central processing. Annu Rev Neurosci 26:1–30. https://doi.org/10.1146/annurev.neuro.26.041002.131022

Craig AD (2003b) Interoception: the sense of the physiological condition of the body. Curr Opin Neurobiol 13:500–505

Craig AD (2004a) Lamina I, but not lamina V, spinothalamic neurons exhibit responses that correspond with burning pain. J Neurophysiol 92:2604–2609. https://doi.org/10.1152/jn.00385.2004

Craig AD (2004b) Distribution of trigeminothalamic and spinothalamic lamina I terminations in the macaque monkey. J Comp Neurol 477:119–148. https://doi.org/10.1002/cne.20240

Craig AD (2006) Retrograde analyses of spinothalamic projections in the macaque monkey: input to ventral posterior nuclei. J Comp Neurol 499:965–978. https://doi.org/10.1002/cne.21154

Craig AD (2009a) How do you feel–now? The anterior insula and human awareness. Nat Rev Neurosci 10:59–70. https://doi.org/10.1038/nrn2555

Craig ADB (2009b) Emotional moments across time: a possible neural basis for time perception in the anterior insula. Philos Trans R Soc Lond Ser B Biol Sci 364:1933–1942. https://doi.org/10.1098/rstb.2009.0008

Craig AD (2010) The sentient self. Brain Struct Funct 214:563–577. https://doi.org/10.1007/s00429-010-0248-y

Craig AD (2011) Significance of the insula for the evolution of human awareness of feelings from the body. Ann N Y Acad Sci 1225:72–82. https://doi.org/10.1111/j.1749-6632.2011.05990.x

Craig AD (2014) Topographically organized projection to posterior insular cortex from the posterior portion of the ventral medial nucleus (VMpo) in the long-tailed macaque monkey. J Comp Neurol. https://doi.org/10.1002/cne.23425

Craig AD (2015) How do you feel? An interoceptive moment with your neurobiological self. Princeton University Press

Craig AD (2018) Central neural substrates involved in temperature discrimination, thermal pain, thermal comfort, and thermoregulatory behavior. Handb Clin Neurol 156:317–338. https://doi.org/10.1016/B978-0-444-63912-7.00019-9

Craig AD, Blomqvist A (2002) Is there a specific lamina I spinothalamocortical pathway for pain and temperature sensations in primates? J Pain 3:95–101; discussion 113–4. https://doi.org/10.1054/jpai.2002.122953

Craig AD, Kniffki KD (1985) Spinothalamic lumbosacral lamina I cells responsive to skin and muscle stimulation in the cat. J Physiol 365:197–221. https://doi.org/10.1113/jphysiol.1985.sp015767

Craig AD, Zhang ET (2006) Retrograde analyses of spinothalamic projections in the macaque monkey: input to posterolateral thalamus. J Comp Neurol 499:953–964. https://doi.org/10.1002/cne.21155

Craig AD, Heppelmann B, Schaible HG (1988) The projection of the medial and posterior articular nerves of the cat's knee to the spinal cord. J Comp Neurol 276:279–288. https://doi.org/10.1002/cne.902760210

Craig AD, Bushnell MC, Zhang ET, Blomqvist A (1994) A thalamic nucleus specific for pain and temperature sensation. Nature 372:770–773. https://doi.org/10.1038/372770a0

Craig AD, Krout K, Zhang ET (1995) Cortical projections of VMpo, a specific pain and temperature relay in primate thalamus. Soc Neurosci Abstr 21:1165

Craig AD, Chen K, Bandy D, Reiman EM (2000) Thermosensory activation of insular cortex. Nat Neurosci 3:184–190. https://doi.org/10.1038/72131

Critchley H, Seth A (2012) Will studies of macaque insula reveal the neural mechanisms of self-awareness? Neuron 74:423–426. https://doi.org/10.1016/j.neuron.2012.04.012

DelParigi A, Chen K, Salbe AD et al (2005) Sensory experience of food and obesity: a positron emission tomography study of the brain regions affected by tasting a liquid meal after a prolonged fast. NeuroImage 24:436–443. https://doi.org/10.1016/j.neuroimage.2004.08.035

Di Cesare G, Gerbella M, Rizzolatti G (2020) The neural bases of vitality forms. Natl Sci Rev 7: 202–213. https://doi.org/10.1093/nsr/nwz187

Di Cesare G, Marchi M, Lombardi G et al (2021) The middle cingulate cortex and dorso-central insula: a mirror circuit encoding observation and execution of vitality forms. Proc Natl Acad Sci USA 118:e2111358118. https://doi.org/10.1073/pnas.2111358118

Dijkstra AA, Lin LC, Nana AL et al (2016) Von Economo neurons and fork cells: a neurochemical signature linked to monoaminergic function. Cereb Cortex. https://doi.org/10.1093/cercor/bhw358

Drzezga A, Darsow U, Treede RD et al (2001) Central activation by histamine-induced itch: analogies to pain processing: a correlational analysis of O-15 H2O positron emission tomography studies. Pain 92:295–305

Duerden EG, Arsalidou M, Lee M, Taylor MJ (2013) Lateralization of affective processing in the insula. NeuroImage 78:159–175. https://doi.org/10.1016/j.neuroimage.2013.04.014

Dum RP, Levinthal DJ, Strick PL (2009) The spinothalamic system targets motor and sensory areas in the cerebral cortex of monkeys. J Neurosci 29:14223–14235. https://doi.org/10.1523/JNEUROSCI.3398-09.2009

Duncan J (2010) The multiple-demand (MD) system of the primate brain: mental programs for intelligent behaviour. Trends Cogn Sci 14:172–179. https://doi.org/10.1016/j.tics.2010.01.004

Duong A, Quabs J, Kucyi A et al (2023) Subjective states induced by intracranial electrical stimulation matches the cytoarchitectonic organization of the human insula. Brain Stimul 16: 1653–1665. https://doi.org/10.1016/j.brs.2023.11.001

Eickhoff SB, Amunts K, Mohlberg H, Zilles K (2006a) The human parietal operculum. II. Stereotaxic maps and correlation with functional imaging results. Cereb Cortex 16:268–279. https://doi.org/10.1093/cercor/bhi106

Eickhoff SB, Weiss PH, Amunts K et al (2006b) Identifying human parieto-insular vestibular cortex using fMRI and cytoarchitectonic mapping. Hum Brain Mapp 27:611–621. https://doi.org/10.1002/hbm.20205

Engström M, Karlsson T, Landtblom A-M, Craig ADB (2014) Evidence of conjoint activation of the anterior insular and cingulate cortices during effortful tasks. Front Hum Neurosci 8:1071. https://doi.org/10.3389/fnhum.2014.01071

Evrard HC (2015) Insular cortex. In: Toga AW (ed) Brain Mapping: An Encyclopedic Reference, vol 2. Elsevier, pp 343–347. https://doi.org/10.1016/B978-0-12-397025-1.00060-0

Evrard HC (2018) Von Economo and fork neurons in the monkey insula, implications for evolution of cognition. Curr Op Behav Sci 21:182–190

Evrard HC (2019) The organization of the primate insular cortex. Front Neuroanat 13:43. https://doi.org/10.3389/fnana.2019.00043

Evrard HC (2021) Interoceptive integration in the primate insular cortex. In: Nguyen DK, Isnard J, Kahane P (eds) Insular Epilepsies. Cambridge University Press

Evrard HC, Blomqvist A (2025) Balancing the role of the ventral posterior thalamus in pain and temperature. J Neurophysiol 133:1014–1015. https://doi.org/10.1152/jn.00009.2025

Evrard HC, Forro T, Logothetis NK (2012) Von Economo neurons in the anterior insula of the macaque monkey. Neuron 74. https://doi.org/10.1016/j.neuron.2012.03.003

Evrard HC, Logothetis NK, Craig AD (2014) Modular architectonic organization of the insula in the macaque monkey. J Comp Neurol 522:64–97. https://doi.org/10.1002/cne.23436

Feil K, Herbert H (1995) Topographic organization of spinal and trigeminal somatosensory pathways to the rat parabrachial and Kölliker-Fuse nuclei. J Comp Neurol 353:506–528. https://doi.org/10.1002/cne.903530404

Fischer DB, Boes AD, Demertzi A et al (2016) A human brain network derived from coma-causing brainstem lesions. Neurology 87:2427–2434. https://doi.org/10.1212/WNL.0000000000003404

Fuller PM, Sherman D, Pedersen NP et al (2011) Reassessment of the structural basis of the ascending arousal system. J Comp Neurol 519:933–956. https://doi.org/10.1002/cne.22559

Grady F, Peltekian L, Iverson G, Geerling JC (2020) Direct parabrachial-cortical connectivity. Cereb Cortex 30:4811–4833. https://doi.org/10.1093/cercor/bhaa072

Hakeem AY, Sherwood CC, Bonar CJ et al (2009) Von Economo neurons in the elephant brain. Anatomical Records 292:242–248. https://doi.org/10.1002/ar.20829

Ham T, Leff A, de Boissezon X et al (2013) Cognitive control and the salience network: an investigation of error processing and effective connectivity. J Neurosci 33:7091–7098. https://doi.org/10.1523/JNEUROSCI.4692-12.2013

Han ZS, Zhang ET, Craig AD (1998) Nociceptive and thermoreceptive lamina I neurons are anatomically distinct. Nat Neurosci 1:218–225. https://doi.org/10.1038/665

Harrison NA, Gray MA, Gianaros PJ, Critchley HD (2010) The embodiment of emotional feelings in the brain. J Neurosci 30:12878–12884. https://doi.org/10.1523/JNEUROSCI.1725-10.2010

Hartig R (2019) An examination of the neural correlates of interoceptive processes in the macaque monkey. University of Tuebingen

Hartig R, Battal C, Chavez G et al (2017) Topographic mapping of the primate primary interoceptive cortex. Front Neurosci. Conference Abstract

Hartig R, Battal C, Oeltermann A et al (2019) Thermal activation and thalamocortical connections of the insular cortex in the macaque monkey. Front Neurosci. Conference Abstract

Hartig R, Karimi A, Evrard HC (2023) Interconnected sub-networks of the macaque monkey gustatory connectome. Front Neurosci 16

Hassanpour MS, Yan L, Wang DJJ, et al (2018) How the heart speaks to the brain: Neural activity during cardiorespiratory interoceptive stimulation. Philosophical Transactions of the Royal Society B: Biological Sciences 371. https://doi.org/10.1098/rstb.2016.0004

Henderson LA, Gandevia SC, Macefield VG (2007) Somatotopic organization of the processing of muscle and cutaneous pain in the left and right insula cortex: a single-trial fMRI study. Pain 128: 20–30. https://doi.org/10.1016/j.pain.2006.08.013

Herbert H, Moga MM, Saper CB (1990) Connections of the parabrachial nucleus with the nucleus of the solitary tract and the medullary reticular formation in the rat. J Comp Neurol 293:540–580. https://doi.org/10.1002/cne.902930404

Hermanson O, Blomqvist A (1996) Subnuclear localization of FOS-like immunoreactivity in the rat parabrachial nucleus after nociceptive stimulation. J Comp Neurol 368:45–56. https://doi.org/10.1002/(SICI)1096-9861(19960422)368:1<45::AID-CNE4>3.0.CO;2-K

Hermanson O, Blomqvist A (1997) Subnuclear localization of FOS-like immunoreactivity in the parabrachial nucleus after orofacial nociceptive stimulation of the awake rat. J Comp Neurol 387:114–123

Horn F (2020) Elemental localization of the von Economo neuron and fork neuron in the insular cortex in humans and macaque monkeys. University of Tuebingen

Horn FM, Hertl M, Logothetis NK, Evrard HC (2017) Elemental localization of the von Economo neuron in the macaque monkey insula. Front Neurosci

Hua H, Strigo IA, Baxter LC et al (2005) Anteroposterior somatotopy of innocuous cooling activation focus in human dorsal posterior insular cortex. Am J Physiol Regul Integr Comp Physiol 289:R319–R325. https://doi.org/10.1152/ajpregu.00123.2005

Huang Z, Tarnal V, Vlisides PE et al (2021) Anterior insula regulates brain network transitions that gate conscious access. Cell Rep 35:109081. https://doi.org/10.1016/j.celrep.2021.109081

Iannetti GD, Zambreanu L, Cruccu G, Tracey I (2005) Operculoinsular cortex encodes pain intensity at the earliest stages of cortical processing as indicated by amplitude of laser-evoked potentials in humans. Neuroscience 131:199–208. https://doi.org/10.1016/j.neuroscience.2004.10.035

Ifuku H, Nakamura T, Hirata S, Ogawa H (2006) Neuronal activities in the reward phase in primary and higher-order gustatory cortices of monkeys. Neurosci Res 55:54–64. https://doi.org/10.1016/j.neures.2006.02.001

Ito SI, Craig AD (2008) Thalamocortical projections of the vagus-responsive region of the basal part of the ventral medial nucleus in monkeys. Progam No 36410 2008 Neuroscience meeting planner Washington, DC: Society for Neuroscience, 2008 Online

James W (1894) The physical bases of emotion. Psychol Rev 1:516–529

Janig W (2022) The integrative action of the autonomic nervous system. Cambridge University Press

Jantsch HHF, Kemppainen P, Ringler R et al (2005) Cortical representation of experimental tooth pain in humans. Pain 118:390–399. https://doi.org/10.1016/j.pain.2005.09.017

Johnson JI, Buchanan KJ, Morris JA, Fobbs AJ (2009) Interrelation of gyral formations, cytoarchitectural variations, and sensory regions in human insular cortex

Kaada BR (1951) Somato-motor, autonomic and electrocorticographic responses to electrical stimulation of rhinencephalic and other structures in primates, cat, and dog; a study of responses from the limbic, subcallosal, orbito-insular, piriform and temporal cortex, hippocampus-fornix and amygdala. Acta Physiol Scand Suppl 24:1–262

Kadohisa M, Rolls ET, Verhagen JV (2005) Neuronal representations of stimuli in the mouth: the primate insular taste cortex, orbitofrontal cortex and amygdala. Chem Senses 30:401–419. https://doi.org/10.1093/chemse/bji036

Kaskan PM, Dean AM, Nicholas MA et al (2019) Gustatory responses in macaque monkeys revealed with fMRI: comments on taste, taste preference, and internal state. NeuroImage 184:932–942. https://doi.org/10.1016/j.neuroimage.2018.10.005

Kitamura T, Tsuruhara A, Sakai A (2001) Afferent projections to the parabrachial nucleus from the spinal and medullary dorsal horns in the rat. Neuroscience Research 39:33–43. https://doi.org/10.1016/S0168-0102(00)00198-3

Krockenberger MS, Saleh-Mattesich TO, Evrard HC (2023) Cytoarchitectonic and connection stripes in the dysgranular insular cortex in the macaque monkey. J Comp Neurol 531:2019–2043. https://doi.org/10.1002/cne.25571

Ku SP, Tolias AS, Logothetis NK, Goense J (2011) fMRI of the face-processing network in the ventral temporal lobe of awake and anesthetized macaques. Neuron 70:352–362. https://doi.org/10.1016/j.neuron.2011.02.048

Kurth F, Eickhoff SB, Schleicher A et al (2010) Cytoarchitecture and probabilistic maps of the human posterior insular cortex. Cereb Cortex 20:1448–1461. https://doi.org/10.1093/cercor/bhp208

Lange C (1885) Om Sindsbevaegelser et Psyko-Fysiologisk Studie. Jacob Lunds Forlag, Copenhagen

Langley JN (1921) The autonomic nervous system. W. Heffer & Sons, Ltd, Cambridge
Lenz FA, Gracely RH, Zirh TA et al (1997) Human thalamic nucleus mediating taste and multiple other sensations related to ingestive behavior. J Neurophysiol 77:3406–3409. https://doi.org/10.1152/jn.1997.77.6.3406
Lenz FA, Dougherty PM, Meeker TJ et al (2024) Neuroscience of the human thalamus related to acute pain and chronic "thalamic" pain. J Neurophysiol 132:1756–1778. https://doi.org/10.1152/jn.00065.2024
Mallela AN, Deng H, Gholipour A et al (2023) Heterogeneous growth of the insula shapes the human brain. Proc Natl Acad Sci 120:e2220200120. https://doi.org/10.1073/pnas.2220200120
Mandonnet V, Obaid S, Descoteaux M et al (2024) Electrostimulation of the white matter of the posterior insula and medial operculum: perception of vibrations, heat, and pain. Pain 165:565–572. https://doi.org/10.1097/j.pain.0000000000003069
Mazzola L, Isnard J, Peyron R et al (2009) Somatotopic organization of pain responses to direct electrical stimulation of the human insular cortex. Pain 146:99–104. https://doi.org/10.1016/j.pain.2009.07.014
Mazzola L, Mauguiere F, Isnard J (2019) Functional mapping of the human insula: data from electrical stimulations. Rev Neurol (Paris) 175:150–156. https://doi.org/10.1016/j.neurol.2018.12.003
Melzack R, Wall PD (1965) Pain mechanisms: a new theory. Science 150:971–979. https://doi.org/10.1126/science.150.3699.971
Menon V, Uddin LQ (2010) Saliency, switching, attention and control: a network model of insula function. Brain Struct Funct 214:655–667. https://doi.org/10.1007/s00429-010-0262-0
Mesulam MM, Mufson EJ (1982) Insula of the old world monkey. I. Architectonics in the insulo-orbito-temporal component of the paralimbic brain. J Comp Neurol 212:1–22. https://doi.org/10.1002/cne.902120102
Mesulam MM, Mufson EJ (1985) The insula of reil in man and monkey. In: Peters A, Jones EG (eds) Cerebral cortex. Plenum Press, New York, pp 179–226
Mitchell DJ, Bell AH, Buckley MJ et al (2016) A putative multiple-demand system in the macaque brain. J Neurosci 36:8574–8585. https://doi.org/10.1523/JNEUROSCI.0810-16.2016
Moga MM, Herbert H, Hurley KM et al (1990) Organization of cortical, basal forebrain, and hypothalamic afferents to the parabrachial nucleus in the rat. J Comp Neurol 295:624–661. https://doi.org/10.1002/cne.902950408
Molnar-Szakacs I, Uddin LQ (2022) Anterior insula as a gatekeeper of executive control. Neurosci Biobehav Rev 139:104736. https://doi.org/10.1016/j.neubiorev.2022.104736
Mufson EJ, Sobreviela T, Kordower JH (1997) Chemical neuroanatomy of the primate insula cortex: relationship to cytoarchiteconics, connectivity, function and neurodegeneration. In: VBloom FE, Bjorklund A, Hokfelt T (eds) Handbook of chemical neuroanatomy. Elsevier, New York
Munk MH, Roelfsema PR, König P et al (1996) Role of reticular activation in the modulation of intracortical synchronization. Science 272:271–274. https://doi.org/10.1126/science.272.5259.271
Naidich TP, Kang E, Fatterpekar GM et al (2004) The insula: anatomic study and MR imaging display at 1.5 T. AJNR Am J Neuroradiol 25:222–232
Nakamura Y, Goto TK, Tokumori K et al (2011) Localization of brain activation by umami taste in humans. Brain Res 1406:18–29. https://doi.org/10.1016/j.brainres.2011.06.029
Ngowyang G (1932) Beschreibung einer Art von Spezialzellen in der Inselrinde – zugleich Bemerkungen über die v. Economoschen Spezialzellen. J Physchol Neurol 44:671–674
Nimchinsky EA, Gilissen E, Allman JM et al (1999) A neuronal morphologic type unique to humans and great apes. Proc Natl Acad Sci USA 96:5268–5273
Norgren R, Leonard CM (1973) Ascending central gustatory pathways. J Comp Neurol 150:217–237
Ogawa H, Ito S, Nomura T (1989) Oral cavity representation at the frontal operculum of macaque monkeys. Neurosci Res 6:283–298. https://doi.org/10.1016/0168-0102(89)90021-7

Olausson H, Lamarre Y, Backlund H et al (2002) Unmyelinated tactile afferents signal touch and project to insular cortex. Nat Neurosci 5:900–904. https://doi.org/10.1038/nn896

Ongur D, Ferry AT, Price JL (2003) Architectonic subdivision of the human orbital and medial prefrontal cortex. J Comp Neurol 460:425–449. https://doi.org/10.1002/cne.10609

Palmiter RD (2018) The parabrachial nucleus: CGRP neurons function as a general alarm. Trends Neurosci 41:280–293. https://doi.org/10.1016/j.tins.2018.03.007

Parvizi J, Damasio AR (2003) Neuroanatomical correlates of brainstem coma. Brain J Neurol 126: 1524–1536. https://doi.org/10.1093/brain/awg166

Porcu E, Benz KM, Ball F et al (2020) Macroscopic information-based taste representations in insular cortex are shaped by stimulus concentration. Proc Natl Acad Sci USA 117:7409–7417. https://doi.org/10.1073/pnas.1916329117

Pritchard TC, Hamilton RB, Morse JR, Norgren R (1986) Projections of thalamic gustatory and lingual areas in the monkey, Macaca fascicularis. J Comp Neurol 244:213–228. https://doi.org/10.1002/cne.902440208

Pritchard TC, Hamilton RB, Norgren R (1989) Neural coding of gustatory information in the thalamus of Macaca mulatta. J Neurophysiol 61:1–14

Pritchard TC, Hamilton RB, Norgren R (2000) Projections of the parabrachial nucleus in the old world monkey. Exp Neurol 165:101–117. https://doi.org/10.1006/exnr.2000.7450

Raghanti MA, Spurlock LB, Treichler FR et al (2015) An analysis of von Economo neurons in the cerebral cortex of cetaceans, artiodactyls, and perissodactyls. Brain Struct Funct 220:2303–2314. https://doi.org/10.1007/s00429-014-0792-y

Ramirez-Villegas JF, Besserve M, Murayama Y et al (2020) Coupling of hippocampal theta and ripples with pontogeniculooccipital waves. Nature. https://doi.org/10.1038/s41586-020-2914-4

Ran C, Boettcher JC, Kaye JA et al (2022) A brainstem map for visceral sensations. Nature 609: 320–326. https://doi.org/10.1038/s41586-022-05139-5

Reil JC (1809) Die sylvische Grube. Arch Physiol (Halle) 9:195–208

Remedios R, Logothetis NK, Kayser C (2009) Monkey drumming reveals common networks for perceiving vocal and nonvocal communication sounds. Proc Natl Acad Sci USA 106:18010–18015. https://doi.org/10.1073/pnas.0909756106

Roberts TS, Akert K (1963) Insular and opercular cortex and its thalamic projection in Macaca mulatta. Schweizer Archiv fur Neurologie, Neurochirurgie und Psychiatrie = Archives suisses de neurologie, neurochirurgie et de psychiatrie 92:1–43

Rolls ET (2005) Taste, olfactory and food texture processing in the brain, and the control of food intake. Physiology & Behavior, 85(1), 45–56. https://doi.org/10.1016/j.physbeh.2005.04.012

Rubio A, Van Oudenhove L, Pellissier S et al (2015) Uncertainty in anticipation of uncomfortable rectal distension is modulated by the autonomic nervous system–a fMRI study in healthy volunteers. NeuroImage 107:10–22. https://doi.org/10.1016/j.neuroimage.2014.11.043

Sanides F (1968) The architecture of the cortical taste nerve areas in squirrel monkey (Saimiri sciureus) and their relationships to insular, sensorimotor and prefrontal regions. Brain Res 8:97–124

Saper CB (2002) The central autonomic nervous system: conscious visceral perception and autonomic pattern generation. Annu Rev Neurosci 25:433–469

Schreckenberger M, Siessmeier T, Viertmann A et al (2005) The unpleasantness of tonic pain is encoded by the insular cortex. Neurology 64:1175–1183. https://doi.org/10.1212/01.WNL.0000156353.17305.52

Segerdahl AR, Mezue M, Okell TW et al (2015) The dorsal posterior insula is not an Island in pain but subserves a fundamental role – response to: "evidence against pain specificity in the dorsal posterior insula" by Davis et al. F1000Research 4:1207. https://doi.org/10.12688/f1000research.7287.1

Seth AK, Friston KJ (2016) Active interoceptive inference and the emotional brain. Philos Trans R Soc Lond Ser B Biol Sci 371. https://doi.org/10.1098/rstb.2016.0007

Sherrington CS (1948) The integrative action of the central nervous system. Cambridge University Press

Simone L, Caruana F, Elena B et al (2025) Anatomo-functional organization of insular networks: from sensory integration to behavioral control. Prog Neurobiol 247:102748. https://doi.org/10.1016/j.pneurobio.2025.102748
Small DM (2010) Taste representation in the human insula. Brain Struct Funct 214:551–561. https://doi.org/10.1007/s00429-010-0266-9
Smith R, Braden BB, Chen K et al (2015) The neural basis of attaining conscious awareness of sad mood. Brain Imaging Behav 9:574–587. https://doi.org/10.1007/s11682-014-9318-8
Spetter MS, de Graaf C, Mars M et al (2014) The sum of its parts–effects of gastric distention, nutrient content and sensory stimulation on brain activation. PLoS One 9:e90872. https://doi.org/10.1371/journal.pone.0090872
Sridharan D, Levitin DJ, Menon V (2008) A critical role for the right fronto-insular cortex in switching between central-executive and default-mode networks. Proc Natl Acad Sci USA 105:12569–12574. https://doi.org/10.1073/pnas.0800005105
Steiner S, Zilles K, Amunts K et al (2024) Conserved localization of the von Economo neuron across primate suborders. In: Neuroscience meeting planner. Chicago
Stephani C, Fernandez-Baca Vaca G, Maciunas R et al (2011) Functional neuroanatomy of the insular lobe. Brain Struct Funct 216:137–149. https://doi.org/10.1007/s00429-010-0296-3
Striedter GF, Srinivasan S, Monuki ES (2015) Cortical folding: when, where, how, and why? Annu Rev Neurosci 38:291–307. https://doi.org/10.1146/annurev-neuro-071714-034128
Strigo IA, Craig AD (2016) Interoception, homeostatic emotions and sympathovagal balance. Philos Trans R Soc Lond Ser B Biol Sci 371. https://doi.org/10.1098/rstb.2016.0010
Strigo IA, Andrew D, Simmons AN et al (2024) Interoception and AD (Bud) Craig's paradigm-shifting legacy. Pain. https://doi.org/10.1097/j.pain.0000000000003208
Terasawa Y, Shibata M, Moriguchi Y, Umeda S (2013) Anterior insular cortex mediates bodily sensibility and social anxiety. Soc Cogn Affect Neurosci 8:259–266. https://doi.org/10.1093/scan/nss108
Testard C, Brent LJN, Andersson J et al (2022) Social connections predict brain structure in a multidimensional free-ranging primate society. Sci Adv 8:eabl5794. https://doi.org/10.1126/sciadv.abl5794
Todd RM, Talmi D, Schmitz TW et al (2012) Psychophysical and neural evidence for emotion-enhanced perceptual vividness. J Neurosci 32:11201–11212. https://doi.org/10.1523/JNEUROSCI.0155-12.2012
Touroutoglou A, Bliss-Moreau E, Zhang J et al (2016) A ventral salience network in the macaque brain. NeuroImage 132:190–197. https://doi.org/10.1016/j.neuroimage.2016.02.029
Trulsson M, Francis ST, Bowtell R, McGlone F (2010) Brain activations in response to vibrotactile tooth stimulation: a psychophysical and fMRI study. J Neurophysiol 104:2257–2265. https://doi.org/10.1152/jn.00565.2010
Ture U, Yasargil DC, Al-Mefty O, Yasargil MG (1999) Topographic anatomy of the insular region. J Neurosurg 90:720–733. https://doi.org/10.3171/jns.1999.90.4.0720
Van Essen DC (1997) A tension-based theory of morphogenesis and compact wiring in the central nervous system. Nature 385:313–318. https://doi.org/10.1038/385313a0
Van Essen DC (2020) A 2020 view of tension-based cortical morphogenesis. Proc Natl Acad Sci 117:32868–32879. https://doi.org/10.1073/pnas.2016830117
Vartiainen N, Perchet C, Magnin M et al (2016) Thalamic pain: anatomical and physiological indices of prediction. Brain J Neurol 139:708–722. https://doi.org/10.1093/brain/awv389
Veldhuizen MG, Douglas D, Aschenbrenner K et al (2011) The anterior insular cortex represents breaches of taste identity expectation. J Neurosci 31:14735–14744. https://doi.org/10.1523/JNEUROSCI.1502-11.2011
Verhagen JV, Kadohisa M, Rolls ET (2004) Primate insular/opercular taste cortex: neuronal representation of the viscosity, fat texture, grittiness, temperature, and taste of foods. Journal of Neurophysiology 92:1685–1699. https://doi.org/10.1152/jn.00321.2004
Von Economo C, Koskinas G (1925) Die Cytoarchitectonik der Hirnrinde des erwachsenen Menschen. Springer, Berlin

Wang L, Uhrig L, Jarraya B, Dehaene S (2015) Representation of numerical and sequential patterns in macaque and human brains. Curr Biol 25:1966–1974. https://doi.org/10.1016/j.cub.2015.06.035

Warnaby CE, Seretny M, Ni Mhuircheartaigh R et al (2016) Anesthesia-induced suppression of human dorsal anterior insula responsivity at loss of volitional behavioral response. Anesthesiology 124:766–778. https://doi.org/10.1097/ALN.0000000000001027

Wattendorf E, Westermann B, Lotze M et al (2016) Insular cortex activity and the evocation of laughter. J Comp Neurol 524:1608–1615. https://doi.org/10.1002/cne.23884

Weber EH (1846) Der Tastsinn und das Gemeingefühl. In: Wagner R (ed) Handwörterbuch der Physiologie. Friedrich Vieweg und Sohn, Braunschweig

Westlund KN, Craig AD (1996) Association of spinal lamina I projections with brainstem catecholamine neurons in the monkey. Exp Brain Res 110:151–162

Wilk HA, Ezekiel F, Morton JB (2012) Brain regions associated with moment-to-moment adjustments in control and stable task-set maintenance. NeuroImage 59:1960–1967. https://doi.org/10.1016/j.neuroimage.2011.09.011

Wilson LB, Andrew D, Craig AD (2002) Activation of spinobulbar lamina I neurons by static muscle contraction. J Neurophysiol 87:1641–1645. https://doi.org/10.1152/jn.00609.2001

Wittmann M, Simmons AN, Aron JL, Paulus MP (2010) Accumulation of neural activity in the posterior insula encodes the passage of time. Neuropsychologia 48:3110–3120. https://doi.org/10.1016/j.neuropsychologia.2010.06.023

Wu T, Wang X, Wu Q et al (2019) Anterior insular cortex is a bottleneck of cognitive control. NeuroImage. https://doi.org/10.1016/j.neuroimage.2019.02.042

Wysiadecki G, Malkiewicz A, Rozniecki J et al (2018) Anatomical variations of the insular gyri: a morphological study and proposal of unified classification. Clin Anat 31:347–356. https://doi.org/10.1002/ca.23060

Yang Y-P, Li X, Stuphorn V (2022) Primate anterior insular cortex represents economic decision variables proposed by prospect theory. Nat Commun 13:717. https://doi.org/10.1038/s41467-022-28278-9

Yaxley S, Rolls ET, Sienkiewicz ZJ (1990) Gustatory responses of single neurons in the insula of the macaque monkey. Journal of Neurophysiology 63:689–700. https://doi.org/10.1152/jn.1990.63.4.689

Zaki J, Davis JI, Ochsner KN (2012) Overlapping activity in anterior insula during interoception and emotional experience. NeuroImage 62:493–499. https://doi.org/10.1016/j.neuroimage.2012.05.012

Zhang ZH, Dougherty PM, Oppenheimer SM (1998) Characterization of baroreceptor-related neurons in the monkey insular cortex. Brain Res 796:303–306

Zweynert S, Pade JP, Wüstenberg T et al (2011) Motivational salience modulates hippocampal repetition suppression and functional connectivity in humans. Front Hum Neurosci 5:144. https://doi.org/10.3389/fnhum.2011.00144

Interoception, Insula, and Autonomic Integration: Relevance to the Expression and Treatment of Psychiatric Symptoms

Hugo D. Critchley and Joel Patchitt

Contents

Abstract The mind is embodied, and this is relevant to mental health and psychiatric illness. Interoception is the body-to-brain axis of sensory information flow and its representation at the neural and psychological levels. Interoception is the purported basis for motivation and emotion, and as an inescapable stream of information about the health and functioning of the whole organism, it is proposed to be the foundation to the conscious unitary sense of self. Correspondingly, this central representation of internal state is relevant to understanding the expression of psychological symptoms and behaviours and ultimately psychiatric disorders. Here we review interoception, particularly from a cardiovascular perspective, and how understanding theoretical neural and psychological aspects of interoception relates to perceptions, thoughts, and feelings. We examine how perturbations in interoceptive

H. D. Critchley (✉) and J. Patchitt
Department of Clinical Neuroscience, Brighton and Sussex Medical School, University of Sussex, Brighton, UK
e-mail: h.critchley@bsms.ac.uk

Curr Topics Behav Neurosci (2025) 70: 63–85
https://doi.org/10.1007/7854_2024_518
Published Online: 13 November 2024

processing are expressed in mental symptoms and psychiatric disorders and show how this knowledge may yield new treatment targets.

Keywords Anxiety · Arousal · Autism · Autonomic · Baroreceptor · Biobehavioural · Biofeedback · Cardiac · Craving · Dissociation · Interoception · Psychiatry · Therapy

1 Introduction

Psychiatric research is at a crossroads (Paul and Potter 2024). There have been major recent advances in data science, molecular biology, and in the neuroscientific understanding of the function of brain systems underpinning perception and behaviour. However, big pharmaceutical companies continue to show a reluctance to invest in new treatments for mental illness. Psychological therapies are also showing their limits (NHS Digital 2023). Conventional therapies are very effective for many people but often fail to reach many others. Pressure on mental health services in many countries have led to an increasing reorientation towards public mental health and prevention. Nevertheless, growing recognition that physical and mental illness go together, and the major advances outside of psychiatry and psychology research (Subbiah 2023), in areas such as metabolic health (Furlan and Petrus 2023), vaccinations, and associated fertile drug development, promise much for the future of psychiatry. This includes management of mental illness and associated functional impairments with approaches that transcend established silos of physical and mental health treatments.

Such an integrative approach that understands the relationship between mind, brain, and body, can provide fresh perspectives and novel interventions to support people with, or vulnerable to, psychiatric disorders. To build such insight, it is necessary first to pursue fundamental research into the mechanisms through which mental and social stressors can impact the state of the body and shape physical health and, conversely, determine how the internal physiological state of the body feeds back via representations within the brain to influence thoughts, feelings, and actions (Critchley 2005). Anatomical studies, notably the work of Arthur (Bud) Craig, and functional imaging studies in the beginning of the twenty-first century, have helped illuminate such mechanisms. The role of cortical brain areas such as the anterior cingulate has been shown to have specific visceromotor functions, altering the internal physiological state of the body via the autonomic nervous system to meet changing, emotional, motivational, cognitive, and motor demands (Critchley 2004, 2005; Critchley et al. 2000, 2003; Gianaros et al. 2005; Wager et al. 2009). In parallel, inspired by the writing of Antonio Damasio (2000, 2006) and his reappraisal of the theoretical models of William James and Carl Lange (James 1884; Lange 1895; Lange and James 1922), cortical regions notably the insula (Craig 2002; Critchley et al. 2002, 2004; Paulus and Stein 2006) have been shown to represent and translate bodily information into physiological, motivational, and emotional states that enrich our conscious experience and underpin our behaviours.

These empirical insights into the field of interoceptive science have been further shaped by theoretical advances and frameworks to characterise how changes within the body are encoded at different levels of neural, perceptual, and cognitive representation (Garfinkel and Critchley 2013; Garfinkel et al. 2015; Quadt et al. 2018). Related models also speak to how conscious experience is grounded on a fundamental representation of biological selfhood which arises from the inescapable stream of information from the body (Damasio 2000; Craig 2002) and the brain's capacity to predict and control biological processes for optimal function (Seth 2013; Seth and Tsakiris 2018). This predictive nature of self-representation, grounded ultimately on integrated physiological feeling, also enables an individual to simulate the inner mental experiences of other people (Critchley 2005; Singer et al. 2009).

This combination of fundamental empirical science and theoretical formulation has enabled translational insights into how emotional arousal and mental stress can impact the function of internal organs (and undermine physical health). Moreover, it has provided a basis for understanding how common psychiatric symptoms arise. In this context, anxiety, low mood, craving, fatigue, and dissociation can be understood in terms of feeling states linked to changes within the body evoked by physiological challenges to homeostatic stability, and the brain's efforts to predict and control such signals. By extension, such feeling states characterise specific mental illnesses including anxiety and affective disorders, substance misuse disorders, and psychoses such as schizophrenia, wherein a breakdown in self-representation or compromised self-efficacy plausibly underlies both positive and negative symptoms (Khalsa et al. 2018). The bidirectional mind-brain-body framework also provides direct insight into conditions that underlie the interface of mental and physical health, notably functional somatic and neurological disorders (Bonaz et al. 2021). This integrative approach to understanding clinical symptoms and conditions has also fuelled therapeutic innovation including the development of new interventional approaches through biobehavioural therapies, electroceutical targets, and potential repurposing of pharmacological agents (Chen et al. 2021).

2 Concept of Interoception

Charles Sherrington was an eminent physiologist who, in 1906, wrote *The integrative action of the nervous system* (Sherrington 1911). He won the Nobel Prize for Physiology or Medicine in 1932 in recognition of his 40+ years of scientific discovery. Much of that science was undertaken, not at the University of Oxford where he was employed when he received the Nobel Prize, but at the University of Liverpool in the Department of Physiology. Here, he named the *synapse* and postulated the division of the senses into *interoception*, *exteroception*, and *proprioception*. While exteroception refers to information concerning the external environment, and proprioception refers to the position of the body in space, interoception refers to sensing of information about the inner state. The most accepted current definition is that interoception is *the sense of the internal physiological state of the body* (Craig 2002; Khalsa et al. 2018). Interoception is thus the signalling, central representation, and the integration of viscerosensory neural and humoral (bloodborne) parameters. Interoceptive pathways inform the brain about the function of visceral organs, state of cardiorespiratory arousal, of blood gases, body temperature, nutrients, inflammation, and immune status, and the multiple parameters that reflect what is now known as the gut–brain axis. Nevertheless, the boundaries of interoception are blurred: Taste, affective touch, pain (wherever felt), vestibular and proprioceptive sensations all can claim to convey interoceptive information about physiological state, relevant both to the homeostatic control of bodily functions and to motivational behaviour. In addition, we use the term quasi-interoceptive to refer to exteroceptive senses that, either actively or incidentally, carry information about internal interoceptive processes. Examples might include the tactile somatosensation of air moving through your upper respiratory tract, the proprioceptive movement of one's chest wall when breathing deeply, and the somatosensation of your heart beating against your chest wall or face, and fingers pulsing in states of heightened cardiorespiratory arousal. Such sensations are not viscerosensory yet are correlated with internal bodily changes.

3 Somatic Markers

In the 1990s, Antonio Damasio and colleagues put forward the notion of *somatic markers* by which they meant changes in bodily state that feed back to the brain (often implicitly) to guide cognitive processes and shape adaptive behaviours (Damasio 2006; Bechara et al. 1997, 2005). Observations on patients with damage to medial prefrontal and orbital frontal cortices led Damasio and co-workers to seek a different explanation for deficits in behaviour consequent upon such acquired brain lesions. They observed that such patients were unable to learn from negative experiences, despite apparently knowing how they should act. In an experimental test that modelled learning in the context of risk and uncertainty, the Iowa Gambling Task, patients with damage to ventral and orbital regions of the frontal lobe did not show the typical bodily signatures of arousal before making risky choices that was observed in healthy participants. These bodily responses were proposed to guide adaptive behaviour and act as somatic markers. This set of observations, and related empirical and theoretical work by others, highlighted the importance of body-to-brain communication in the support of emotional and motivational behaviours and of aspects of cognition from memory to decision-making (Damasio 2006; Bechara et al. 1997, 2005). Damasio's work inspired a more general appreciation of mind–brain–body interactions, including the contribution of interoception to subjective aspects of emotion and clinically significant feeling states.

4 Cardiac Signalling

As described above, interoception is the body-to-brain axis of sensory information flow and representation. At a fundamental level, interoception is the afferent limb of homeostatic reflex arcs. For example, in the arterial baroreflex, information about blood pressure is signalled by arterial baroreceptors located in the aortic arch and carotid sinus (Benarroch 2008). Baroreceptors are stretch receptors that fire when blood is pumped into the arteries and stretches the vessel walls. Activation is fed up to the medulla oblongata (in the brainstem) where through a brief series of synapses autonomic output is regulated. A strong response from arterial baroreceptors will inhibit sympathetic outflow, increasing vasodilation of arterioles in the muscular vascular beds, and causing enhanced parasympathetic outflow to the heart via the vagus nerve. This slows down the next heartbeat at the atrioventricular 'pacemaker' node. The consequence of this reflex is to maintain blood pressure at a constant level during states of low or stable activity. However, critical to adaptive behaviour is the ability to change bodily state to meet physical and mental demands that may increase the need for greater blood flow to brain and muscle. This adaptive and predictive regulation is achieved through *allostasis* (Sterling 2012). The purpose of allostasis is to anticipate and mitigate future challenges to bodily physiology to maintain the overall well-being of the organism. Stability is achieved through change in a predicted way (Sterling 2012). This is apparent, for example, in hunger, which occurs before nutritional status is compromised to the point of putting the body at risk and re-orientates perceptions and behaviour accordingly. The same logic operates for other motivational states. Emotional states represent social framing of motivational experience and control and similarly build upon these internal signals. Consequently, both the feelings of motivational need and of emotional arousal have an origin in internal bodily signals. The brain and the mind constantly review and control the internal state of the body. This dynamic aspect of interoception is considered to be the foundation of a sense of self. Damasio and Craig recognised the primacy of this first-order representation of the internal state of the body from which a unitary model of the whole organism can be derived (Damasio 2000; Craig 2002, 2009, 2014). This permits the control of behaviour at the level of the individual through the integration, prioritisation, and coordination of interoceptive signals and organ specific control.

5 Selfhood

Higher order representations of selfhood can build upon this 'neurobiological self' as an agent and within its external environment in a predictive allostatic way to navigate motivational, social, and other behavioural needs and challenges. Craig identified posterior insular cortex as the neural substrate for a primary representation of interoceptive state, upon which the self is built (Craig 2002, 2009, 2014).

Projections along the posterior to anterior axis of the insular cortex contextualise the self as a motivational entity, terminating within anterior insular cortex where the self-representation becomes consciously accessible as feelings states and becomes the substrate for self-awareness within cognitive, motivational social conditions (Craig 2009, 2014). Craig's contribution to understanding anatomy interoception and the functional role of insular cortex has been foundational. His description of the lamina 1 spinothalamic tract as the principle spinal axis for conveying motivationally salient physiological information to the brain challenged previous models of somatosensory representation and coding. Within the brain, he further recognised the fast tracking of interoceptive information that appears as a special evolutionary characteristic of primates, through which posterior insular cortex receives and integrates signals concerning internal states of the body with high granularity. In a series of review articles from 2002, through to his publication of the monograph *'How do you feel? An interoceptive moment with your neurobiological self'* published in 2015, Craig presented a model of somatic markers and emotional states that reinforced notions of embodied cognition and self-representation for the benefit of consciousness science (Craig 2002, 2009, 2014).

6 Growth in Interoception Research

Craig, along with Antonio Damasio and colleagues, and experimental researchers (including Critchley and colleagues), brought interoceptive neuroscience into the foreground in the 2000s. The first interoceptive summit was held in Tulsa, Oklahoma, in 2016, bringing together affective scientists and clinicians to consider the status of interoception and its potential wider impact for basic and clinical neuroscience. From this meeting, *Roadmap paper on interoception and mental health* was published, drawing attention to the current state of the art and the potential for interoceptive science to advance understanding and treatment of mental health problems (Khalsa et al. 2018). In 2019, the National Institutes of Health (NIH) held a first Blueprint meeting on the *Science of interoception and its roles in nervous system disorders*. This meeting considered whether interoception could bridge NIH funding initiatives, notably the Brain Project (focusing on advancing central nervous system neuroscience) and the SPARC Project (focusing on bodily organs and peripheral neural mechanisms and disorders). Critchley's keynote lecture on disorders of interoception was mirrored by a keynote from Ardem Patapoutian on peripheral transduction and signalling of changes in visceral organs. Patapoutian, with David Julius, won the Nobel Prize in 2021, 2 years after this meeting for work on PIEZO and TRPV receptors (Andolfo et al. 2022). The NIH subsequently released a funding call (Notice of Special interest) across all the Institutes of Health for 'projects on interoception to improve understanding of brain body function'. By 2024, there have been two rounds of awards and two investigator meetings at NIH. Other major funding agencies also recognise the importance of interoception. In

2023, the UK research charity Wellcome Trust highlighted interoception as one of its key 'transdiagnostic' research targets across mental health diagnoses.

7 Sensory Anatomy

Sherrington, whose work won him the Nobel Prize in 1943, referred to *interoceptors* as the sensory end-organs that capture and transduce information about the internal state of the body (Sherrington 1911). They comprise general and specialised chemosensors, mechanoceptors, and thermosensors that respond to changes in the internal environment over time. As noted above, Patapoutian and Julius identified such interoceptors (also mediating touch, temperature, and pain perception) notably PIEZO stretch-sensitive ion channels and TRPV temperature sensitive ion channels. Interoception was thus mentioned in recent Nobel speeches, David Julius referring to the communication of core body temperature, inflammatory, neuropathic, and visceral pain and associated protective reflexes; Ardem Patapoutian referring to mechanical pain and the signals that control urination, breathing, and blood pressure.

The activation of interoceptors triggers neural impulses along fine, visceral afferent fibres. These afferent axons track back along pathways shared with autonomic efferents, for example, along sympathetic nerves to the spinal cord, where the afferents terminate in spinal lamina 1 (Craig 2002). Visceral afferents also track up cranial nerves, notably the vagus nerve, to synapse in the nucleus of the solitary tract located in the medulla oblongata (Critchley and Harrison 2013).

In addition to these neural pathways of interoceptive signalling, the brain contains circumventricular organs, which represent points at which the blood brain barrier is modified to enable the contents of the blood to be directly sampled, and for large molecules (Critchley and Harrison 2013), including metabolic hormones such as leptin, to be taken up into the brain via specialised cells known as tanycytes (Prevot et al. 2021). Sampling at these sites of immune mediators, such as cytokines, and blood oxygenation levels also provide information relevant to the physiological state and health of the body (Critchley and Harrison 2013).

8 Forebrain Representation of Viscerosensory Signals

Within the primate brain, the pathways through which interoceptive information is conveyed and represented have been mapped by functional neuroanatomists, notably Craig, and more recently Evrard (Craig 2002, 2009; Evrard 2019): Ascending information from the medulla oblongata branches to synapses in areas such as the parabrachial nucleus and hypothalamus, but also travels directly to specific thalamic nuclei. These nuclei, particularly the posterior part of the ventral media nucleus (VMpo), project to the posterior insular cortex, a region identified as the primary interoceptive cortex. There is also an accessory projection from the ventrocaudal part of the dorsomedial thalamus (VMvc) to dorsal medial anterior cingulate cortex (Critchley and Harrison 2013).

Functional neuroimaging experiments, reinforced by observations in patients with autonomic disorders and those with brain lesions, elicited a more detailed understanding of the proposed human Central Autonomic Network. While cognitive neuroscience tended to leave consideration of brain body interactions at the level of the brainstem, these imaging experiments implicated cortical areas previously ascribed to cognitive functions, in the control and representation of internal bodily state. Importantly, activity changes within the anterior cingulate cortex, particularly the supra-genual 'dorsal' region (including anterior mid cingulate), were found to predict changes in autonomic arousal, particularly those mediated by the sympathetic nervous system (Critchley 2005; Critchley et al. 2000, 2003; Critchley 2004; Gianaros et al. 2005; Wager et al. 2009). The nature of the tasks that elicited such changes, together with the particular axis of autonomic control measured, appeared to determine where specific activation was localised within this region (Critchley 2004, 2005; Critchley et al. 2003). Nevertheless, lesions to this dorsal anterior cingulate area impaired the integrative generation of physiological arousal, during both effortful cognitive and motor behaviours (Critchley et al. 2003). In contrast, activity within the subgenual region of the anterior cingulate cortex typically shows anti-sympathetic and/or parasympathetic responses, such that decreases in activity here predict autonomic arousal characterised by a shift in sympathetic to parasympathetic balance (Critchley 2004, 2005; Wager et al. 2009; Nagai et al. 2004). Thus, our conclusion was that one important function of the anterior cingulate cortex is the visceromotor generation of states of physiological arousal that match and putatively facilitate behavioural demands (Critchley 2004).

In addition, brain areas such as amygdala are also implicated by animal experiments as a forebrain generator of autonomic sympathetic drive during affective behaviours, including responses to threats. Human neuroimaging experiments similarly linked amygdala activation to the generation of sympathetic skin responses during fear conditioning (Knight et al. 2005). However, the amygdala also seems to support interoceptive representations: A decrease in amygdala activation is observed in patients with peripheral autonomic failure, during threat processing (Critchley et al. 2002). Such a pattern is also seen in the insular cortex and reflects the absence of evoked physiological change in the body during threat. Thus, some regions of the amygdala are viscerosensory rather than purely visceromotor.

Consistent with the automatic engagement of the amygdala in threat reactions, studies on intact humans show that the amygdala responds to subliminal as well as supraliminal threats. In the fear conditioning study involving autonomic failure patients (mentioned above), subliminal ('masked') and explicit ('unmasked') conditioned fear stimuli were used to modify awareness of threat: In contrast to amygdala responses, activity within regions of right insula were shown to be sensitive to both conscious awareness of the (associated) threat and the presence and absence of a change in physiological (autonomic) arousal (Critchley et al. 2002). The right insula thus seemed to integrate the occurrence of a change in body state with conscious awareness of the threatening stimulus (Fig. 1c). This was compatible

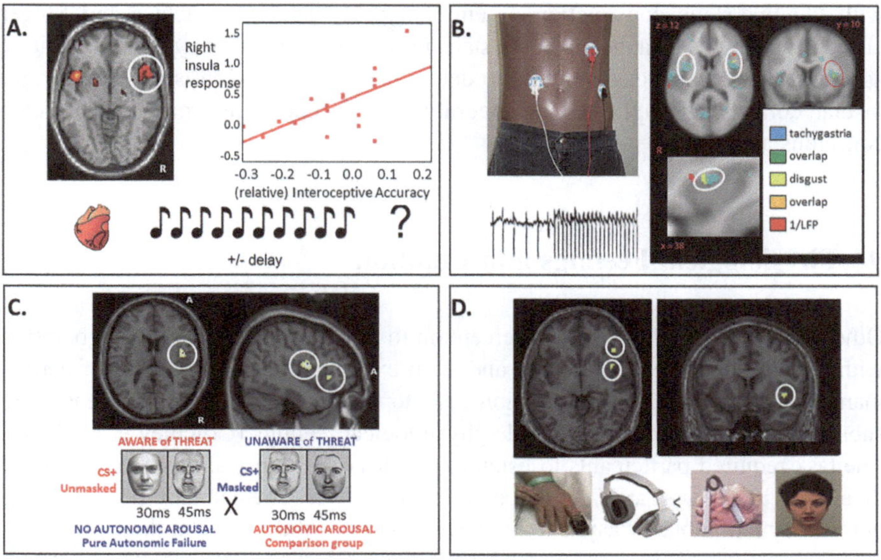

Fig. 1 Role of anterior insula in translating interoceptive representations into affective feelings. (**a**) Heartbeat discrimination[18]. During a heartbeat discrimination task performed concurrently with functional magnetic resonance imaging (fMRI)[18], increased activity during interoceptive attention to timing of heartbeats relative to quality of exteroceptive notes (task illustrated in lower half of figure) increased activity in regions including insular cortex (upper left panel), where activity in right anterior insular cortex (RAI) predicted individual difference in accuracy of task performance (relative to exteroceptive task). Additional data from this study showed RAI activation also predicted daily experiences of mood symptoms and was also sensitive to interoceptive/exteroceptive timing mismatch between heartbeat timing exteroceptive stimuli in an attention-dependent way[18]. (**b**) Representation of disgust[43]. Participants in an fMRI experiment viewed, then rated, brief videos designed to evoke different degrees of nauseating (core) and 'bloody' (body boundary violation, bbv) disgust, during simultaneous electogastrography (EGG) (electrode position shown upper left) and electrocardiography (ECG). Nauseating disgust provoked a tachygastria response from the 3 cycles/min normogastric pattern (lower left panel)[43]. Brain activity correlating with the degree of tachygastria overlapped within anterior insular cortex with activity correlated with the rated experience of disgust. Similar findings were found for bloody/bbv disgust in other anterior insular regions where activity overlapped between correlates of evoked change in heart rate variability (e.g. low frequency power 1/LFP) and rated experience of disgust (right panel)[43]. (**c**) Constructed fear[17]. In a fear conditioning study during fMRI, the responses to overt and subliminal threat was compared in healthy individuals and patients with pure autonomic failure (PAF) who are unable to generate (hence feedback interoceptively) autonomic arousal responses. Conscious awareness of the conditioned threat stimuli (CS+ anger face) was manipulated using backward masking – rapidly presenting the face before a covering mask (neutral face) – see bottom section of figure[17]. Both conscious and unconscious threat CS+ engaged activity in right amygdala, which was also sensitive to presence and absence of autonomic arousal (somatomotor expressions of threat were preserved in PAF patients). However, the right insular cortex (mid and RAI) (illustrated in upper panel) showed activity sensitive to the presence and absence of both awareness of threat and presence of physiological autonomic arousal[17], consistent with the region being a putative substrate for 2 stage models of constructed emotion wherein 'automatic' physiological responses require cognitive appraisal to define the emotional experience[42]. (**d**) Emotional attribution of unexplained arousal[48] Participants in an fMRI study made emotion intensity judgements of each of a series of emotional and neutral faces while listening, through headphones, to notes they were told reflected their heartbeats. During the task, they also performed block of handgrip exercise and had their pulse

with the two-stage construction of emotion proposed by Schachter and Singer (1962), wherein a change in body state may trigger emotion, yet the cognitive appraisal of the context is required to determine which emotion is felt. Here, right insular cortex was implicated in integrating interoceptive information with such conscious awareness.

9 Physiological Feelings and Emotion

Other studies also link conscious perception to physiological state: In collaboration with Öhman and Wiens, Critchley and team examined neural correlates of performance on a heartbeat discrimination task to determine which brain areas were substrates for consciously accessible physiological feelings (Critchley et al. 2004). The task required participants to listen to a series of 10 musical notes, triggered by their own heartbeats, and after each series to answer whether they perceived the notes as synchronous or asynchronous with their heartbeats. When the participants attended to their heartbeat, several brain areas, including somatosensory, anterior cingulate cortex, and insular cortex, were activated. Nevertheless, using task accuracy and a proxy for reported awareness, the right anterior insular cortex showed the strongest relationship with people's conscious perception of their heart beating (Fig. 1a). Activation and even the size of this brain region also predicted day-to-day experiences of anxiety. This study therefore suggested that the right anterior insular cortex is an important substrate for the conscious readout of signals from the body and its translation into physiological and affective feeling states (Critchley et al. 2004), consistent with the inferences made by Craig (2009, 2014). Subsequent studies have linked regions of the anterior insular cortex to different aspects of body brain coupling during emotion. In one study from our lab, in which participants saw films that were designed to evoke disgust, changes in stomach electromyography (electrogastrography), corresponding to induction of nausea, predicted the magnitude of activity change within regions of left and right insular cortex that also correlated with self-reported feelings of disgust evoked by these videos (Harrison et al. 2010) (Fig. 1b). When the same participants viewed videos of bloody disgust

Fig. 1 (continued) measured using finger oximetry (lower panel). The pulsatile auditory feedback through the headphones was veridical for half task, i.e. accurately reflecting the heart rate when the participant was exercising or not exercising. However, false feedback was also delivered where the auditory stimuli reflected exercise condition when the participant was not exercising and vice versa. This mismatch/false feedback was found to influence the emotional intensity rating of ambiguous neutral faces (more than intrinsically arousing happy or fearful faces). Activity within right insula (mid and RAI) represented this mismatch and the RAI activity predicted the resulting degree of change in the emotional rating of the face stimuli (upper panels). These findings are consistent with the role of right insula in ascribing salience to external stimuli through embodied higher order representation of interoceptive prediction errors[48] (more likely to make them faint), evoked changes in heart rate variability predicted changes in left anterior insular cortex where activity also correlated with the intensity of self-reported *bloody* (body boundary violation) disgust (Harrison et al. 2010). Such studies further reinforced the notion that the anterior insular cortex is a site for conscious readouts of internal bodily information underpinning emotional states (Fig. 1b).

Interoceptive pathways were also implicated in responses to induced systemic inflammation. Early neuroimaging studies in this field of neuro psycho immunology examined the central responses to inflammatory challenges. Activity changes within brainstem, midbrain, insula, and subgenual cingulate cortex were linked to specific symptoms of fatigue induced psychomotor slowing and low mood induced by typhoid vaccine evoked inflammation (Brydon et al. 2008; Harrison et al. 2009a, b). These observations provided the basis for the consideration of mechanisms through which depressive illness may hijack usually adaptive (physiological and behavioural) responses known as sickness behaviours that occur when inflammation/infection challenges the physical health of an individual. This field, of immunopsychiatry has grown considerably, exploring the potential for anti-inflammatory medications to be used in management of depressive illness, particularly if treatment resistant and associated with increased inflammatory markers.

10 Autonomic and Interoceptive Prediction

These studies, mapping visceromotor and viscerosensory regions of the human brain, during different types of behavioural and physiological challenges, typically show the conjoint activation of anterior cingulate and anterior insular cortices often with brainstem regions such as dorsal pons. This pattern of activity is commonly now ascribed to engagement of the salience network, attesting, by reverse inference to the emotional and behavioural significance of stimuli or behaviours that typically evokes such correlated patterns of activity. Nevertheless, our insights into the differential visceromotor and viscerosensory functions of cingulate and insular cortex provide more mechanistic understanding of such brain activity patterns that engage embodied responses.

Drawing on the emerging field of computational neuroscience, it was proposed that an efference copy (forward model) of the visceromotor drive from anterior cingulate cortex is mapped in insular and related orbital frontal cortex, to potentially allow for predictive and stimulatory modelling of internal body state, and by extension, the psychophysiological states of other people (Critchley 2005). This proposal represents an early instantiation of interoceptive predictive coding, a framework that gained explanatory power for understanding symptoms of anxiety (Paulus and Stein 2006) and which has found widespread application to in fields such as consciousness science and psychosomatic medicine (Seth 2013; Seth and Tsakiris 2018; Bonaz et al. 2021; Edwards 2016).

In fact, interoceptive prediction and the role of insular cortex as a comparator for the generation of prediction errors have previously been recognised: For example, in an attention-dependent way, mismatch between exteroceptive (notes) and interoceptive stimuli (heartbeats) during the heartbeat discrimination task engaged regions such as right, anterior insula and adjacent frontal operculum, along with anterior cingulate cortex and medial dorsal thalamus (Critchley et al. 2004). Moreover, in a study of false physiological feedback, in which heartbeat-like auditory tones were delivered to individuals when they appraised the emotional intensity of faces, activity of right anterior insular cortex predicted the enhanced ratings of emotional intensity of faces, particularly neutral (emotionally ambiguous) expressions (Gray et al. 2007) (Fig. 1d). Thus, when people experience interoceptive prediction errors (especially when provided with feedback) a representation of 'unattributed arousal' is generated, which appears to engage the insular cortex in the role of ascribing salience to external stimuli to explain the cause of the unattributed feedback. Similar mechanisms may underlie abnormal threat processing, paranoia, and aberrant salience in clinical groups such as patients with schizophrenia.

However, as it relates to tactile false feedback, a mere mismatch does not explain the observed effects. For example, a reduction in subjective and objective measures of anxiety occurs in those receiving feedback of a decreased HR when anticipating a public speech (Azevedo et al. 2017). Another investigation found that when tactile false feedback is presented at a supraliminal level, the falsified signals are progressively embodied resulting in the engagement of right posterior, middle and anterior insula (Patchitt et al. 2024: in press). This in turn alters body-centred parietal representations, biasing how affective salience is ascribed to external stimuli in a bidirectional manner, such that increases and decreases to affective salience are dependent on increased and decreased (relative to actual HR) false feedback, respectively. Such effects occur in the absence of evidence for physiological entrainment to the falsified stimuli (Patchitt et al. 2024: in press) (Fig. 2).

11 Baroreceptor Afferents and Cardiac Timing

In considering how the brain represents physiological states, particularly of cardiovascular arousal, the activation of arterial baroreceptors is an important interoceptive channel that is the subject of focused research. As noted earlier, at cardiac systole when blood is ejected from the ventricles, the walls of the aortic arch and carotids are dilated activating baroreceptors which signal via left vagus and bilateral glossopharyngeal nerves to the brainstem nucleus of the solitary tract. These signals, concerning the strength and timing of each heartbeat, inform the homeostatic baroreflex to control blood pressure[32], and communicate the state of cardiovascular arousal to the brain to impact other processes. We can examine the effects of this interoceptive signalling pathway by presenting stimuli at different phases of the cardiac cycle: A stimulus presented at ventricular systole occurs in the context of enhanced afferent cardiac signalling, in contrast to stimuli presented at ventricular

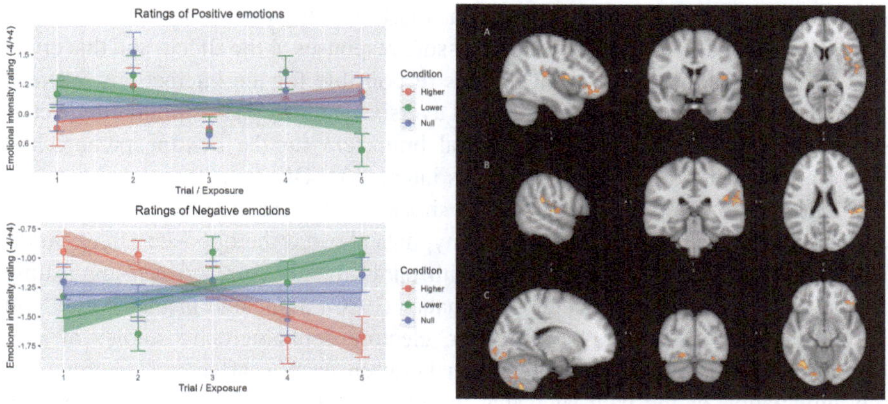

Fig. 2 Right insular engagement by false somatosensory feedback of cardiovascular arousal. (**a**) Behavioural effects of somatosensory false physiological feedback. Following on from Fig. 1d[48], a study was undertaken using somatosensory false feedback (FFB) of heart rate (HR), in which each participant rated emotional faces during 20s blocks of heartbeat-like pulsatile tactile stimulation at rates slower, faster, or same as their veridical HR. The panel illustrates the bidirectional exposure effect of tactile false physiological feedback on mean subjective intensity ratings of emotionally ambiguous faces over time (95% confidence interval error bars). FFB higher than veridical HR exerted increases in intensity ratings over time for both positive (upper graph) and negative emotions (lower graph). FFB lower than veridical HR exerted the opposite effect. (**b**) Neural substrates mediating effects of somatosensory false physiological feedback. Functional magnetic resonance imaging analysis revealed brain regions activated more by FFB-implied arousal (faster than veridical HR) than lower HR FFB: notably: (upper 3 brain figures 'A') a swathe of activity across right posterior, middle, and anterior insula through to right orbitofrontal cortex (arguably reflecting interoceptive representation). Also (middle 3 brain figures 'B'), right supramarginal gyrus, parietal operculum, and superior temporal cortex (regions implicated in proprioceptive embodiment and emotion recognition). Lastly (lower 3 brain figures 'C'), bilateral fusiform cortex and cerebellum and right orbitofrontal cortex (regions implicated in representation of visual salience (esp. faces)), and affective priors. Colour bars represent significant cluster-corrected z-statistics

diastole when baroreceptors are largely quiescent. Interactions with exteroceptive stimulation are demonstrable: Reflex responses to pain and perception of pain are also inhibited by baroreceptor activation at cardiac systole, in contrast to diastole when baroreceptor firing is much weaker (Wilkinson et al. 2013; Droste et al. 1994). If an electro-cutaneous shock is delivered at cardiac systole, when baroreceptors are firing, this inhibits the activity of sympathetic nerves to muscle vascular beds (but not skin sympathetic responses) (Donadio et al. 2002a). For the majority of people, this effect usually habituates with repetition, but not in the case of people with phobic syncope: These individuals who may faint at the sight of blood (body boundary violation phobia) show more enduring inhibition of muscle sympathetic activity (Donadio et al. 2002b). This observation attests to the fact that cardiac afferent signalling interacts with the presence of exteroceptive aversive signals in a way that also captures individual differences in emotional reactivity and affective behaviour. A neuroimaging study was conducted to understand further such interactions (Gray et al. 2009): Electro-cutaneous shock presented at systole was compared to similar shocks presented at diastole. This revealed an inhibition at systole of evoked (beat to beat) blood pressure responses to the shock, and this effect was tracked to identify the brain regions responsible for the interactions between interoceptive signals (cardiac timing) and aversive exteroceptive stimulation. Activity within bilateral insular cortex, dorsal brainstem (in the pontine parabrachial region), and right amygdala reflected this interaction. Of particular interest was the fact that systole, in conjunction with the shock, inhibited activity within the insula and brainstem yet the response of the amygdala showed the opposite effect. Additional analyses revealed relationships with parasympathetic responses implicating regions including periaqueductal grey matter (PAG) alongside insular cortex and amygdala (Gray et al. 2009). Moreover, electroencephalography studies of pain evoked potentials showed an interaction between cardiac afferent signalling and the anticipation of the pain such that the exaggeration of the neural response to anticipated pain was blocked by these systolic cardiac afferent signals (Gray et al. 2010). Together, these experiments provided some insights into how physiological states arousal, modelled at the level of heartbeats, can impact the processing of emotionally salient stimuli.

A follow-up study tested whether there was an effect on the processing of emotional face stimuli and found similar effects (Gray et al. 2012): Evoked changes in heart rate (inter-beat interval) evoked by different emotions were modulated as a function of cardiac timing (systole vs diastole). This was particularly the case for faces portraying disgusted and happy facial expressions, emotions typically linked to parasympathetic responses. Correspondingly, this effect was linked to changes in the activation of periaqueductal grey matter and medial orbital prefrontal/subgenual cingulate cortex (Gray et al. 2012). A much stronger effect was observed for the effects of cardiac timing on fear processing (Garfinkel et al. 2014). In contrast to pain processing, presenting fear stimuli at cardiac systole greatly enhanced subjective ratings of fear intensity and this was distinct from other emotions (Garfinkel and Critchley 2016). The effect predicted amygdala activation and correlated with individual differences in trait anxiety, indicating that more anxious people showed amplification of this effect (Garfinkel et al. 2014). This observation led to a programme of research examining the *cardiac control of fear in the brain*, which provided evidence attesting to the clinical value of exploiting this mechanism.

In a study of fear conditioning, for example, the effects of presenting either CS+ (associated with aversive shock stimulation) or CS− ('safe' stimuli with no risk of shock) stimuli at cardiac systole overshadowed the initial phase of fear learning: Using electrodermal activity as an objective measure of perceived threat, systole generalised fear responses such that CS+ and CS− stimuli presented at systole evoked greater fear reactivity[58]. Even when fear conditioning had been fully established and participants in this study had learned the stimulus contingencies, i.e. which were the threat stimuli (CS+), predicting shock and which were the safe (CS−) stimuli predicting no shock, the effects of cardiac timing persisted: CS− stimuli presented at systole were rated less pleasant than those presented at diastole. Next, the conditioned fear was extinguished by repeatedly presenting all stimuli without shock until well after no differences in fear responses were present. The next day participants returned to the test room and fear was reinstated by delivering two random shocks. This typically brings back the extinguished fear memory. However, a dominance of responses to stimuli presented previously at systole was also observed, such that even (safe) CS− stimuli originally presented at systole gave a stronger fear response (measured electrodermally) than (threat) CS+ stimuli presented at diastole (Garfinkel et al. 2021).

These observations, relating to cardiac afferent signals, pain, and fear processing, make intrinsic sense when escaping from threats (Garfinkel and Critchley 2016). The inhibition of pain is obviously adaptive over shorts periods of 'fight or flight' behaviour where we do not want to be slowed by injury. Moreover, during escape behaviour, it is necessary to maintain alertness for ongoing and other potential threats. Interestingly this work on cardiac interaction and amplification of threat has been recently 'back-translated' into animal models (Klein et al. 2021; Hsueh et al. 2023). For example, optogenetic acceleration of a mouse heart drives threat-related behaviours in the mice exposed to anxiogenic contexts (Hsueh et al. 2023). Functional analysis of the brain mechanisms supporting this change in behaviours highlights the role of cardiac interoceptive representation within posterior insular cortex.

12 Interoceptive Signals and Therapy for Anxiety

This heart-brain mechanism has potential clinical application: If the 'threat-value' of stimuli can be modified as a function of physiological state, and cardiac timing, to impact the learning and unlearning of fear, then this may be exploited in therapies for anxiety. A study of patients with spider phobia tested whether exposure therapy (a key component of treatment approaches to extinguish phobic fear) could be enhanced by harnessing cardiac timing effect. A computerised task was designed in which images of spiders could be shown at different phases of the cardiac cycle, such that one group of phobic individuals might see spiders only at cardiac systole, while another only at diastole (Watson et al. 2019). Initial pilot work showed that presenting stimuli at cardiac systole indeed enhanced the subjective fear response (which led to several people dropping out from this group). Those individuals that remained showed the biggest change in fear responses due to this enhanced exposure effect. This pilot work led to a larger clinical trial, in which incentives were added to avoid dropouts and ensure that membership of each group was retained. There were three groups of spider phobic participants in this study: People who saw spiders only at cardiac systole, those that saw spiders only at cardiac diastole, and a third group seeing spiders randomly over the course of the cardiac cycle. Since this was an exposure therapy, all groups improved (Watson et al. 2019). However, individuals who saw spider stimuli during their exposure therapy at cardiac systole improved the most on a subjective measure of spider phobia (spider phobia questionnaire score). Interestingly, this effect was strongest for people who also performed poorly on a test of cardiac interoceptive perception. Also, after treatment, the 'random' group was significantly worse than both systole and diastole on a behavioural test of spider phobia (how close one would go towards and potentially even touch a moult of a tarantula spider). People who performed better on tests of cardiac interoceptive perception appeared to benefit most. This observation suggests that the 'self-context' (in which phasic visceral cardiac signals are linked to external stimuli) lead to more effective extinction of fear behaviour. Additionally, the greater one's capacity to perceive such self-relatedness increases this therapeutic exposure effect (Watson et al. 2019).

Such observations are consistent with the notion that, from an autonomic/interoceptive perspective, the brain is interested in unexplained feedback, i.e. sensory surprise, while cancelling-out the salience of self-generated, self-related (even self-synchronous) information. Actions are matched against internal sensory feedback and the mismatch equates to prediction errors. The brain attempts to resolve prediction error by (1) changing the prediction, including the model of the self/world belief, or (2) by taking action to correct and change the feedback. Both responses to prediction error can be seen as learning. However, actions potentially gain more knowledge and allow for better future prediction; this process can be described as active inference. In bodily control, autonomic efferent responses (coupled to behaviour) represent actions. Here, learning to avoid bigger interoceptive prediction errors by taking anticipatory *autonomic action* ultimately underpins the prospective control of allostasis. There have been various accounts proposed to account for how predictive coding of internal state interacts with exteroceptive perception and behaviour. Stephan and Petschzner proposed a hierarchical model in which low-level homeostatic reflexes are yoked to allostatic predictive coding in the generation of allostatic predictions within areas such as anterior cingulate and insular cortices (Stephan et al. 2016). At a superordinate level, metacognitive appraisal of the system as a whole enables the representational modelling of *self-efficacy* through which one can determine whether or not the strategies (behavioural and physiological) that adaptively maintain well-being are working. Excessive inescapable allostatic burden may be appraised though this metacognitive model of self-eficacy as being insurmountable, and trigger a change in strategy. This is proposed by Stephan, Petschzner and colleagues to be the basis of fatigue states and depression (Stephan et al. 2016). Relatedly, anxiety itself has been proposed to be based on a hypersensitivity to interoceptive prediction-error signals represented in insular cortex[19]. Based on such a predictive model and prior work – unsatisfactorily relating intense emotional experience (including anxiety) to increased interoceptive sensitivity – the proposed mismatch between actual and predicted interoceptive states was operationalised within our hierarchical framework of interoception (Garfinkel and Critchley 2013; Garfinkel et al. 2015). This framework distinguishes between objective measures interoceptive perception (e.g. behavioural performance scores on heartbeat detection tasks) and subjective measures from self-reports (e.g. scores on interoceptive questionnaires) that probe an individual's subjective representation of their experience (hence predicted experience) of interoceptive sensations. Observations across different populations showed that anxiety symptoms correlated with the mismatch between 'objective' heartbeat detection task performance and 'subjective' questionnaire measures (such as the Porges Body Perception questionnaire score) and referred to this mismatch as interoceptive trait prediction error. This was demonstrated to be the case particularly for autistic individuals (Garfinkel et al. 2016). These observations led to a clinical trial of an interoceptive training intervention, labelled as *ADIE*; Aligning Dimensions of Interoceptive Experience. Pilot work in nonclinical individuals allowed us to optimise the design of the therapy and undertake a clinical trial in 121 autistic adults who were vulnerable to anxiety symptoms (Quadt et al. 2021). The group was divided into those receiving interoceptive therapy ADIE, and those receiving an exteroceptive training therapy, aiming to improve the emotional recognition, and understanding of non-verbal speech (i.e. prosody training) to help decrease social anxiety. The ADIE therapy itself involved biofeedback practice on heartbeat detection tasks, combined with periods of light physical activity designed to increase transiently the strength of heartbeats, thus amplifying the target sensations. Six sessions of ADIE training were associated with improvements in anxiety score; this improvement was observed to be sustained up to a year later. Over 30% or patients in this active group met criteria for recovery, in contrast to around 15% of the comparison group who received the exteroceptive training (Quadt et al. 2021). Moreover, computational modelling of the learning over the course of training indicated that low-level increases in signal precision of the cardiac afferent information underpinned the improvements in performance and clinical symptoms (Suksailp, Smith, and Garfinkel personal communication). This finding mirrored some of the feedback from patients in the ADIE treatment group who stated that the 'body channel became clearer'. The primary outcome (reduced anxiety) was accompanied by a decrease in scores on the 'Body Perception Questionnaire', which can be viewed as subjective rating of day-to-day experience of interoceptive surprise, and an improved score on heartbeat perception task performance, consistent with a reduction in interoceptive trait prediction error (Quadt et al. 2021). Neuroimaging showed the ADIE training was associated with changes in connectivity between dorsal and genual cingulate cortex with left and right insular cortex, respectively.

This validation of an interoceptive training therapy in a fully powered superiority trial is important since it offers increased access to treatments for mitigating and managing anxiety. Of the 1.8M people referred in the UK for psychological therapies via the Improved Access to Psychological Therapies (IAPT)/Talking Therapies programme (bridging primary care and secondary services) only half complete a full course of therapy (NHS Digital 2023). While there are several reasons for this, autistic individuals, while very prone to anxiety, often struggle to engage with therapies that require reflection on emotional states due to alexithymia. The body focused nature of the ADIE therapy provides an effective alternative. A second (feasibility) trial, *ADAPT*, was undertaken over the COVID lockdown period(s) in people who experienced anxiety in the context of pain and fatigue (Davies et al. 2021). The ADIE approach was implemented for home therapy using a platform in which a tablet computer was linked to a pulse oximeter. Sessions were overseen by a therapist online. Equivalence was shown with online psychological (compassion) therapy in this group, but importantly this study showed a means of digitising the therapy for home use to improve accessibility further. There is now an ongoing work, including an additional clinical trial, to develop and compare a stand-alone digital platform for ADIE (software or therapist guided) therapy. This platform (HeartRater:Clinical) also incorporates an ECG based sensor to replace optical heartbeat sensing (photoplethysmography/pulseoximetry) which underperforms in people with poor peripheral circulation and/or dark skin ('HeartRater | Clinical' Anon n.d.-a).

13 Aberrant Interoception as a Transdiagnostic Mechanism

Beyond anxiety, abnormal interoception is implicated in the expression of psychiatric symptoms and stress-sensitive medical disorders. Mood symptoms, and fatigue, craving, compulsions and urges, and dissociative phenomena can be linked to inner physiological sensations. Correspondingly, interoception is relevant to affective disorders, substance misuse, tic disorders, obsessive-compulsive disorder (OCD), neurodevelopmental conditions, and schizophrenia, as well as 'functional' somatic and neurological conditions (Bonaz et al. 2021; Gray and Critchley 2007; Betka et al. 2018; Rae et al. 2019; Koreki et al. 2020; Elkommos et al. 2023; Hoyos et al. 2019). In these contexts, symptoms may arise from the interoceptive signalling which are linked to differences in organ-level function and control (e.g. in syncope and postural orthostatic tachycardiac syndrome, bladder instability, irritable bowel syndrome (Bonaz et al. 2021; Owens et al. 2018)) as well as through unstable interoceptive underpinnings of self-representation and/or a drive to increase precision in sensory representation of one's body through attention-dependent *quasi-interoceptive* (e.g. somatosensory sensations that are intermittently correlated with some interoceptive signals) rather than viscerosensory (interoceptive) pathways (Quadt et al. 2018; Edwards 2016; Gray et al. 2007; Wilkinson et al. 2013).

Interoceptive signals (e.g. of physiological arousal) may also, through misattribution, underlie aberrant salience or be more fundamentally impaired in disorders like schizophrenia (evidence suggests issues with both low-level signalling and interoceptive metacognition, the latter linked to dissociation). In functional (dissociative) seizures, often also present in people who experience epileptic seizures, psychological aspects of interoception, including mismatch between subjective (questionnaire) and objective (heartbeat perception task performance) measures, correlate with functional seizure frequency (Koreki et al. 2020), while attenuation of the amplitude of heartbeat-evoked potentials (an attention-dependent neural signature of cardiac afferent signalling) precedes functional seizures, thus providing an electroencephalographic index of non-epileptic attacks (Elkommos et al. 2023).

Together a focus on interoception is providing a fresh transdiagnostic framework to reconceptualise difficult-to-manage conditions and offers opportunities to optimise exciting interventions and provide novel therapeutic approaches.

Pharmacological approaches to modulate interoceptive neural signals include obvious interventions such as analgesics (local and systemic), and capsaicin (e.g. into bladder) and perhaps medications like pregabalin, betablockers (to attenuate physiological components of anxiety), and botulinum toxin (Botox) which also blocks autonomic responsivity hence afferent feedback. Even established psychiatric medicines such as the Selective Serotonin Reuptake Inhibitors (SSRIs, e.g. citalopram) act on central interoceptive responses and enhance interoceptive metacognition (Livermore et al. 2022, 2024). The new generation of metabolic medicines (e.g. GLP1 agonists such as semaglutide) are fundamentally acting via interoceptive pathways; their impact likely extends well beyond their licences targeting of diabetes and obesity to impact other aspects of motivational and affective behaviours. Immune modulators from probiotics to biologics similarly target immune–brain interactions mediated interoceptively.

Such pharmacological modulation of interoceptive pathways and representations are complemented by bioelectric approaches, with vagus nerve, pelvic nerve, and spinal (dorsal root ganglion) stimulation already established. Brain stimulation of interoceptive targets may emerge in due course as these are mechanistically mapped and characterised. Biobehavioural interoceptive training protocols, including those validated in the ADIE clinical trial (Quadt et al. 2021; Davies et al. 2021;) and the heartbeat timing study of spider phobics (Watson et al. 2019) provide further therapeutic tools. Relatedly, biofeedback therapy (Nagai et al. 2018) has a long history and, in its classical form, uses exteroceptive information to bypass consciously inaccessible interoceptive pathways to enable volitional control of physiological processes. In practice, this means that interoceptive regulation is achieved aided by exteroceptive yoking in the early phases of establishing control. These medicinal, immune, and behavioural strategies to intervene through interoceptive mechanisms for the mitigation control and alleviation of psychiatric symptoms and disorders can of course be used in combination.

14 Conclusion

The growing science of interoception revitalised theoretical and empirical understanding, from Damasio, Craig, and others over the first decade of this century, of the links between physical and mental well-being, health, and illness with a focus on interoceptive mechanisms and measures that reflect function. This work is offering multiple new targets for helping people manage their lives, with the potential for rapid translation of promising therapies into practice, provided that interdisciplinarity, including co-production with the people that we seek to help, is at the foreground of such innovation. With major mechanistic advances in basic neuroscience mapping onto treatment innovations across medical disciplines, interoception provides both a cross-disciplinary channel and framework for a new integrated, even holistic, culture to support health and well-being and presents biopsychosocial approaches to reversing the burden of disease, particularly that originating from mental illness and psychological disorders.

References

Andolfo I et al (2022) Nobel prize in physiology or medicine 2021, receptors for temperature and touch: implications for hematology. Am J Hematol 97(2):168–170

Anon (n.d.-a) HeartRater | Clinical [online]. Available from: https://www.heartrater.co.uk/heartrater-clinical. Accessed 19 July 2024

Azevedo R et al (2017) The calming effect of a new wearable device during the anticipation of public speech. Sci Rep 7(1):2285

Bechara A et al (1997) Deciding advantageously before knowing the advantageous strategy. Science 275(5304):1293–1295

Bechara A et al (2005) The Iowa gambling task and the somatic marker hypothesis: some questions and answers. Trends Cogn Sci 9(4):159–162

Benarroch EE (2008) The arterial baroreflex: functional organization and involvement in neurologic disease. Neurology 71(21):1733–1738

Betka S et al (2018) How do self-assessment of alexithymia and sensitivity to bodily sensations relate to alcohol consumption? Alcohol Clin Exp Res 42(1):81–88

Bonaz B et al (2021) Diseases, disorders, and comorbidities of Interoception. Trends Neurosci 44(1):39–51

Brydon L et al (2008) Peripheral inflammation is associated with altered Substantia Nigra activity and psychomotor slowing in humans. Biol Psychiatry 63(11):1022–1029

Chen WG et al (2021) The emerging science of interoception: sensing, integrating, interpreting, and regulating signals within the self. Trends Neurosci 44(1):3–16

Craig AD (2009) How do you feel – now? The anterior insula and human awareness. Nat Rev Neurosci 10(1):59–70

Craig AD (2014) How do you feel? An interoceptive moment with your neurobiological self. Princeton University Press. Available from: https://academic.oup.com/princeton-scholarship-online/book/23787. Accessed 24 May 2024

Craig AD (2002) How do you feel? Interoception: the sense of the physiological condition of the body. Nat Rev Neurosci 3(8):655–666

Critchley HD et al (2000) Cerebral correlates of autonomic cardiovascular arousal: a functional neuroimaging investigation in humans. J Physiol 523(1):259–270

Critchley HD et al (2002) Fear conditioning in humans: the influence of awareness and autonomic arousal on functional neuroanatomy. Neuron 33(4):653–663

Critchley HD et al (2003) Human cingulate cortex and autonomic control: converging neuroimaging and clinical evidence. Brain 126(10):2139–2152

Critchley HD (2005) Neural mechanisms of autonomic, affective, and cognitive integration. J Comp Neurol 493(1):154–166

Critchley HD et al (2004) Neural systems supporting interoceptive awareness. Nat Neurosci 7(2): 189–195

Critchley HD (2004) The human cortex responds to an interoceptive challenge. Proc Natl Acad Sci 101(17):6333–6334

Critchley HD, Harrison NA (2013) Visceral influences on brain and behavior. Neuron 77(4): 624–638

Damasio AR (2006) Descartes' error: emotion, reason and the human brain, Review edn with a new preface edn. Vintage, London

Damasio AR (2000) The feeling of what happens: body and emotion in the making of consciousness. A harvest book, vol 1, Harvest edn. Harcourt, San Diego

Davies G et al (2021) Altering dynamics of autonomic processing therapy (ADAPT) trial: a novel, targeted treatment for reducing anxiety in joint hypermobility. Trials 22(1):645

Donadio V, Kallio M et al (2002a) Inhibition of human muscle sympathetic activity by sensory stimulation. J Physiol 544(1):285–292

Donadio V, Karlsson T et al (2002b) Interindividual differences in sympathetic and effector responses to arousal in humans. J Physiol 544(1):293–302

Droste C et al (1994) Baroreceptor stimulation: pain perception and sensory thresholds. Biol Psychol 37(2):101–113

Edwards MJ (2016) Chapter 12 – neurobiologic theories of functional neurologic disorders. In: Hallett M et al (eds) Handbook of clinical neurology, Functional neurologic disorders. Elsevier, pp 131–137. Available from: https://www.sciencedirect.com/science/article/pii/B9780128017722000126. Accessed 29 May 2024

Elkommos S et al (2023) Changes in the heartbeat-evoked potential are associated with functional seizures. J Neurol Neurosurg Psychiatry 94(9):769–775

Evrard HC (2019) The organization of the primate insular cortex. Front Neuroanatomy:1343

Furlan A, Petrus P (2023) Brain–body communication in metabolic control. Trends Endocrinol Metab 34(12):813–822

Garfinkel SN et al (2016) Discrepancies between dimensions of interoception in autism: implications for emotion and anxiety. Biol Psychol:114117–114126

Garfinkel SN et al (2014) Fear from the heart: sensitivity to fear stimuli depends on individual heartbeats. J Neurosci 34(19):6573–6582

Garfinkel SN et al (2021) Interoceptive cardiac signals selectively enhance fear memories. J Exp Psychol Gen 150(6):1165–1176

Garfinkel SN et al (2015) Knowing your own heart: distinguishing interoceptive accuracy from interoceptive awareness. Biol Psychol:10465–10474

Garfinkel SN, Critchley HD (2013) Interoception, emotion and brain: new insights link internal physiology to social behaviour. Commentary on: social cognitive and affective. Neuroscience 8(3):231–234

Garfinkel SN, Critchley HD (2016) Threat and the body: how the heart supports fear processing. Trends Cogn Sci 20(1):34–46

Gianaros PJ et al (2005) Anterior cingulate activity correlates with blood pressure during stress. Psychophysiology 42(6):627–635

Gray MA et al (2010) Baroreceptor activation attenuates attentional effects on pain-evoked potentials. Pain 151(3):853

Gray MA et al (2012) Emotional appraisal is influenced by cardiac afferent information. Emotion 12(1):180–191

Gray MA et al (2009) Following one's heart: cardiac rhythms gate central initiation of sympathetic reflexes. J Neurosci 29(6):1817

Gray MA et al (2007) Modulation of emotional appraisal by false physiological feedback during fMRI. PLoS One 2(6):e546

Gray MA, Critchley HD (2007) Interoceptive basis to craving. Neuron 54(2):183–186

Harrison NA, Brydon L, Walker C, Gray MA, Steptoe A, Critchley HD (2009a) Inflammation causes mood changes through alterations in subgenual cingulate activity and mesolimbic connectivity. Biol Psychiatry 66(5):407–414

Harrison NA, Brydon L, Walker C, Gray MA, Steptoe A, Dolan RJ et al (2009b) Neural origins of human sickness in interoceptive responses to inflammation. Biol Psychiatry 66(5):415–422

Harrison NA et al (2010) The embodiment of emotional feelings in the brain. J Neurosci 30(38): 12878–12884

Hoyos C et al (2019) The role of dissociation and abuse among adolescents who self-harm. Aust N Z J Psychiatry 53(10):989–999

Hsueh B et al (2023) Cardiogenic control of affective behavioural state. Nature 615(7951):292–299

James W (1884) What is an emotion? Mind:19188–19205

Khalsa SS et al (2018) Interoception and mental health: a roadmap. Biol Psychiatry Cogn Neurosci Neuroimaging 3(6):501–513

Klein AS et al (2021) Fear balance is maintained by bodily feedback to the insular cortex in mice. Science 374(6570):1010–1015

Knight DC et al (2005) The role of the human amygdala in the production of conditioned fear responses. Neuroimage 26(4):1193–1200

Koreki A et al (2020) Trait and state interoceptive abnormalities are associated with dissociation and seizure frequency in patients with functional seizures. Epilepsia 61(6):1156–1165

Lange CG (1895) Les émotions; étude psychophysiologique. Alcan. Available from: https://books.google.co.uk/books?id=qyPGd-dUtNcC

Lange CG, James W (1922) Open library ID: OL6645245M. The emotions. Williams & Wilkins Company, Baltimore

Livermore JJA et al (2022) A single oral dose of citalopram increases interoceptive insight in healthy volunteers. Psychopharmacology (Berl) 239(7):2289–2298

Livermore JJA et al (2024) General and anxiety-linked influence of acute serotonin reuptake inhibition on neural responses associated with attended visceral sensation. p 2024.01.23.24301647. [online]. Available from: https://www.medrxiv.org/content/10.1101/2024.01.23.24301647v1. Accessed 29 May 2024

Nagai Y et al (2004) Activity in ventromedial prefrontal cortex covaries with sympathetic skin conductance level: a physiological account of a "default mode" of brain function. Neuroimage 22(1):243–251

Nagai Y et al (2018) Epileptic seizures are reduced by autonomic biofeedback therapy through enhancement of Fronto-limbic connectivity: a controlled trial and neuroimaging study. EBioMedicine:27112–27122

NHS Digital (2023) Psychological therapies: reports on the use of IAPT services, England, May 2023 final including a report on the IAPT employment advisers pilot. Available from: https://digital.nhs.uk/data-and-information/publications/statistical/nhs-talking-therapies-monthly-statistics-including-employment-advisors/performance-may-2023-provisional-june-2023#resources

Owens AP et al (2018) Investigating the relationship between cardiac interoception and autonomic cardiac control using a predictive coding framework. Auton Neurosci Basic Clin:21065–21071

Patchitt J, Garfinkel S, Strawson WH, Miller M, Tsakiris M, Clarke A, Critchley HD (2024) Somatosensory false feedback biases emotional ratings through interoceptive embodiment. Sci Rep. Manuscript in press

Paul SM, Potter WZ (2024) Finding new and better treatments for psychiatric disorders. Neuropsychopharmacology 49(1):3–9

Paulus MP, Stein MB (2006) An insular view of anxiety. Biol Psychiatry 60(4):383–387

Prevot V et al (2021) Tanycytes in the infundibular nucleus and median eminence and their role in the blood–brain barrier. In: Handbook of clinical neurology. Elsevier, pp 253–273. Available from: https://linkinghub.elsevier.com/retrieve/pii/B9780128201077000161. Accessed 28 May 2024

Quadt L et al (2021) Interoceptive training to target anxiety in autistic adults (ADIE): a single-center, superiority randomized controlled trial. eClinicalMedicine:39. Available from: https://www.thelancet.com/journals/eclinm/article/PIIS2589-5370(21)00322-9/fulltext. Accessed 29 May 2024

Quadt L et al (2018) The neurobiology of interoception in health and disease. Ann N Y Acad Sci 1428(1):112–128

Rae CL et al (2019) Dimensions of interoception predict premonitory urges and tic severity in Tourette syndrome. Psychiatry Res:271469–271475

Schachter S, Singer J (1962) Cognitive, social, and physiological determinants of emotional state. Psychol Rev 69(5):379–399

Seth AK (2013) Interoceptive inference, emotion, and the embodied self. Trends Cogn Sci 17(11):565–573

Seth AK, Tsakiris M (2018) Being a beast machine: the somatic basis of selfhood. Trends Cogn Sci 22(11):969–981
Sherrington CS (1911) The integrative action of the nervous system. Yale University Press, New Haven. Available from: https://content.apa.org/books/13798-000. Accessed 28 May 2024
Singer T et al (2009) A common role of insula in feelings, empathy and uncertainty. Trends Cogn Sci 13(8):334–340
Stephan KE et al (2016) Allostatic self-efficacy: a metacognitive theory of dyshomeostasis-induced fatigue and depression. Front Hum Neurosci:10550
Sterling P (2012) Allostasis: a model of predictive regulation. Physiol Behav 106(1):5–15
Subbiah V (2023) The next generation of evidence-based medicine. Nat Med 29(1):49–58
Wager TD et al (2009) Brain mediators of cardiovascular responses to social threat. Neuroimage 47(3):821–835
Watson DR et al (2019) Computerized exposure therapy for spider phobia: effects of cardiac timing and interoceptive ability on subjective and behavioral outcomes. Psychosom Med 81(1):90–99
Wilkinson M et al (2013) Electrocutaneous pain thresholds are higher during systole than diastole. Biol Psychol 94(1):71–73

The Touched Body and the Experience of Self

Rebecca Böhme

Contents

Abstract Slow caressing of the skin activates C-tactile fibers in the periphery and the posterior insula cortex. Because of this, Bud Craig considered affective touch to be an interoceptive modality. Through the tactile sense, we perceive the border of our own body and the largest of our organs, the skin. Whether or not C-targeted touch is considered interoceptive, it contributes fundamentally to the development and maintenance of the bodily self. This is supported by experimental data from cases, where somatosensory processing is altered, and the other way around, i.e., when the bodily sense of self is changed either pharmacologically or in psychiatric conditions. Self-touch can be seen as a special case contributing to the bodily self-model by providing high fidelity signals within a closed feedback loop. Social touch, especially between parents and children and between romantic partners, plays a crucial role in social allostasis and the co-regulation of physiology and emotions. Touch, both self-touch and social touch, should therefore be considered foundational for the bodily self and essential for mental and physiological well-being. Through touch, we perceive the self in its most basic form, as a social body.

Key words Self · Bodily self · Touch · Social touch · Interoception · Self-other-distinction

R. Böhme (✉)
Center for Social and Affective Neuroscience, Linköping University, Linköping, Sweden
e-mail: Rebecca.Bohme@liu.se

Curr Topics Behav Neurosci (2025) 70: 87–106
https://doi.org/10.1007/7854_2025_577
Published Online: 12 March 2025

1 Affective Touch – An Interoceptive Modality?

When touching something or being touched by someone else, we always perceive ourselves as a body, the borders of this body, and the distinction between "self" and "non-self." The skin is our largest organ which both separates and insulates us from the outside world, but also mediates intensive interactions with it and with other beings (Böhme 2019; McGlone et al. 2014). Therefore, the sense of touch might be considered a special case when distinguishing intero- and exteroception.

Interoception is the perception of signals and sensations from within the body, typically the viscera, whereas exteroception is the perception of signals from outside the body. While this distinction is simple in its definition, it can become difficult to draw a clear line between these two concepts. The classification of the "far"-senses vision and audition as exteroception is easy. Also the chemical senses, taste and smell, are usually activated by external stimuli (although one might potentially consider the case where they are activated by stimuli arising from within, e.g., one's own gastrointestinal system, as interoceptive). Similarly, sensations from within like feeling one's heartbeat, the need for air, or the need to go to the toilet are clearly interoceptive. But what about nociception, i.e., the perception of pain? Pain can be evoked by causes both from the inside and from the outside. Also, the perception of temperature can be triggered by cold air on a winter morning, but one can also feel cold "from the inside" when getting sick. I will now discuss why the sense of touch also lies at the intersection between intero- and exteroception.

Exploratory touch of surfaces and shapes can easily be understood as exteroception, but even in this case one might argue: when touching a surface, while we perceive the surface structure, we always also perceive our own body and one of its organs – the skin – in this interaction. The percept of the specific surface structure is mediated not only by activation of the different touch receptors but also by the deformation of the skin (Sakaguchi et al. 2023; Sanzeni et al. 2019). However, humans do not only use their sense of touch for exploration of the environment, but we also receive touch from other humans. For humans, like for many other mammals, touch is a form of pair-bonding. We hug, kiss, and caress each other – these types of touch are often referred to as "social touch." The slow, caressing touch is further specified as "affective touch." From a psychological or systems perspective, the reception of social and affective touch lets us sense not only the other person providing the touch, but also always ourselves. When touched, we clearly perceive the borders of our own body (Boehme and Olausson 2022).

In this sense, touch might be considered to be right at the border between extero- and interoception: we perceive the boundaries of the inside through the outside. In his opinion article "How do you feel?," Bud Craig suggests "that interoception should be redefined as the sense of the physiological condition of the entire body, not just the viscera" (Craig 2002). He bases this statement on anatomical and physiological findings, specifically on the spinothalamocortical system that begins in the dorsal horn of the spinal cord and projects to the brain stem and then to the thalamus and the cortex. Signals that travel along this pathway originate in small-diameter primary afferents. These afferents are the slow Aδ- and C-type fibers which innervate all tissues of the body, including the skin. They respond to a variety of physiologically relevant stimuli, for example, temperature and painful stimuli, metabolic changes, e.g., of pH, or immune system activity. The slowest fiber type, the C-fiber, had been termed C-nociceptor, because these fibers were known to respond to pain stimuli. All tactile receptors – those receptors that respond to, e.g., vibration or pressure – relay their signals via the fast, myelinated fibers, which project to the primary somatosensory cortex – or so it was thought. However, in 1939, Yngve Zotterman discovered a special sub-class of C-fibers, the so-called C-tactile (Ct) fibers, which respond not to painful stimulation but to a type of touch: slow stroking of the skin (Zotterman 1939).

In a series of experiments, it has been shown that Ct-fibers specifically respond to slow stroking of their receptive fields. The responsiveness of these nerves, i.e., their spiking frequency, follows an inverted U-shape with the peak activity at a preferred speed of 3 cm/s (Ackerley et al. 2014a; McGlone et al. 2014; Olausson et al. 2010; Vallbo et al. 1999; Wessberg and Norrsell 1993). Both slower and faster speeds can still activate the C-fibers, but to a lesser degree. Interestingly, this inverted U-shape corresponds to the average subjective rating of the pleasantness of the stroking stimulus, i.e., on average, study participants reported the stroking at 3 cm/s to be the most pleasant kind of stroking compared to both slower and faster stroking (Löken et al. 2009). Ct-fibers, which are located in the hairy skin, also respond differently depending on stimulus temperature. While most receptors react to environmental or internal changes away from the norm, e.g., in the case of thermoreceptors to temperatures warmer or colder than body temperature, the Ct-fibers show an optimal firing response to stimuli of around 32°C (Ackerley et al. 2014b). This is a bit below body temperature and approximately the temperature of human fingertips. Taken together, these findings make a strong case for Ct-fibers to be specifically tuned to interpersonal caress.

Do we humans indeed have a receptor that evolved specifically to let us perceive when we are stroked in a pleasant way by another human? This beautiful conclusion might seem too good to be true. And as in most cases of scientific discoveries, the actual nature underlying the phenomena might be more complex than first thought. Follow-up studies paint a more diffuse picture on both the subjective experience of the preferred stroking speed (there are large interindividual differences (Croy et al. 2021)) and the physiology of the Ct fibers (Ackerley 2022). But while the details might prove more complex, the role of Ct fibers in mediating slow mechanostimulation of the skin at body-temperature or slightly below still speaks for them as a receptor for the emotional and affective aspects of social touch.

In "How do you feel?," Craig then takes his discussion from the spinal cord laminae fibers and their initial relay stations in the brain stem and the thalamus to higher-order cortical areas. In both humans and monkeys, there is good evidence that suggests that the signals from these pathways terminate in the insula cortex, leading to the terminology "interoceptive cortex." While the projections from the thalamus in monkeys reach the entire insular cortex, in humans they appear to terminate in the posterior parts of it. Imaging studies in humans show activity in the posterior parts of the insula in response to a variety of typical interoceptive stimuli, including pain, temperature, itch, muscle ache, gastric distention, gustatory percepts, etc. (Craig 2011). It is suggested that in humans – very broadly speaking – these signals are integrated with each other and with signals from other brain areas in the mid-insula. This integrated information is then relayed to the anterior insula, which relates it to subjective conscious perceptions. Craig proposed "that the human anterior insula substantializes what can be called 'the sentient self', based on a progressive integration of all neural activity across all networks of the brain onto the homeostatically relevant representation of inner and outer salience formed in the mid-insula, formed on the basis of the interoceptive template for a 'feeling'" (Craig 2011).

If affective, social touch should be considered interoceptive, one would expect it to also elicit activation of the posterior insula as has been described for other interoceptive stimuli. This was not easy to prove, since stimulation with slow, affective stroking would never only activate the Ct-fibers, but always also the faster Aβ afferents, which are associated with the discriminative aspects of the tactile sense and activate primary somatosensory cortex. First support for the specific involvement of the insular cortex in the processing of Ct-fiber signals arose from studies with rare patients who had lost their myelinated, i.e., large diameter, fibers. The two patients with this condition can perceive slow stroking on the arm; however, their subjective percepts are inconsistent and vague (McGlone et al. 2014). McGlone and Olausson write in their review article: "Although the two neuronopathy subjects are not able to give a concise or detailed description of the sensation evoked by CT stimulation, some consistent features of the quality emerged. The sensation has no quality of pain, tickle, or itch. Further, they report that the sensation elicited by CT stimulation is slightly or moderately pleasant." A functional magnetic resonance imaging (fMRI) study with one of these patients revealed indeed selective activity in the insular cortex, while no activation in the somatosensory cortex was observed (Olausson et al. 2008). These results have been further corroborated by a large body of work in healthy participants comparing slow (Ct-optimal) versus fast (non-Ct-optimal) stroking or stroking of the hairy skin (where the Ct-fibers are located) to stroking of the non-hairy skin (where no Ct-fibers have been found). These types of comparisons allow for a separation of the Ct-fiber-evoked brain signals, as fast stroking and stroking of the non-hairy skin activate the fast, myelinated fibers, but not or to a lesser degree the Ct-fibers. By subtracting the signal evoked in the fast-stroking conditions or during stroking of the non-hairy skin from the Ct-optimal conditions, the brain signal can be "cleared" from the Aβ-related activity, allowing for investigation of the Ct-fiber-related activations. A recent meta-analysis of such studies demonstrated that posterior insula does indeed respond specifically to slow stroking of the hairy skin (Morrison 2016).

Whether these findings should lead to the classification of affective social touch as an interoceptive stimulus or as a stimulus that is both intero- and exteroceptive continues to be a matter of debate. Bud Craig thought so as can been read in many of his articles.

2 Touch and the Bodily Self

Touch perceptions depend on a variety of factors including who the interaction partner is, what type of relationship exists between touching partners, which state an individual currently is in, emotionally and physiologically, and what the surrounding environment looks like (is it a private or a public space?). This further indicates the close entanglement with the physiological state of the body and its homeo- and allostatic needs (more on this below), and it complicates research on these topics immensely. Because of the similarity of the Ct-fibers' anatomical pathways and functional neural processing with the other interoceptive modalities, I will focus on those types of touch interactions that activate Ct-fibers and are therefore based on their physiological processing potentially to be considered interoceptive – or at least seem to involve different processing pathways than classic exploratory touch behaviors.

In addition to the physiologically and anatomically distinct processing of Ct-fiber targeted touch, i.e., soft, slow stroking, the already mentioned perception of one's own bodily boundaries supports the case for touch to play a special role in the establishment and maintenance of the bodily self. Already prenatally in the mother's womb, tactile sensations can be perceived and might contribute to the development of an understanding where the own body ends, and the outside world begins (Ciaunica and Fotopoulou 2017). The amniotic fluid causes movement of the thin hairs which cover the whole body of an embryo at 21 weeks called lanugo hair. Ct-fibers are typically located around hair follicles and might be mediating these early tactile sensations. Embryos have been observed to touch themselves as well as the other body that is present in their environment – in most cases this can be the wall of the mother's womb, and in the case of twin-pregnancies, this can also be the other twin (Castiello et al. 2010). Through such early self- and other-touch, a basic self-other-differentiation might be established which can be understood as a minimal bodily self-percept.

This understanding of self versus non-self is essential for any interaction with the outside world and any advanced social interaction with other (human) beings. Only when an organism knows where its limbs are and how far they extend into space, can it use them effectively to interact with this space. Only if we understand that we are separate beings from the other humans around us, can we have effective and meaningful interactions with them. Many different sensory and motorsensory perceptions contribute to this understanding: proprioception, i.e., the sense of where and in which position our limbs are; interoceptive percepts from within the different organ systems, including our heartbeat and breathing; the sense of agency, i.e., understanding when we are the one acting and causing an effect; and potentially more. This perception of the self as a body typically only receives our conscious attention, when it either dysfunctions, in the case of discomfort, pain, or illness, or when we experience high bodily pleasure, mostly in the form of sexual pleasure (Böhme 2008). Sensing our bodies during pain and pleasure, but especially if we are able to experience this bodily self also outside these extremes, provides a unique opportunity to perceive ourselves as nature (Böhme 2010). Our felt body is crucial for our human experience, for the experience of being in the here and now, for the experience of reality and our connection to it (Böhme 2016). There are many different ways how one can evoke or increase the bodily self-percept. Typical exercises involve the increased focus on interoceptive sensations, e.g., in meditative practices where the mediator focuses on their breathing or heart rate or lets their attention travel through their body when doing a "body scan." Conscious touch perception, both of self-touch and of social touch, might be another way of directing attention toward our bodily sense of self.

Craig famously defined interoception as the sense of the physiological state of the body (Craig 2003). The conscious perception of our physiological state can easily be evoked by all senses, e.g., suddenly noticing that one is feeling hungry when smelling or seeing a tasty food or having to go to the bathroom when hearing the sound of running water. However, in these cases, the exteroceptive perceptions only call the physiological state to our attention. In the case of touch, we might sense and directly alter aspects of our physiology directly. The philosopher Edmund Husserl, the founder of phenomenology, describes the intricate connection between touch and the sensing of one's own state (Husserl 1952). When we feel pressure, warmth, or cold, we do not only perceive objects and surface temperatures but also sense "ourselves as a body" – and ourselves in relation to these things. Whether we perceive something as warm or cold also depends on our own body temperature and our internal state (for instance, we are more likely to feel cold when we are tired). In addition to temperature, we can perceive other aspects of our state when we are touched by another person: we can feel pain or that we are stressed, tense, or anxious. The latter sensations are often only identified through touch releasing them (more on how touch alters physiology is discussed below and in part 4 "Touch as a co-regulator of homeostasis"). On the positive side, sexual desire, relaxation, feeling safe, supported, comforted, or consoled are often experienced when being touched by others. Whether these sensations are to be considered a "physiological state" might be debated. All of them do entail physiologically measurable components. For instance, relaxation and feeling safe are associated with the activity of the parasympathetic nervous system. Another example is how we use self-touch to check ourselves for a fever or an injury. We can also use self-touch to guide our attention to a certain body part or more generally to call our attention into our body (more on self-touch below) or even to ensure us of being awake (although typically not through Ct-targeted caress but by pinching ourselves).

In order to be able to study touch and its functions, we can try to differentiate types of touch: touch can be active or passive, and a touch interaction can occur with the outside world and other living beings, or it can be self-directed or self-evoked. In our daily life, these types often co-occur, and they cannot be fully separated. When I touch someone else, I will always also perceive the touch on my own skin. When someone else touches me, I will in many cases actively touch them back. Regarding Ct-fiber directed touch, most research has used passive touch, where a robot or another human provides stroking at slow speeds to the forearm of the participant, who in turn is to sit still and not react in any way, a type of interaction far from our typical daily life touch interaction, rendering these experiments low on their ecological validity. Most of these studies have investigated emotional or affective experiences of this type of touch (Is it pleasant? How intense is it?) and the neural correlates (Which brain areas respond to this type of touch?).

Research has shown that Ct-targeted touch and social touch interactions with our close others have physiological and psychological consequences. For example, touch can reduce heart rate, pain perception and processing, subjective stress level, and cortisol. These effects have been even shown using ecologically valid paradigms, for example, hand holding with the romantic partner, massage, or skin-to-skin caress (Ditzen et al. 2007; Holt-Lunstad et al. 2008; Liljencrantz et al. 2017; Savallampi et al. 2023). Using skin-to-skin touch, we have shown that social touch is associated with a decrease of sympathetic and an increase of parasympathetic nervous system activity (Candia-Rivera et al. 2024a) and that these changes are coupled to changes in brain activity (Candia-Rivera et al. 2025).

Whether touch has such positive effects depends on a variety of factors, as listed above. This leads to an interesting consideration: even if slow stroking of the skin activates Ct-fibers and is generally associated with a pleasant sensation, something else must happen along the processing of this basic peripheral signal, which could modulate its evaluation, but even its unconscious effects on physiology, depending on context, interaction partner, and current physiological state. At which processing state this happens is currently unknown. Recent work in animals suggests that the brain sends a large amount of modulatory signals to the spinal cord (Abraira et al. 2017; Liu et al. 2018) and the spinal cord integrates signals from different sensory fibers within intricate local networks (Emanuel et al. 2021). It might therefore be possible that these different contributing factors can modulate the processing of a touch sensation already at the spinal cord level, e.g., upregulate the signal transduction if we are touched by a romantic partner and downregulate it if we need to focus, for example, to finish a difficult mental or physical task (Bohic and Abraira 2022). We have demonstrated differences between self- and social touch processing latencies already at spinal cord level, which might indicate a modulation of tactile signals within spinal cord circuitry (Boehme et al. 2019). Such an early integration and modulation of tactile sensations, which fundamentally contribute to our bodily sense of self, further corroborate the essential contribution of the body and the localization of this sense of self at least in part within this felt body.

Like touch can be understood as something between intero- and exteroception, self-touch can also be considered a special case at an exciting intersection: during self-touch, we are both the toucher and the touched, we perform the touch actively and receive the touch passively. Within this closed-loop, we can perfectly predict how the touch is going to feel. Because of this high predictability, it has been suggested that the self-evoked sensation will be attenuated. The original model by Helmholtz calls this prediction the "efference copy"- a copy of the motor command which would be used to cancel out the predicted sensory outcome. The modern models of predictive processing suggest a similar mechanism (Friston 2009). While the neural implementation of these models is still under debate, the attenuation of the self-evoked sensations has been shown both behaviorally and neurally (Blakemore et al. 1998; Boehme et al. 2019; Kilteni et al. 2020).

An important question arises from these observations: if the self-evoked sensation is attenuated, which could be interpreted as this sensation being treated as non-relevant by the brain, why do we keep touching ourselves? As became very clear during the recent covid-pandemic, humans touch their own face very often – numbers up to 50 times an hour have been reported (Rahman et al. 2020). Interestingly, it has been observed, that self-touch, specifically to the face, is increased when we find ourselves in a social and/or challenging situation (Spille et al. 2021). It is important to note that the specifics of these spontaneous self-touches have not been analyzed, i.e., we currently do not know if these can be understood as Ct-fiber targeted or not. Different explanations for the occurrence of spontaneous self-touch are suggested: nonverbal communication of negative affect, emotion regulation, attention regulation, and development of the bodily self. The latter, our main interest here, was specifically discussed with regard to infants, who show spontaneous self-touch, which can be interpreted as a way of "tuning" the predictive model (DiMercurio et al. 2018; Moszkowski and Stack 2007; Rochat and Hespos 1997). However, we do not simply stop touching ourselves when we have developed functional models of the bodily self. In adults, self-evoked sensations might continue to fine-tune, reset, and maintain the bodily self-model (Boehme and Olausson 2022). In support of this idea, it has been shown experimentally that self-touch can restore a disruption of the bodily self-model, both if it is induced experimentally through a body illusion (Cataldo et al. 2024; Cataldo et al. 2021) and if it occurs as the consequence of neural injury (Van Stralen et al. 2011). It might also be possible that the high predictability of the self-evoked tactile sensations has a calming or grounding effect. This could be the reason that self-touch is observed to occur more frequently in highly unpredictable situation, like social gatherings or emotionally challenging situations. Through their high predictability, self-touches might reduce stress and arousal in an unpredictable environment, while at the same time re-enforcing the bodily sense of self.

3 Alterations of the Sense of Self Coincide with Altered Touch Processing

So far, only a few studies have tried to understand experimentally how touch relates to the bodily self. Often these studies approached this question from a deficit perspective: What happens to the bodily self if touch processing or perceptions are altered? Is an alteration of the sense of self, e.g., in depersonalization, schizophrenia, or anorexia nervosa, related to differences in touch processing and perception?

The first question is not easily addressed, as somatosensory processing seems to be so fundamentally important that humans are rarely born without it functioning properly. In cases where the somatosensory system shows a developmental loss of function, the consequences are far reaching, which does not allow for drawing conclusions specific to the bodily self. There are of course cases where a whole limb is lost or where spinal or cortical injuries lead to loss of sensation. Investigations with these patients point toward changes of the bodily sense of self, e.g., by higher degrees of depersonalization experiences (Lenggenhager et al. 2012). Interestingly, there is also some support for a relationship between touch-frequency in daily life in the general population and depersonalization experiences (Ciaunica et al. 2022; Salamone et al. 2024). However, also these situations do not allow for the investigation of the specific contribution of affective, social, or Ct-targeted touch.

Neurodiverse individuals, especially autistic people and people with attention-deficit hyperactivity disorder (ADHD), regularly report altered touch perceptions (Cascio 2010). This can be both hyper- and hyposensitivity. Work with these groups provides some experimental support for the idea that alterations of somatosensory processing relate to bodily self-processing. We tested the neural processing of self-touch and other-touch as well as somatosensory integration in a group of adults who had a diagnosis of ADHD. Using both Ct-fiber-targeted touch (i.e., slow, caress-like skin-to-skin) and Aβ-fiber targeted touch (i.e., short electric pulses), we showed that ADHD participants were more sensitive to somatosensory stimulation (Frost-Karlsson et al. 2024) and that the distinctive processing of self- and other-touch was enhanced (Boehme et al. 2020). This was accompanied by a reduced susceptibility to the rubber hand illusion (Boehme et al. 2020), a body illusion thought to rely on bodily self-malleability. However, these findings do not allow for conclusions about causality between alteration of somatosensory processes and the distinctiveness of the bodily self, and it cannot be excluded that underlying attention-related processes mediate both these findings.

Another recent study provided further support for the suggested association. It demonstrated a relationship of left insular cortex response to interpersonal touch with individual sensitivity to somatosensory stimuli, when controlling for personality traits (Schaefer et al. 2022). Considering the role of the insula in interoception and the bodily sense of self, this finding supports that basic somatosensation could be tightly coupled to aspects of the self. Interestingly, this study used touch to the palm, which is non-hairy skin and, therefore, this touch is not expected to activate Ct-fibers.

Another point worth our attention is that the correlation was found specifically for the left insula. There is some evidence for potential lateralization in insula function (Craig 2011). The left insula seems to be specifically involved in the processing of positive emotions (Duerden et al. 2013), for example it is activated more when experiencing intense love (Bartels and Zeki 2004). When stimulated or if in the center of an epileptic seizure, activity of the left insula has been reported to often result in sensations of intense joy, euphoric and ecstatic sensations (Gschwind and Picard 2016). If this is the case, the relationship of the left insula with sensory sensitivity suggests an intriguing hypothesis: sensory hypersensitivity is especially common in neurodiverse populations, i.e., autistic people and people with ADHD. While neurodiverse individuals suffer from being overstimulated (Taels et al. 2023), they also report what is referred to as "sensory joy" (Boehme in press; Sibeoni et al.

2022). This experience of sensory joy (sometimes also referred to as "glimmer" by the neurodiverse community) could be related to left insular hyperresponsivity. This might be mediated at any stage of sensory processing; however, there is some evidence that tactile peripheral thresholds are unaltered in neurodiverse populations (Boehme et al. 2020; Parush et al. 1997), suggesting the hyperresponsivity to occur later in the processing stream, e.g., in the spinal cord (which is supported by animal work (Orefice et al. 2016)) or at cortical levels. Whether the experience of sensory hypersensitivity and sensory joy is accompanied with a heightened sense of bodily self or presentness has not been investigated yet.

The insula activation in relation to sensory sensitivity is also in line with another recent study, which found insula to also respond to deep pressure touch; however, in this case, the activation was localized more medially (Case et al. 2021). The authors speculate that the deep pressure sensation is at least in part mediated by afferents in deeper tissues than the skin, in muscle and connective tissues. This further supports the idea that touch should be considered to be a sensation somewhere between intero- and exteroception if deep pressure, like during a massage or a tight hug, is perceived by deeper bodily tissues and activates the insular cortex.

Instead of investigating altered touch processing, one can also begin from the other side of the hypothesized association and investigate somatosensation in the case of an altered bodily self. This can either be studied when experimentally evoking an alteration of the sense of self or in psychiatric conditions associated with such an alteration. Before discussing such studies, it should be noted that both approaches are limited regarding their support for a causal relationship between touch processing and sense of self. In the case of an experimental intervention to change the sense of self, there will typically be additional physiological changes (e.g., of heartrate and breathing) which could introduce moderating or mediating effects. In the case of a psychiatric condition, there are many additional symptoms which persevere for years or a lifetime, limiting the interpretation of a direct relationship between the sense of self and somatosensation.

Experimentally, an alteration of the sense of self can be evoked through a pharmacological intervention. Using this approach, we have shown that a change of neural touch processing did indeed accompany a reduced self-perception (Kaldewaij et al. 2023). We administered ketamine to our healthy participants, while they were in the fMRI scanner and were instructed to either touch their own arm or receive touch from the experimenter. Ketamine at sub-anesthetic doses causes an alteration of the sense of self: dissociation, out-of-body perceptions, altered perceptions of time and space, etc., are common. We found that the neural difference between self-touch and touch by another person was reduced during the ketamine-session compared to a placebo-session. This effect was driven by lower activation during touch by the other person in a secondary processing area, which included temporal and parietal cortex and also the posterior insula. Similarly, administration of psilocybin, a psychedelic substance evoking an altered state of self, also elicited reduced neural responses to surprising tactile stimuli (Duerler et al. 2022).

A different approach to altering the sense of bodily self are body illusions, like the rubber hand illusion and out-of-body illusions, or in a modern adaptation the embodiment of an avatar in virtual reality. The vast literature on studies employing these methods cannot be summarized here comprehensively. In short, these illusions rely on multisensory integration, involving vision and touch. For example, in the out of body illusion, the participant sees a mannequin in front of them being stroked on the back while being themselves stroked on their back in a synchronous fashion (Aspell et al. 2012). This can cause the sensation of actually being located in or at least closer to the mannequin. However, it is important to note that this does not mean that the participant really loses their bodily sense of self and fully believes to be located at the mannequin's position. While it is interesting that the combination of these sensory stimuli can cause body illusions, I would argue that these illusions cannot provide causal evidence regarding the relationship between the actual bodily self and touch processing, if the manipulation to evoke an illusory, altered self involves touch. However, that touch contributes to evoke these illusions can be seen as support for its contribution to integrating other objects, like prostheses or tools, into the body schema (Miller et al. 2019).

A different, also somewhat limited, approach is the investigation of psychiatric conditions characterized by self-dysfunctions. In anorexia, own body perception is heavily altered. While anorexia patients are often described as having a distorted body "image" and there is a strong focus on visual processing of the own body, affected patients typically report that they do not perceive the border of their own body. This suggests that dysfunctions in self-other-distinction and/or bodily sense of self, including interoception, might underlie this distortion. Studies have indeed reported interoception to be affected in anorexia (Jacquemot and Park 2020; Lucherini Angeletti et al. 2022). This includes also affective touch processing: anorexia patients report reduced pleasantness for slow stroking stimuli (Crucianelli et al. 2016) and alterations in brain responses to Ct-targeted touch have been found (Davidovic 2016). These effects seem to be related to deeper, underlying alterations in body- or self-perception, as they persist even after remission (Bischoff-Grethe et al. 2018; Crucianelli et al. 2021). In order to test whether a reduced self-other-distinction might underlie the dissolution of body boundaries in anorexia, we performed a study comparing self-touch and touch by others during fMRI (Frost-Karlsson et al. 2022). We did not find support for the hypothesis that self-touch and other-touch are processed more similarly in anorexia. However, we found that anorexia patients demonstrated an increased neural response to both types of touch. We interpret this finding as follows: previous work suggests that anorexia patients might be "stuck" in a misrepresentation of their own body as perceived from the outside ("allostatic lock"), which precludes them from updating their body perception based on veridical sensory and interoceptive signals (Riva 2012). This would evoke prediction errors when the body is touched – both by others but also by oneself, as the actual percept does not match the percept predicted by the incorrect bodily self-model. These prediction errors render the signal more salient. However, this might be an unpleasant salience in the case of anorexia patients, as it draws the attention toward the body without evoking an update of the body-model. Based on these findings, training of veridical interoception and somatosensation might prove beneficial for strengthening the perception of bodily boundaries. A case study did indeed find improved body image and weight gain in anorectic patient regularly wearing a tight-fit neoprene diving suit which provided clear and continued tactile signals of body boundaries (Grunwald and Weiss 2005). Others tried to alter the bodily sense of self using posture (Weineck et al. 2021) and floating in a warm water tank with high salt concentration (Choquette et al. 2023) with some success. Considering the allostatic lock hypothesis, additional support for unfreezing prior expectations might further improve such treatment approaches. Such a state might be evoked using psilocybin, a psychedelic substance considered to increase neural plasticity and model-updating (Carhart-Harris and Friston 2019; Spriggs et al. 2021).

Also schizophrenia patients display dysfunctional processing and perceptions related to the self (Haug et al. 2012; Hur et al. 2014; Parnas and Handest 2003): They often believe that events in their environment are directed toward them (Menon et al. 2011), like believing a newspaper article was written specifically for or about them. This can be understood as heightened self-referential thinking (Pankow et al. 2015). At the same time, schizophrenia patients can have problems assigning self-generated sensations to themselves (Blakemore et al. 2000). For example, they might misidentify their own voice as alien (Allen et al. 2004). Such altered interpretation of self-generated perceptions is possibly associated with more general self-dysfunctions, especially with respect to the bodily self (Gallese and Ferri 2014). Deficits in the sense of body ownership and the sense of agency have been reported (Hur et al. 2014). A few studies have investigated heartbeat interoception in schizophrenia and reported reduced interoceptive accuracy (Yao and Thakkar 2022).

While heightened self-referential thinking might underlie paranoid delusions or delusions of grandeur, interpretation of self-generated sensations as alien might underlie delusions of alien control and could relate to social cognitive deficits (Haug et al. 2014). A possible mechanistic explanation for the perception of self-generated sensations as alien might be that schizophrenia patients exhibit aberrant salience, i.e., irrelevant stimuli appear meaningful to them (Deserno et al. 2017; Heinz and Schlagenhauf 2010; Pankow et al. 2015). A consequence of this aberrant salience hypothesis would be that self-touch appears salient, leading to a reduced difference between self- and non-self-generated sensations. Indeed, schizophrenia patients experience self-produced touch as more ticklish than healthy people (Blakemore et al. 2000). Touch by others evoked reduced activation in the posterior insula (Ebisch et al. 2013). However, it is unknown whether schizophrenia patients exhibit a reduced neural and behavioral differentiation between self-touch and other-touch, and how such an alteration relates to symptoms in the self-function domain. Unpublished data from my group demonstrates an overall increased neural response to Ct-targeted touch in schizophrenia in the posterior insula among other regions. This increase was also present for self-touch, indicating again a dysfunction of the underlying predictive model, which might relate to specific aspects of bodily self-perception. These findings suggest that training the predictive model, e.g., by targeted self-touch exercises, could improve predictions about self-evoked sensation and thereby sharpen self-other-distinction (given that model-updating processes are intact). While there are many therapeutic approaches that employ somatosensation and self-touch as parts of their clinical practice, high-quality research, e.g., randomized controlled studies, on the specific contributions of social, Ct-targeted, or self-touch and the underlying mechanisms is lacking. When looking at touch-interventions more broadly, there is now good evidence that they in general have beneficial effects on health and well-being, especially with regard to stress, pain, depression, and anxiety (Packheiser et al. 2024).

4 Touch as a Co-regulator of Homeostasis

We need to sense the state of our current physiology, e.g., whether we are hungry or thirsty, tired, or in the need of oxygen, in order to regulate it. To react to the body's physiological condition and balance it out, i.e., bring it back to a somehow internally defined set value or level, is called homeostasis. An even more "advanced" form of homeostasis is called allostasis. Allostasis means to not only react to physiological changes, e.g., to feel hungry and eat in response to a drop in blood sugar level, but also to predict these changes and adapt to them preemptively, if possible, e.g. packing a snack when one assumes to be out longer and get hungry (Schulkin 2011). In order to perform such anticipatory regulatory acts, the individual needs to have built some kind of predictive model about their own physiology (e.g., "I typically get hungry at lunch time") and about the world (e.g., "there won't be food available where I am going"). My examples here are highly cognitive, but it is assumed that these anticipatory adaptations occur also on an unconscious level. For instance, the excretion of the so-called "hunger hormone" ghrelin (which we cannot control consciously) is highly tuned to expected feeding times (LeSauter et al. 2009), thereby preparing for the predicted food intake.

A newborn child still has to tune their internal models to the world around them and can therefore not or only very rudimentary perform allostasis. They therefore rely heavily on their caregiver for regulation of their physiological needs and it is through this parental co-regulation that they learn and build their model of the world and of the (bodily) self (Fotopoulou and Tsakiris 2017). The concept of social allostasis suggests that the parent regulates the child's homeostasis through predictive processes (Atzil and Barrett 2017). This means that the sense of the own body and of its regulation will from the beginning on include, and in the beginning heavily rely on, the social other who plays a key role in the allostatic loop. Within this loop, interpersonal touch, and potentially specifically Ct-targeted touch, seems to play a crucial role.

An extensive body of work has documented the importance of interpersonal touch and skin-to-skin contact for the healthy development of children. If newborn babies receive close bodily, skin-to-skin contact from their mother, they show better adaptation along many different physiological parameters: body temperature, blood glucose levels, heart rate and breathing, blood oxygenation level, energy conservation, acid–base balance (for review see Winberg (2005)). Positive effects of bodily contact during the first hour after birth have been reported still 3 years later (Wiberg et al. 1989), and the positive consequences of increased bodily contact for prematurely born children were even still observed after 10 years (Feldman et al. 2014). The latter study found improved emotional self-regulation and increased physiological regulation (e.g., in the form of having an easier time going to sleep) in the group who received early bodily contact. This could be explained within the framework of social allostasis as a more finely tuned allostatic loop during early development which then allows for improved self-regulation of one's physiology.

Regarding specifically Ct-targeted touch as an interoception-like modality, studies have found that Ct-optimal touch decreases heart rate in babies (Püschel et al. 2022), when they are (or believe to be) stroked by their caregiver (Aguirre et al. 2019). Ct-targeted touch has also been shown to have further reaching behavioral consequences in babies and children, including faster learning and higher social orientation (Croy et al. 2022).

These findings give rise to the hypothesis that touch plays a crucial role in the parental co-regulation of the infant's physiology. Through providing touch, the parent can purposefully or non-purposefully direct the child's arousal and attention. We purposefully provide comforting touch to children who are hurt physically or emotionally. It has been shown that Ct-targeted touch reduces pain in adults both subjectively (Liljencrantz et al. 2017) and neurally (Habig et al. 2017; Savallampi et al. 2023). While measuring pain in infants is challenging, there is good support that suggests that this pain reduction can also already be observed in newborn babies (Gursul et al. 2018). However, parents and caregivers also co-regulate their children unconsciously. For example, a mothers' heartbeat is correlated to stroking speed (Bytomski et al. 2020), i.e., mothers whose heart beats faster also stroke their children faster, a way how they can transfer their own arousal level. The consequences of this parental regulation of the infant's physiology might have far-reaching consequences. Animal experiments even suggest alterations in epigenetics, a potential mechanism for inter-generational transmission of, e.g., stress sensitivity (Champagne and Meaney 2007; Hellstrom et al. 2012).

Taken together, these results paint a picture of touch as a crucial player in the highly complex development of physiological regulation. Early-life touch interactions shape our ability to sense our body and its borders and to adaptively regulate its physiological condition. This physiological regulation is learned in social interaction with caregivers and probably continues to be socially informed through close interpersonal relationships. As described above, tactile interaction between romantic partners reduces pain perception and processing, subjective stress, heart rate, blood pressure, and biomarkers of stress (Ditzen et al. 2007; Holt-Lunstad et al. 2008; Light et al. 2005; Liljencrantz et al. 2017; Savallampi et al. 2023). In addition, tactile interactions also modulate adult physiological state by contributing to pair bonding through eliciting oxytocin release (Yu et al. 2022) and physiological synchronization between romantic partners (Chatel-Goldman et al. 2014). Also emotional co-regulation seems in part to be mediated by touch (Debrot et al. 2013). Considering these far-reaching effects of social touch, it should not come as a surprise that loneliness and social isolation have been associated with increased mortality risk (Holt-Lunstad et al. 2015). As a social species, humans rely on each other. From an evolutionary perspective, we can understand social touch as a safety signal: being touched by another human means physical proximity, and the proximity of other humans, at least those that belong to our tribe, means safety (Eckstein et al. 2020).

5 Conclusion

Touch, including social and self-touch, lies at the intersection between extero- and interoception. Specifically slow, soft caressing touch, that activates Ct-fibers and is processed in the posterior insula, had been suggested by Bud Craig to be considered an interoceptive modality (Craig 2003). It becomes therefore an especially interesting modality to study the bodily self with regard to its development but also to its maintenance throughout the lifespan. For many psychiatric conditions, which demonstrate an alteration or dysfunction of the sense of self, touch might offer a unique non-invasive, non-pharmacological intervention or therapy, which has not been explored enough in currently available treatment options. Touch allows us to feel ourselves and to be in close relationship and connection to others and the surrounding world. It conveys proximity of others and contributes to the interpersonal co-regulation of our physiology and emotions. In a world that is becoming more isolating and where technology begins to mediate all of our interactions – with others, but even with our own bodies – the tactile sense could be a unique opportunity to still experience ourselves in our most basic form, as a body.

References

Abraira VE, Kuehn ED, Chirila AM, Springel MW, Toliver AA, Zimmerman AL, Orefice LL, Boyle KA, Bai L, Song BJ (2017) The cellular and synaptic architecture of the mechanosensory dorsal horn. Cell 168(1):295–310. e219

Ackerley R (2022) C-tactile (CT) afferents: evidence of their function from microneurography studies in humans. Curr Opin Behav Sci 43:95–100

Ackerley R, Carlsson I, Wester H, Olausson H, Backlund Wasling H (2014a) Touch perceptions across skin sites: differences between sensitivity, direction discrimination and pleasantness. Front Behav Neurosci 8:54. https://doi.org/10.3389/fnbeh.2014.00054

Ackerley R, Wasling HB, Liljencrantz J, Olausson H, Johnson RD, Wessberg J (2014b) Human C-tactile afferents are tuned to the temperature of a skin-stroking caress. J Neurosci 34(8): 2879–2883

Aguirre M, Couderc A, Epinat-Duclos J, Mascaro O (2019) Infants discriminate the source of social touch at stroking speeds eliciting maximal firing rates in CT-fibers. Dev Cogn Neurosci 36: 100639

Allen PP, Johns LC, Fu CH, Broome MR, Vythelingum GN, McGuire PK (2004) Misattribution of external speech in patients with hallucinations and delusions. Schizophr Res 69(2–3):277–287

Aspell JE, Lenggenhager B, Blanke O (2012) Multisensory perception and bodily self-consciousness: from out-of-body to inside-body experience. In: The neural bases of multisensory processes. CRC Press/Taylor & Francis, Boca Raton

Atzil S, Barrett LF (2017) Social regulation of allostasis: commentary on “Mentalizing homeostasis: the social origins of interoceptive inference” by Fotopoulou and Tsakiris. Neuropsychoanalysis 19(1):29–33

Bartels A, Zeki S (2004) The neural correlates of maternal and romantic love. NeuroImage 21(3): 1155–1166

Bischoff-Grethe A, Wierenga CE, Berner LA, Simmons AN, Bailer U, Paulus MP, Kaye WH (2018) Neural hypersensitivity to pleasant touch in women remitted from anorexia nervosa. Transl Psychiatry 8(1):161

Blakemore S-J, Wolpert DM, Frith CD (1998) Central cancellation of self-produced tickle sensation. Nat Neurosci 1(7):635–640

Blakemore S-J, Smith J, Steel R, Johnstone E, Frith CD (2000) The perception of self-produced sensory stimuli in patients with auditory hallucinations and passivity experiences: evidence for a breakdown in self-monitoring. Psychol Med 30(5):1131–1139

Boehme R (in press) Leib-Sein und betroffene Selbstgegebenheit im Kontext der Neurodiversitätsbewegung. In: Boehme R, Gahlings U, Mersch D, Nordmann A, Schiemann G (eds) Philosophie als Kritik und Lebenspraxis. Herder Verlag

Boehme R, Olausson H (2022) Differentiating self-touch from social touch. Curr Opin Behav Sci 43:27–33

Boehme R, Hauser S, Gerling GJ, Heilig M, Olausson H (2019) Distinction of self-produced touch and social touch at cortical and spinal cord levels. Proc Natl Acad Sci USA 116:2290

Boehme R, Karlsson MF, Heilig M, Olausson H, Capusan AJ (2020) Sharpened self-other distinction in attention deficit hyperactivity disorder. NeuroImage: Clin 27:102317

Bohic M, Abraira VE (2022) Wired for social touch: the sense that binds us to others. Curr Opin Behav Sci 43:207–215

Böhme G (2008) Ethik leiblicher Existenz. Über unseren Umgang mit der eigenen Natur, Frankfurt aM

Böhme G (2010) The concept of body as the nature we ourselves are. J Specul Philos 24(3): 224–238

Böhme G (2016) Gut Mensch sein. Anthropologie als Proto-Ethik. Die Graue Edition

Böhme R (2019) Human Touch: Warum körperliche Nähe so wichtig ist, vol 6305. CH Beck

Bytomski A, Ritschel G, Bierling A, Bendas J, Weidner K, Croy I (2020) Maternal stroking is a fine-tuned mechanism relating to C-tactile afferent activation: an exploratory study. Psychol Neurosci 13(2):149

Candia-Rivera D, Boehme R, Salamone PC (2024a) Autonomic modulations to cardiac dynamics in response to affective touch: differences between social touch and self-touch. bioRxiv, 2024.2002. 2020.581146

Candia-Rivera D, de Vico Fallani F, Boehme R, Salamone PC (2025) Linking heartbeats with the cortical network dynamics involved in self-social touch distinction. Communications Biology, 8(1), 52.

Carhart-Harris RL, Friston KJ (2019) REBUS and the anarchic brain: toward a unified model of the brain action of psychedelics. Pharmacol Rev 71(3):316–344

Cascio CJ (2010) Somatosensory processing in neurodevelopmental disorders. J Neurodev Disord 2(2):62–69. https://doi.org/10.1007/s11689-010-9046-3

Case LK, Liljencrantz J, McCall MV, Bradson M, Necaise A, Tubbs J, Olausson H, Wang B, Bushnell MC (2021) Pleasant deep pressure: expanding the social touch hypothesis. Neuroscience 464:3–11

Castiello U, Becchio C, Zoia S, Nelini C, Sartori L, Blason L, D’Ottavio G, Bulgheroni M, Gallese V (2010) Wired to be social: the ontogeny of human interaction. PLoS One 5(10):e13199

Cataldo A, Dupin L, Gomi H, Haggard P (2021) Sensorimotor signals underlying space perception: an investigation based on self-touch. Neuropsychologia 151:107729

Cataldo A, Crivelli D, Bottini G, Gomi H, Haggard P (2024) Active self-touch restores bodily proprioceptive spatial awareness following disruption by ‘rubber hand illusion. Proc R Soc B 291(2015):20231753

Champagne FA, Meaney MJ (2007) Transgenerational effects of social environment on variations in maternal care and behavioral response to novelty. Behav Neurosci 121(6):1353

Chatel-Goldman J, Congedo M, Jutten C, Schwartz JL (2014) Touch increases autonomic coupling between romantic partners. Front Behav Neurosci 8:95. https://doi.org/10.3389/fnbeh.2014.00095

Choquette EM, Flux MC, Moseman SE, Chappelle S, Naegele J, Upshaw V, Morton A, Paulus MP, Feinstein JS, Khalsa SS (2023) The impact of floatation therapy on body image and anxiety in anorexia nervosa: a randomised clinical efficacy trial. eClinicalMedicine 64:102173

Ciaunica A, Fotopoulou A (2017) The touched self: psychological and philosophical perspectives on proximal intersubjectivity and the self. In: Embodiment, enaction, and Culture investigating the constitution of the shared world, pp 173–192

Ciaunica A, McEllin L, Kiverstein J, Gallese V, Hohwy J, Woźniak M (2022) Zoomed out: digital media use and depersonalization experiences during the COVID-19 lockdown. Sci Rep 12(1):3888

Craig AD (2002) How do you feel? Interoception: the sense of the physiological condition of the body. Nat Rev Neurosci 3(8):655–666. https://doi.org/10.1038/nrn894

Craig AD (2003) Interoception: the sense of the physiological condition of the body. Curr Opin Neurobiol 13(4):500–505. https://doi.org/10.1016/S0959-4388(03)00090-4

Craig A (2011) Significance of the insula for the evolution of human awareness of feelings from the body. Ann N Y Acad Sci 1225(1):72–82

Croy I, Bierling A, Sailer U, Ackerley R (2021) Individual variability of pleasantness ratings to stroking touch over different velocities. Neuroscience 464:33–43

Croy I, Fairhurst MT, McGlone F (2022) The role of C-tactile nerve fibers in human social development. Curr Opin Behav Sci 43:20–26

Crucianelli L, Cardi V, Treasure J, Jenkinson PM, Fotopoulou A (2016) The perception of affective touch in anorexia nervosa. Psychiatry Res 239:72–78. https://doi.org/10.1016/j.psychres.2016.01.078

Crucianelli L, Demartini B, Goeta D, Nisticò V, Saramandi A, Bertelli S, Todisco P, Gambini O, Fotopoulou A (2021) The anticipation and perception of affective touch in women with and recovered from anorexia nervosa. Neuroscience 464:143–155

Davidovic M (2016) Brain mechanisms for processing of gentle touch in healthy subjects and patients with anorexia nervosa. Thesis, Göteborg University

Debrot A, Schoebi D, Perrez M, Horn AB (2013) Touch as an interpersonal emotion regulation process in couples' daily lives: the mediating role of psychological intimacy. Personal Soc Psychol Bull 39(10):1373–1385. https://doi.org/10.1177/0146167213497592

Deserno L, Boehme R, Mathys C, Katthagen T, Kaminski J, Stephan KE, Heinz A, Schlagenhauf F (2017) Overestimating environmental volatility increases switching behavior and is linked to activation of dorsolateral prefrontal cortex in schizophrenia. bioRxiv. https://doi.org/10.1101/227967

DiMercurio A, Connell JP, Clark M, Corbetta D (2018) A naturalistic observation of spontaneous touches to the body and environment in the first 2 months of life. Front Psychol 9:2613

Ditzen B, Neumann ID, Bodenmann G, Von Dawans B, Turner RA, Ehlert U, Heinrichs M (2007) Effects of different kinds of couple interaction on cortisol and heart rate responses to stress in women. Psychoneuroendocrinology 32(5):565–574

Duerden EG, Arsalidou M, Lee M, Taylor MJ (2013) Lateralization of affective processing in the insula. NeuroImage 78:159–175

Duerler P, Brem S, Fraga-González G, Neef T, Allen M, Zeidman P, Stämpfli P, Vollenweider FX, Preller KH (2022) Psilocybin induces aberrant prediction error processing of tactile mismatch responses—a simultaneous EEG–FMRI study. Cereb Cortex 32(1):186–196

Ebisch SJ, Salone A, Ferri F, De Berardis D, Romani GL, Ferro FM, Gallese V (2013) Out of touch with reality? Social perception in first-episode schizophrenia. Soc Cogn Affect Neurosci 8(4):394–403. https://doi.org/10.1093/scan/nss012

Eckstein M, Mamaev I, Ditzen B, Sailer U (2020) Calming effects of touch in human, animal, and robotic interaction—scientific state-of-the-art and technical advances. Front Psych 11:555058

Emanuel AJ, Lehnert BP, Panzeri S, Harvey CD, Ginty DD (2021) Cortical responses to touch reflect subcortical integration of LTMR signals. Nature 600(7890):680–685

Feldman R, Rosenthal Z, Eidelman AI (2014) Maternal-preterm skin-to-skin contact enhances child physiologic organization and cognitive control across the first 10 years of life. Biol Psychiatry 75(1):56–64

Fotopoulou A, Tsakiris M (2017) Mentalizing homeostasis: the social origins of interoceptive inference. Neuropsychoanalysis 19(1):3–28

Friston K (2009) The free-energy principle: a rough guide to the brain? Trends Cogn Sci 13(7): 293–301. https://doi.org/10.1016/j.tics.2009.04.005

Frost-Karlsson M, Capusan AJ, Perini I, Olausson H, Zetterqvist M, Gustafsson PA, Boehme R (2022) Neural processing of self-touch and other-touch in anorexia nervosa and autism spectrum condition. NeuroImage: Clin 36:103264

Frost-Karlsson M, Capusan AJ, Olausson H, Boehme R (2024) Altered somatosensory processing in adult attention deficit hyperactivity disorder. BMC Psychiatry 24(1):558

Gallese V, Ferri F (2014) Psychopathology of the bodily self and the brain: the case of schizophrenia. Psychopathology 47(6):357–364

Grunwald M, Weiss T (2005) Inducing sensory stimulation in treatment of anorexia nervosa. QJM 98(5):379–380

Gschwind M, Picard F (2016) Ecstatic epileptic seizures: a glimpse into the multiple roles of the insula. Front Behav Neurosci 10:21

Gursul D, Goksan S, Hartley C, Mellado GS, Moultrie F, Hoskin A, Adams E, Hathway G, Walker S, McGlone F (2018) Stroking modulates noxious-evoked brain activity in human infants. Curr Biol 28(24):R1380–R1381

Habig K, Schänzer A, Schirner W, Lautenschläger G, Dassinger B, Olausson H, Birklein F, Gizewski ER, Krämer HH (2017) Low threshold unmyelinated mechanoafferents can modulate pain. BMC Neurol 17:1–11

Haug E, Lien L, Raballo A, Bratlien U, Øie M, Andreassen OA, Melle I, Møller P (2012) Selective aggregation of self-disorders in first-treatment DSM-IV schizophrenia spectrum disorders. J Nerv Ment Dis 200(7):632–636

Haug E, Øie M, Andreassen OA, Bratlien U, Raballo A, Nelson B, Møller P, Melle I (2014) Anomalous self-experiences contribute independently to social dysfunction in the early phases of schizophrenia and psychotic bipolar disorder. Compr Psychiatry 55(3):475–482

Heinz A, Schlagenhauf F (2010) Dopaminergic dysfunction in schizophrenia: salience attribution revisited. Schizophr Bull 36(3):472–485. https://doi.org/10.1093/schbul/sbq031

Hellstrom IC, Dhir SK, Diorio JC, Meaney MJ (2012) Maternal licking regulates hippocampal glucocorticoid receptor transcription through a thyroid hormone–serotonin–NGFI-A signalling cascade. Philos Trans R Soc B: Biol Sci 367(1601):2495–2510

Holt-Lunstad J, Birmingham WA, Light KC (2008) Influence of a "warm touch" support enhancement intervention among married couples on ambulatory blood pressure, oxytocin, alpha amylase, and cortisol. Psychosom Med 70(9):976–985

Holt-Lunstad J, Smith TB, Baker M, Harris T, Stephenson D (2015) Loneliness and social isolation as risk factors for mortality: a meta-analytic review. Perspect Psychol Sci 10(2):227–237

Hur J-W, Kwon JS, Lee TY, Park S (2014) The crisis of minimal self-awareness in schizophrenia: a meta-analytic review. Schizophr Res 152(1):58–64

Husserl E (1952) Ideen zu einer reinen Phänomenologie und phänomenologischen Philosophie: Die Phänomenologie und die Fundamente der Wissenschaften. Springer

Jacquemot AMMC, Park R (2020) The role of interoception in the pathogenesis and treatment of anorexia nervosa: a narrative review. Front Psych 11:281

Kaldewaij R, Salamone P, Enmalm A, Östman L, Pietrzak M, Karlsson H, Löfberg A, Gauffin E, Samuelsson M, Gustavson S (2023) Ketamine reduces the neural distinction between self- and other-produced affective touch-a double-blind placebo-controlled study. Neuropsychopharmacology. https://doi.org/10.1038/s41386-024-01916-0

Kilteni K, Engeler P, Ehrsson HH (2020) Efference copy is necessary for the attenuation of self-generated touch. IScience 23(2):100843

Lenggenhager B, Pazzaglia M, Scivoletto G, Molinari M, Aglioti SM (2012) The sense of the body in individuals with spinal cord injury. PLoS One 7(11):e50757

LeSauter J, Hoque N, Weintraub M, Pfaff DW, Silver R (2009) Stomach ghrelin-secreting cells as food-entrainable circadian clocks. Proc Natl Acad Sci 106(32):13582–13587

Light KC, Grewen KM, Amico JA (2005) More frequent partner hugs and higher oxytocin levels are linked to lower blood pressure and heart rate in premenopausal women. Biol Psychol 69(1): 5–21

Liljencrantz J, Strigo I, Ellingsen D, Krämer H, Lundblad L, Nagi S, Leknes S, Olausson H (2017) Slow brushing reduces heat pain in humans. Eur J Pain 21(7):1173–1185

Liu Y, Latremoliere A, Li X, Zhang Z, Chen M, Wang X, Fang C, Zhu J, Alexandre C, Gao Z (2018) Touch and tactile neuropathic pain sensitivity are set by corticospinal projections. Nature 561(7724):547–550

Löken LS, Wessberg J, McGlone F, Olausson H (2009) Coding of pleasant touch by unmyelinated afferents in humans. Nat Neurosci 12(5):547–548

Lucherini Angeletti L, Innocenti M, Felciai F, Ruggeri E, Cassioli E, Rossi E, Rotella F, Castellini G, Stanghellini G, Ricca V (2022) Anorexia nervosa as a disorder of the subcortical–cortical interoceptive-self. Eat Weight Disord 27(8):3063–3081

McGlone F, Wessberg J, Olausson H (2014) Discriminative and affective touch: sensing and feeling. Neuron 82(4):737–755. https://doi.org/10.1016/j.neuron.2014.05.001

Menon M, Schmitz TW, Anderson AK, Graff A, Korostil M, Mamo D, Gerretsen P, Addington J, Remington G, Kapur S (2011) Exploring the neural correlates of delusions of reference. Biol Psychiatry 70(12):1127–1133. https://doi.org/10.1016/j.biopsych.2011.05.037

Miller LE, Fabio C, Ravenda V, Bahmad S, Koun E, Salemme R, Luauté J, Bolognini N, Hayward V, Farne A (2019) Somatosensory cortex efficiently processes touch located beyond the body. Curr Biol 29(24):4276–4283. e4275

Morrison I (2016) ALE meta-analysis reveals dissociable networks for affective and discriminative aspects of touch. Wiley-Blackwell

Moszkowski RJ, Stack DM (2007) Infant touching behaviour during mother–infant face-to-face interactions. Infant Child Dev Int J Res Pract 16(3):307–319

Olausson HW, Cole J, Vallbo A, McGlone F, Elam M, Kramer HH, Rylander K, Wessberg J, Bushnell MC (2008) Unmyelinated tactile afferents have opposite effects on insular and somatosensory cortical processing. Neurosci Lett 436(2):128–132. https://doi.org/10.1016/j.neulet.2008.03.015

Olausson H, Wessberg J, Morrison I, McGlone F, Vallbo A (2010) The neurophysiology of unmyelinated tactile afferents. Neurosci Biobehav Rev 34(2):185–191. https://doi.org/10.1016/j.neubiorev.2008.09.011

Orefice LL, Zimmerman AL, Chirila AM, Sleboda SJ, Head JP, Ginty DD (2016) Peripheral Mechanosensory neuron dysfunction underlies tactile and behavioral deficits in mouse models of ASDs. Cell 166:299. https://doi.org/10.1016/j.cell.2016.05.033

Packheiser J, Hartmann H, Fredriksen K, Gazzola V, Keysers C, Michon F (2024) A systematic review and multivariate meta-analysis of the physical and mental health benefits of touch interventions. Nat Hum Behav 8:1–20

Pankow A, Katthagen T, Diner S, Deserno L, Boehme R, Kathmann N, Gleich T, Gaebler M, Walter H, Heinz A (2015) Aberrant salience is related to dysfunctional self-referential processing in psychosis. Schizophr Bull 42(1):67–76

Parnas J, Handest P (2003) Phenomenology of anomalous self-experience in early schizophrenia. Compr Psychiatry 44(2):121–134

Parush S, Sohmer H, Steinberg A, Kaitz M (1997) Somatosensory functioning in children with attention deficit hyperactivity disorder. Dev Med Child Neurol 39(7):464–468

Püschel I, Reichert J, Friedrich Y, Bergander J, Weidner K, Croy I (2022) Gentle as a mother's touch: C-tactile touch promotes autonomic regulation in preterm infants. Physiol Behav 257: 113991

Rahman J, Mumin J, Fakhruddin B (2020) How frequently do we touch facial T-zone: a systematic review. Ann Glob Health 86(1):75

Riva G (2012) Neuroscience and eating disorders: the allocentric lock hypothesis. Med Hypotheses 78(2):254–257

Rochat P, Hespos SJ (1997) Differential rooting response by neonates: evidence for an early sense of self. Infant Child Dev 6(3–4):105–112

Sakaguchi S, Saito K, Arakawa N, Konyo M (2023) The dynamic behavior of skin in response to vibrating touch stimuli affects tactile perception. Skin Res Technol 29(3):e13295

Salamone PC, Dowzuk NA, Gallo MN, Boehme R (2024) From bodily numbness to social connection: unveiling the relationship between depersonalization, self-concept, and social touch. https://doi.org/10.21203/rs.3.rs-4923651/v1

Sanzeni A, Katta S, Petzold B, Pruitt BL, Goodman MB, Vergassola M (2019) Somatosensory neurons integrate the geometry of skin deformation and mechanotransduction channels to shape touch sensing. elife 8:e43226

Savallampi M, Maallo AM, Shaikh S, McGlone F, Bariguian-Revel FJ, Olausson H, Boehme R (2023) Social touch reduces pain perception—an fMRI study of cortical mechanisms. Brain Sci 13(3):393

Schaefer M, Kühnel A, Gärtner M (2022) Sensory processing sensitivity and somatosensory brain activation when feeling touch. Sci Rep 12(1):12024

Schulkin J (2011) Social allostasis: anticipatory regulation of the internal milieu. Front Evol Neurosci 2:111

Sibeoni J, Massoutier L, Valette M, Manolios E, Verneuil L, Speranza M, Revah-Levy A (2022) The sensory experiences of autistic people: a metasynthesis. Autism 26(5):1032–1045

Spille JL, Grunwald M, Martin S, Mueller SM (2021) Stop touching your face! A systematic review of triggers, characteristics, regulatory functions and neuro-physiology of facial self touch. Neurosci Biobehav Rev 128:102–116

Spriggs MJ, Douglass HM, Park RJ, Read T, de Magalhães FJ, Lafrance A, Nicholls DE, Erritzoe D, Nutt DJ, Carhart-Harris RL (2021) Study protocol for "psilocybin as a treatment for anorexia nervosa: a pilot study". Front Psych 12:735523

Taels L, Feyaerts J, Lizon M, De Smet M, Vanheule S (2023) 'I felt like my senses were under attack': an interpretative phenomenological analysis of experiences of hypersensitivity in autistic individuals. Autism 27(8):2269–2280

Vallbo Å, Olausson H, Wessberg J (1999) Unmyelinated afferents constitute a second system coding tactile stimuli of the human hairy skin. J Neurophysiol 81(6):2753–2763

Van Stralen H, Van Zandvoort M, Dijkerman H (2011) The role of self-touch in somatosensory and body representation disorders after stroke. Philos Trans R Soc B: Biol Sci 366(1581): 3142–3152

Weineck F, Hauke G, Lindemann H, Lachenmeir K, Schnebel A, Karačić M, Meule A, Voderholzer U, Pollatos O (2021) Using bodily postures in the treatment of anorexia nervosa: effects of power posing on interoception and affective states. Eur Eat Disord Rev 29(2):216–231

Wessberg J, Norrsell U (1993) A system of unmyelinated afferents for innocuous mechanoreception in the human skin. Brain Res 628(1–2):301–304

Wiberg B, Humble K, De Chateau P (1989) Long-term effect on mother-infant behaviour of extra contact during the first hour post partum. V. Follow-up at three years. Scand J Soc Med 17(2): 181–191

Winberg J (2005) Mother and newborn baby: mutual regulation of physiology and behavior—a selective review. Dev Psychobiol 47(3):217–229

Yao B, Thakkar K (2022) Interoception abnormalities in schizophrenia: a review of preliminary evidence and an integration with Bayesian accounts of psychosis. Neurosci Biobehav Rev 132: 757–773

Yu H, Miao W, Ji E, Huang S, Jin S, Zhu X, Liu M-Z, Sun Y-G, Xu F, Yu X (2022) Social touch-like tactile stimulation activates a tachykinin 1-oxytocin pathway to promote social interactions. Neuron 110(6):1051–1067. e1057

Zotterman Y (1939) Touch, pain and tickling: an electro-physiological investigation on cutaneous sensory nerves. J Physiol 95(1):1

Sensing the Self: The Role of the Insula and Interoception in Body Image

Emily M. Choquette and Sahib S. Khalsa

Contents

Abstract This chapter explores the insula's role in shaping body image through interoception – the neural process of sensing, interpreting, and integrating internal bodily signals to facilitate a coherent sense of self. Drawing on A.D. (Bud) Craig's hierarchical model, which emphasizes the insula's integration of sensory input into higher-order self-representations, we examine how disruptions in interoceptive processing may contribute to body image disturbance. This framework is applied to mental health conditions such as anorexia nervosa, body dysmorphic disorder, as

---

E. M. Choquette
Laureate Institute for Brain Research, Tulsa, OK, USA

S. S. Khalsa (✉)
Department of Psychiatry and Biobehavioral Sciences, Semel Institute for Neuroscience and Human Behavior, David Geffen School of Medicine, University of California at Los Angeles, Los Angeles, CA, USA

Laureate Institute for Brain Research, Tulsa, OK, USA
e-mail: skhalsa@mednet.ucla.edu

Curr Topics Behav Neurosci (2025) 70: 107–125
https://doi.org/10.1007/7854_2025_581
Published Online: 04 April 2025

well as physical health conditions including phantom limb pain and cancer, highlighting evidence for impaired interoceptive signaling and altered insular cortex function. We propose a mechanistic model describing how such disruptions can affect both sensory processing and the subjective experience of the body, leading to distorted body perception. We conclude with a discussion of future research directions and the potential for interoception-based therapeutic interventions targeting body image disturbance.

Keywords Body image · Brain-body connection · Insula · Interoception · Neural integration

"The Eye Sees Only What the Mind Is Prepared to Comprehend"
– Robertson Davies

1 Interoception, the Insula, and the Foundations of Body Image

It is now well established that our perception of the external world (i.e., what we see, hear, touch) is not a simple reflection of reality. Instead, it depends on how the brain interprets and comprehends incoming sensory data, in the context of past experiences, expectations, and cognitive frameworks. Interoception, which can be viewed as the nervous system's capacity to sense, interpret, and integrate signals from within the body (Khalsa et al. 2018a), plays a critical role in shaping how we perceive ourselves as distinct entities existing in the world. Body image is a component of self-perception that involves both external and internal information, as we experience our body not only through visual and tactile signals but also through interoceptive cues. Here we will present the argument that both exteroceptive and interoceptive processes are required for shaping how humans perceive their body, and in particular, how the insular cortex is essential for bridging visceral and visual experiences to form a bodily image of the self.

A.D. "Bud" Craig's trailblazing conceptualization of interoception as the "sense of the physiological condition of the entire body, not just the viscera" helped to revitalize interest in this concept through a neuroanatomical focus on the ascending lamina I spinothalamocortical pathway and a concise delineation of the complex integration of interoceptive information in the insular cortex (Craig 2002). Building upon Antonio Damasio's groundbreaking descriptions of how interoceptive signals are fundamental ingredients for emotional experience, feelings, and consciousness (Damasio 1994, 1999), Craig was able to root the multifaceted nature of interoception within an emerging series of functional neuroimaging studies of interoceptive processing, demonstrating common involvement of the insula in various conditions such as temperature sensation, cutaneous itch, isometric exercise, and chronic pain (Craig 2002). A key aspect of Craig's version of interoception was that this unique form of sensory information was dynamically represented in the human body via a hierarchical ascending neurobiological process meant to inform the optimal balance of the physiological condition of the body. In turn, a bidirectional process comprised of feedback and feedforward loops between the brain and body could continuously update the internal model and support homeostasis (Khalsa et al. 2018a).

Craig's interest in interoception grew from his original work in the neurobiology of pain and thermal processing. Craig understood that pain and temperature sensations challenged the physiological state of the organism, and therefore its survival. He further asserted that these challenges were inextricably tied to affect and motivated behaviors to maintain homeostasis, the reactive regulatory process by which organisms maintain a stable internal environment (Cannon 1929; Ramsay and Woods 2014). Craig defined the role of the lamina I spinothalamic neurons and associated pathways across numerous mammalian species (Craig et al. 2001; Dostrovsky and Craig 1996; Han et al. 1998; Westlund and Craig 1996). He linked this pathway to insular activation in humans (Craig et al. 1996, 2000), and postulated that this functional neuroimaging evidence supported the causal role associating damage to the insula with the central pain syndrome that has been described following a stroke. Based on his work with pain and thermal processing, Craig even theorized that primates have distinct cortical "images" that reflect all physiological conditions including pain, temperature, itch, muscle burn, visceral sensations, hunger, thirst, taste, and even, sensual touch, in an effort to maintain homeostatic conditions of the body. According to this view, the progression of interoceptive information started with small-diameter primary afferent fibers reporting the physiological status to body tissues (Craig 2008). These fibers, terminating in lamina I of the dorsal spinal cord, putatively project their information to homeostatic integration areas including the thalamus and insula (Craig 2002, 2008, 2009). The central notion was that the lamina I neuroanatomical pathway could transmit interoceptive signals from all over the body (via C and A delta fibers). This distinction from traditional nociceptive and visceroceptive pathways catalyzed interest in the study of interoception.

Craig's views on the processing of interoceptive signals evolved further through the proposal that interoceptive information is modified in a hierarchical fashion within the insular cortex from the posterior to anterior regions (Craig 2002, 2009, 2010). In his view, the dorsal posterior insula served as the primary sensory cortex for homeostatic or "interoceptive" feelings such as hunger, pain, graded temperature, touch, and thirst (Craig 2002, 2003). These feelings served as "sensory homeostatic emotions" and were directly dependent on the physiological state of the body. The example often presented by Craig was that of contextual thermal regulation, in which a cold shower feels pleasant on a hot day but is unbearable or even painful when the body is cooler than normal (or hypothermic), due to a discrepancy in the body's state-dependent thermoregulatory needs. Within this hierarchical structure, homeostatic sensory information was projected forward from the posterior to the mid-insula, where its processing converged with emotionally salient activity from the amygdala and subcortical homeostatic control regions (e.g., hypothalamus). The combined physiological/affective information at this level was often considered a percept (i.e., a conscious perception that could be localized to regional neural processing in the posterior/mid insula). But it was the progression to the anterior insula that Craig considered to be of primary importance to human interoception. By converging with signals arriving from a broader array of areas such as the anterior cingulate, orbitofrontal, and dorsolateral prefrontal cortices, he argued that the combined integration of this set of information created not only a mental model of the body that was continuously updated, but did so based on acquired internal models (e.g., from the posterior/mid insula). He incorporated predictive representations of the external and internal world and self by drawing upon Damasio's concept of the "as-if body loop" (Damasio 2000), a neural circuit within the brain whose configuration of activity could mirror the peripherally triggered percept (the "body loop") without requiring a corresponding change in peripheral physiological activity. Pointing toward the anterior insula (on the right in particular) as a key area for human-specific functional integration, he proposed that, in humans, the ability to consciously experience interoceptive awareness and distinguish one's bodily self as distinct from other physical beings began in the anterior insula, through the continuous integration of all salient bodily information (Craig 2002).

Craig's opinion paper (Craig 2002) was so popular (7,101 Google Scholar citations as of this writing), that he was invited to write another perspective for the same journal, which was published in 2009 (Craig 2009). In that article, he went so far as to argue that the signal processing occurring in the anterior insular cortex created a neural meta-representation of what he called the "global emotional moment" (Craig 2009, 2010), stating that the anterior insular cortex's "function in the re-representation of interoception offers one possible basis for its involvement in *all* subjective feelings" (emphasis added). That paper (7,724 Google Scholar citations as of this writing) was subsequently announced as one of the ten most cited articles in *Nature Reviews Neuroscience* (Luo et al. 2010). In reflecting on the article's impact, Craig correctly envisioned that it "will certainly guide new imaging and morphometry studies on the neural bases of mood disorders (anxiety and depression), schizophrenia, the forebrain asymmetry of positive and negative emotions, subjective time perception, music appreciation, meditation, somatization syndromes, focal attention, risk and error processing, and so on" (Luo et al. 2010).

Craig's work on interoceptive processing and the "sentient self" reflects a meaningful contribution to our understanding of the integration of homeostatic and emotional information in healthy individuals. While he did not expand his thinking to the role of interoception in individuals with psychopathological disorders, and never explicitly theorized on the topic of body image, he laid elements of conceptual groundwork that could be seen as relevant to these processes. For example, Craig (2008, p 281) stated "the metarepresentations of the state of the body in the right anterior insula seem to differentiate inner and outer conditions (self from nonself) and provide a subjective mental image of the material self as a feeling (sentient) entity." If we assume the creation of these complex mental self-representations to be true, one implication is that the integration of interoceptive, emotional, and exteroceptive components with the subjective experiences of one's body would be essential to the formation of body image. From there, it might be plausible to postulate that altered body representations and more specifically, disturbances of body image, such as those reported in numerous mental and physical health conditions, could have a basis in the disruption of interoceptive neural pathways. In the ensuing sections we elaborate upon this concept.

2 The Complex Dimensions of Body Image

Body image can be conceptualized as a multidimensional construct encompassing the cognitive, affective, and perceptual representations of one's physical appearance (Cash 2004). The cognitive and affective components of body image are primarily related to subjective thoughts and emotions associated with the experience of one's body, and when disrupted, are also referred to as body dissatisfaction. The cognitive and affective components of body image have been traditionally conceptualized as either negative or positive. Body dissatisfaction is most often discussed as a subcomponent of negative body image, referring to distressing beliefs or feelings about one's own body size, shape, or appearance. It can be measured in several ways. For example, the Body Dissatisfaction subscale from the Eating Disorder Inventory (Garner 2004) asks individuals to indicate the frequency of negative cognitions (such as "I think my stomach is too big" or "I think my thighs are too large"), whereas the body shame scale of the Objectified Body Consciousness Scale (McKinley and Hyde 1996) attempts to quantify the shame one may feel when their body does not conform to societal beauty standards (for example, "When I'm not the size I think I should be, I feel ashamed"). Positive body image measures assess the appreciation of one's body size, shape, or weight. For example, the Body Appreciation Scale (Avalos et al. 2005; Tylka and Wood-Barcalow 2015) assesses favorable thoughts and feelings regarding one's body and acceptance of their body regardless of whether it conforms to societal beauty standards (questions such as "I appreciate the different and unique characteristics of my body" and "I am comfortable in my body"). Other measures explicitly ask individuals how satisfied (or unsatisfied) they are with specific parts of their body or appearance (for example, the Body Areas Satisfaction Scale of the Multidimensional Body-Self Relations Questionnaire (MBSRQ; Cash 2000) presents individuals with nine body parts and asks them to evaluate each part on a satisfaction scale ranging from "very dissatisfied" to "very satisfied").

The perceptual component of body image is typically conceptualized as a distinct mental representation of body size and shape. It may be healthy (or normative) when an individual's mental representation aligns with their actual body size and shape, and distorted when there is a mismatch between their perceived and actual physical appearance. This is most commonly measured using figure rating scales (e.g., the Stunkard Figure Rating Scale; Stunkard et al. 1983 and the Photographic Figure Rating Scale; Swami et al. 2008, 2012), likely due to their ease of use. Participants simply look at line drawings or photographs of bodies to make their self-referential selection in terms of their current body, as well as their ideal body. More nuanced and personalized measures have been developed in recent years. These typically involve the interactive tasks in which individuals adjust virtual avatars to reflect their perceived body size and shape, which can then be compared to their actual body measurements and their "ideal" body image. One such measure, Somatomap, has shown utility in capturing individual variability in body image—both in professional fashion models (Ralph-Nearman et al. 2019) and individuals with eating disorders (Ralph-Nearman et al. 2021), and it represents a promising digital tool for assessing and longitudinally tracking body dissatisfaction in clinical populations.

3 Natural Variability, Cultural Influences, and Health Conditions in Shaping Body Image

Body image varies naturally between individuals, and can fluctuate even in those without clinically significant body image disturbance. Mental representations of the self may differ based on naturally occurring differences in body height, proportions, shape, or physical features such as skin tone and hair color. Perceptual representations of the self typically update when physical appearance changes, for example following weight fluctuations or alterations of physical appearance (e.g., changes in hair color, tattoo application, or body piercing). The cognitive and affective components of body image can also be influenced by natural variations in body characteristics, but these subjective aspects tend to be determined by cultural standards of beauty and individual body size and shape preferences. For example, the ideal body size and shape that an individual seeks or strives for has been shown to be driven by specific cultural influences such as family/peer/media exposures, racial/ethnic/national identity, socioeconomic factors, and individual preferences (e.g., Frederick et al. 2022; Keigan et al. 2024; Ordaz et al. 2018; Swami et al. 2010). Personalized activities may also influence ideal body size/shape and associated body image ideals. For example, an individual engaged in an endurance sport such as marathon running may desire to have a smaller body size than one who is a body builder or power lifter. When an individual's actual or perceived body is at odds with the mental representation they hold for their ideal body, it can create body dissatisfaction, anxiety, and other forms of psychological distress (Grogan 2021). This concern may range from slight dissatisfaction to severe distress with appearance, and can create ego-dystonic conditions related to body image that, in some individuals, may motivate them to adopt health-promoting lifestyle changes. Body dissatisfaction and psychological distress may also cause an individual to undertake maladaptive approaches to body modification such as food or water restriction (to reduce body weight), anabolic steroid use (to build body muscle), or plastic surgery utilization (to reshape body parts).

Body image can also vary in relation to health conditions that may or may not directly affect the composition of the body or associated internal organs. For example, individuals with body dysmorphic disorder (BDD) report extreme body dissatisfaction and perceptual misrepresentation of one or more body parts (e.g., judging the body part as too large or too small, or to be hideously disfigured in some way) without objective evidence of such malformations (American Psychiatric Association [APA] 2022). Body image disturbance may occur in transient or persistent forms when the body is severely under weight. For example, individuals with anorexia nervosa (AN) typically view themselves as larger than their objective body habitus and often hold negative beliefs about their body weight and shape (APA 2022). Other examples of body image disturbance may be related to a chronic state — for example, increased body weight due to obesity, jaundiced appearance from hepatic failure, or normative changes associated with aging. Disturbances can also occur following the loss of a limb (e.g., amputation), which may disrupt body image at both perceptual and cognitive-affective levels due to persistent alterations in the representation of body form.

Building on Craig's normative perspective on the sentient self, it seems plausible that individuals without body image disturbance maintain a self-representation that adaptively updates in response to afferent bodily signals—even as the status or appearance of the body changes—thereby supporting an accurate internalization of body perception. In contrast, body image disturbance may arise when there are impairments in the interoceptive circuits responsible for sensing, interpreting, or integrating these signals. In the remainder of this work, we formulate how abnormalities of interoceptive processing may contribute to the development or maintenance of maladaptive body image across mental and physical health conditions characterized by body image disturbance.

4 A Role for Interoception and Exteroception in Multisensory Integration and Body Image Disturbance

Previous perspectives have posited that body image disturbance occurs as a result of the faulty integration of multisensory inputs affecting perception and cognition, at several levels of integration of afferent visual stimuli, prior cognitive schemas, and incoming or anticipated body signals (Gadsby 2017; Riva and Gaudio 2012; Stanghellini et al. 2019). According to these arguments, the inability or diminished ability to accurately sense, interpret, or integrate the physiological state of the body leads to an increased reliance on exteroceptive signals. In the case of visual body signals, this leads a person to view their own body from a third-person perspective (i.e., to "self-objectify"; Gadsby 2017, Riva and Gaudio 2012, Stanghellini et al. 2019). As a consequence of reduced interoceptive abilities or a reduced connection to state-dependent internal bodily readouts, individuals may have difficulty adaptively integrating external and internal signals, fostering inaccurate mental representations of their own body, body image distortion, and ultimately body dissatisfaction (Riva and Gaudio 2012, Stanghellini et al. 2019; see Fig. 1 for an illustrative model applied to a model disorder). The following sections examine specific clinical conditions in which disrupted interoceptive processing contributes to disintegration

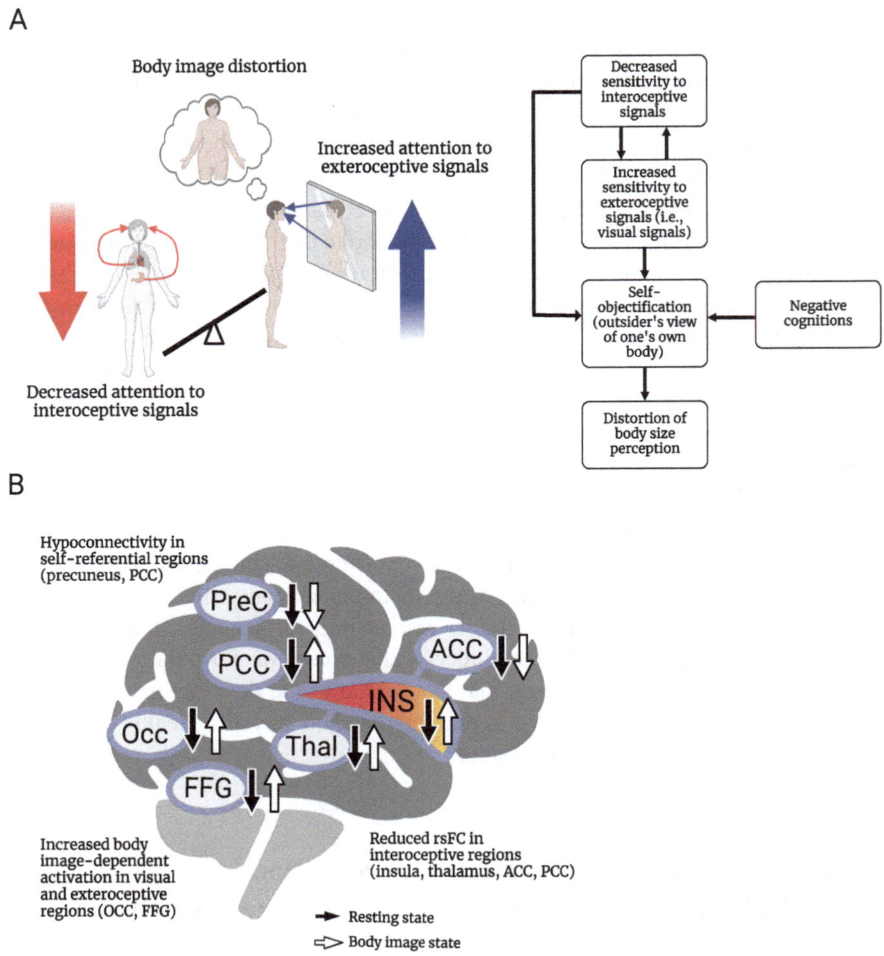

Fig. 1 A neuroanatomically informed model of body image distortion in anorexia nervosa. (**a**) Conceptual model illustrating how decreased sensitivity to interoceptive signals and heightened sensitivity to exteroceptive signals (e.g., visual input) results in an imbalance in body signal processing. This imbalance promotes self-objectification—viewing the body from an outsider's perspective—which contributes to distorted body size perception and negative body-related cognitions. (**b**) Neurobiological correlates of this imbalance. Reduced resting state functional connectivity (rsFC) in interoception-related regions—including the insula (INS), thalamus (Thal), and anterior cingulate cortex (ACC)—may impair the brain's ability to accurately integrate bodily signals, contributing to diminished interoceptive sensitivity. During body image-related tasks, increased activation is observed in visual and exteroceptive regions (e.g., occipital cortex [Occ], fusiform gyrus [FFG]) suggesting compensatory reliance on visual feedback. Hypoconnectivity in self-referential regions (precuneus [PreC], posterior cingulate cortex [PCC]) may further disrupt the integration of body-related information into a coherent self representation. Together, these alterations lead to distorted neural representations of the body, body dissatisfaction, and psychological distress. (Created with Biorender.com)

between internal body signal and body image, illustrating how such mechanisms manifest across both mental and physical health disorders.

5 The Insula's Role in Altered Body Perception and Interoception in Anorexia Nervosa

The primary diagnostic criteria of anorexia nervosa (AN) are (1) caloric restriction leading to significantly low body weight; (2) an intense fear of gaining weight or becoming fat; and (3) a disturbance in body image, overemphasis on body weight or shape in self-evaluation, or lack of recognition of the seriousness of low body weight (APA 2022). However, early depictions of patients with AN also described individuals who were anxious not just about eating food or gaining weight but also about sensations stemming from their own bodies (Pearce 2004). Bodily preoccupations in AN often focus on abnormal perceptions of appearance; however, other preoccupations frequently pertain to internal sensations (Abbate-Daga et al. 2013; Skarderud 2007). For example, classic case reports emphasized a failure to recognize body states (Bruch 1962) and "vague feelings of fullness" (Vandereycken and van Deth 1989), which were complemented by findings demonstrating that individuals with AN experience premature gastric fullness (Bluemel et al. 2017). Other studies suggest individuals with AN often have difficulty accurately sensing and integrating interoceptive signals, via evidence of (1) abnormal insula activation during attention to stomach and heart sensations in weight-restored AN (Kerr et al. 2016), abnormal insula activation during breathing perturbation in recovered AN (Berner et al. 2018), heightened suffocation fear during inspiratory breath holding (Lapidus et al. 2020), and abnormal pre-meal anxiety and heightened cardiorespiratory interoception during adrenergic perturbation (Khalsa et al. 2015, 2018b).

Consistent with both Craig's work and the proposed model (Fig. 1), functional neuroimaging findings in AN appear to support alterations in the insula and its association with body image through the mechanism of aberrant interoception. A review of available studies indicated that, overall, there was reduced functional connectivity in patients with AN (compared to healthy individuals) in areas of self-referential and interoceptive processing (Angeletti et al. 2022). In particular, altered resting state functional connectivity (rsFC) was identified across numerous regions involved in interoception, exteroceptive processing, and self-referential processing, with the most alterations appearing in regions of interoceptive processing. For example, reduced rsFC was found in areas such as the insula, thalamus, anterior cingulate cortex, and postcentral gyrus in individuals with AN compared with healthy individuals. Interestingly, the pattern of lower resting state connectivity was opposed by task-based increases in the activity of many of the same regions preferentially in AN individuals (as compared to HC) during tasks involving state modulations of body image, such as during perceptions of self-body images (unmodified or modified to look thinner/fatter) or other-body images (Angeletti et al.

2022). More recently, using a peripheral adrenergic perturbation of cardiorespiratory signals via isoproterenol, we observed pre- to post-stimulation decreases in rsFC within the bilateral anterior insula, left amygdala, posterior cingulate cortex, and ventromedial prefrontal cortex in individuals with AN (De la Cruz et al. 2022). These regions are part of the central autonomic network (CAN), which is putatively tasked with regulation of peripheral autonomic tone (Benarroch 1993; Ferraro et al. 2022). Most strikingly, these rsFC reductions in between several CAN regions were significantly associated with the severity of body image concerns, providing empirical evidence that disrupted interoceptive processing contributes to distorted perceptual and neural representations of the body in AN. We speculate that the discrepancies in connectivity between resting and body image states may reflect diminished integration of interoceptive and exteroceptive information, resulting in reduced awareness or impaired differentiation of these inputs. This mismatch may disrupt the perception of self and body representation. Furthermore, the consistently reduced rsFC in self-referential regions suggests enduring alterations in subjective experience that persist across internal and body-relevant states. Together, these findings point to a dual disruption: both in the integration of afferent body signals and in the continuity of self-related experience. Such disruptions may give rise to distorted neural and mental representations of the body, which, when incorporated into one's self-schema, could drive maladaptive metacognitions and reinforce negative affective feelings towards the body.

Increased task-dependent activation in individuals with AN (in comparison to hypoconnectivity at rest) may reflect a heightened responsively to external, particularly self-relevant visual stimuli. Given an already disrusted mental representation of the body and impaired interoceptive processing, these individuals may rely more heavily on visual input to update their body representations. One possible therapeutic scenario in this case would be to explore whether the simultaneous modulation of exteroceptive and interoceptive signaling could improve perceptual BID in AN. We recently utilized a non-pharmacologic intervention called floatation-REST (Reduced Environmental Stimulation Therapy) to accomplish this aim (Choquette et al. 2023; Khalsa et al. 2020). During floatation-REST, input from visual, auditory, olfactory, gustatory, thermal, tactile, vestibular, and proprioceptive channels are minimized, and interoceptive input is enhanced (Feinstein et al. 2018a, b). Across two clinical trials in AN (Choquette et al. 2023; Khalsa et al. 2020), we have found that floatation-REST acutely reduces perceptual body image disturbance, indexed by the body dissatisfaction score on the PFRS, after one session (Khalsa et al. 2020) and reliably after multiple sessions (Choquette et al. 2023). A long-term goal of this work is to understand how body image disturbance in AN is associated with dysfunctions of interoceptive processing with the aim of providing a mechanistic framework for developing novel interventions based on concurrent modulation of interoceptive and exteroceptive body signals.

While the therapeutic modulation of interoceptive signals to target body image disturbance seems like a viable avenue of investigation, this perspective must be balanced by research demonstrating that AN individuals also exhibit abnormal neural activation patterns in early visual processing regions when viewing basic visual patterns (Li et al. 2015a, b; Madsen et al. 2013). Such an aberrant form of early visual processing in the brain has prompted the suggestion that abnormal perceptions of the body might also be therapeutically targeted through interventional perturbations of either visual processing or through the modulation of neural circuits that regulate visual perception (Feusner et al. 2017). Thus, whether via modulation of interoceptive or exteroceptive signaling, or both, much empirical work lies ahead to determine the optimal modes of therapies for body image distortion in AN.

6 Visual Processing, Interoception, and Insular Dysfunction in Body Dysmorphic Disorder

Body dysmorphic disorder (BDD) is another psychiatric condition characterized by body image concerns. The primary diagnostic feature of BDD is extreme distress and preoccupation related to one or more perceived defects in appearance which are unobservable to others (APA 2022). From the perspective of Craig's extended framework and our theoretical model, distorted self-representation and heightened emotional arousal to perceived defects in BDD may be linked to abnormal integration of bodily signals within the insula. However, research on BDD has primarily focused on visual processing deficits rather than interoception (Beilharz et al. 2017; Diaz-Fong and Feusner 2024). A recent review of visual processing in BDD highlighted key deficits, including an overemphasis on detail, global integration deficiencies, and selective attentional bias toward minute features (Diaz-Fong and Feusner 2024). It seems plausible to speculate that BDD involves an overreliance on visual signals due to faulty integration of interoceptive signals, similar to the model proposed for AN. Specifically, reduced attention to interoceptive signals may lead individuals with BDD to focus excessively on external stimuli, particularly visual details. When paired with distorted prior beliefs about certain body parts, this heightened focus may exacerbate a distorted neural representation of the body.

Once a neural representation is already skewed, hyperfixation on the perceived flaw may reinforce cognitive biases, further intensifying dissatisfaction with appearance. Although studies on interoceptive processing in BDD are limited, research has linked muscle dysmorphic symptoms (a subtype of BDD; APA 2022) with self-reported interoceptive deficits (Grunewald et al. 2023a, b). More empirical studies are needed to explore the role of interoception in BDD. While the role of the insula and interoception in BDD has largely been theoretical (Jenkinson and Rossell 2024), given the neurobiological similarities in visual processing between AN and BDD (Li et al. 2015a, b; Madsen et al. 2013), it seems likely that both disorders share some mechanisms of abnormal processing within the insula. Emerging research is beginning to explore these connections more deeply (Diaz-Fong and Feusner 2024).

7 Illness, Homeostasis, and Body Image: Disruptions of Self-Representation in Physical Illness or Injury

Physical illness, such as viral or bacterial infection, acutely disrupts the body's natural homeostatic state. In Craig's parlance, if the primary function of afferent pathways from lamina I to cortical regions, particularly the insula, is to integrate sensory information throughout the body with other inputs to maintain homeostasis, then when physical illness alters or damages the body corpora, it should disrupt this balance. For example, physical alterations in the body may occur, when jaundice caused by acute liver dysfunction leads to yellowing of the skin and eyes due to a buildup of bilirubin. Similarly, sudden weight gain or loss may signal conditions like hypothyroidism, cancer, or diabetes. These bottom-up changes in physical appearance create a mismatch between the current physical state and its normative mental representation. This misalignment can be adaptive, such as in the case of triggered awareness of illness, motivating help-seeking behaviors aimed at restoring homeostasis. But it can also lead to subjective distress (e.g., body dissatisfaction) and perceptual disturbances if the mental representation is not updated.

Some treatments for illness, such as cancer, can chronically alter physical appearance through side effects like weight changes, hair loss, or surgical excisions (e.g., mastectomy or amputation). Research on cancer-related body image has primarily focused on cognitive and affective components (Byrne et al. 2023; Kołodziejczyk and Pawłowski 2019; Vani et al. 2021). To date, little research has explored the relationship between body image and illness through the lens of disrupted interoceptive homeostasis, representing a promising area for further study to better understand the mind-body connection.

Limb amputation represents a special circumstance that can uniquely impact body image. In typical recovery, the mental image of one's own body will update and account for the missing limb. When mental imagery fails to update, this may result in phantom limb syndrome, a condition where individuals experience sensations such as pain, tingling, temperature changes, or movement in the limb that has been amputated. These phenomenological symptoms raise the possibility of faulty integration of signals between the body and brain. For example, although afferent interoceptive signals from the limb are no longer conveyed to the brain, it has been demonstrated that the brain still retains the sensory representation of the amputated limb (Kikkert et al. 2018, 2019b; Makin et al. 2013; Preißler et al. 2012). The persistence of pain and the sense of ownership over the missing limb also suggest a mismatch between the brain's perceptual representation of body image and the absence of afferent sensory input. Given the insula's role in integrating sensory information with prior body schemas and predictive models, it is likely that distortions in body representation occur in this region. Functional neuroimaging studies have shown that upper-limb amputees with phantom limb pain activate the same somatosensory regions as healthy individuals when performing phantom limb movements (Kikkert et al. 2018, 2019a, b; Makin et al. 2013), suggesting a preserved representation of the missing limb. Additionally, limb amputees exhibit greater bilateral insula activation during phantom limb movement compared to non-amputees (Kikkert et al. 2018, 2019b; Makin et al. 2013). Kikkert et al. (2018, 2019b) provided a bottom-up hypothesis whereby damaged nerves in amputees persistently send aberrant signals to the somatosensory cortex, preventing the brain from fully updating its body image. However, it is also well known that the loss of sensory input leads to degeneration in the somatosensory cortex representation of the missing limb (Boucard et al. 2009; Preißler et al. 2013). This degeneration has even been shown to occur after prolonged periods of attenuated sensation (e.g., limb immobilization; Travagli and Anselmi 2016, Zanchi et al. 2017) in the absence of amputation, underscoring the dynamic process that underlies the brain's mapping of the body.

In phantom limb pain, ongoing dysfunctional input to the brain may hinder the insula's ability to update the mental representation of the body. As a result, individuals continue to experience sensations from and perceive bodily ownership over the missing limb. This mismatch between the brain's mental image and the physical body could explain the effectiveness of mirror therapy in reducing phantom limb pain (MacLachlan et al. 2004; Ramachandran and Rogers-Ramachandran 1996; Ramachandran et al. 1995). In mirror therapy, a mirror is placed to reflect the intact limb in the space where the phantom limb is perceived, creating the visual illusion of the missing limb. Although the exact mechanism is not fully understood, mirror therapy may help reconcile the mismatch between visual (exteroceptive) signals and interoceptive signaling, temporarily restoring a sense of ownership and alleviating pain (Herrador Colmenero et al. 2018). Perhaps not surprisingly, individuals with phantom limb pain also report higher levels of body image dissatisfaction and hypervigilance regarding the missing limb compared to amputees without phantom limb pain (Beisheim-Ryan et al. 2023). Collectively, these findings suggest a link between disturbed interoception and altered body representation in phantom limb syndrome.

In summary, disruptions in the body's homeostatic state, whether due to illness, injury, or other physical changes, highlight the critical role of interoceptive integration in maintaining an accurate body representation. Our current understanding of the function of the insula underscores its significance in integrating sensory input into our mental body representations to ensure alignment between the physical and perceived self. Conditions such as jaundice, cancer-related appearance changes, and phantom limb syndrome demonstrate the profound effects of disrupted interoception on body image, mental health, and behavior. These findings suggest that treatments targeting the restoration of interoceptive and exteroceptive balance hold promise for alleviating distress and improving quality of life. Expanding research on the interplay between interoception and body image offers a promising avenue to deepen our understanding of the mind-body connection and develop innovative interventions for individuals facing these challenges.

8 Interoception and Body Image: Future Directions

Craig's contributions to the field of neurophysiology, particularly his theories on interoception and the insula's role in homeostasis and affective processes, have significantly influenced current scientific interest in the role of interoception in relation to human consciousness, emotional experience, and self-perception. Since interoception plays a critical role across all major biological systems that regulate homeostasis, its disruption may contribute to both physical and mental illnesses. While Craig's work did not extend into clinical applications, it continues to provide a conceptual basis for generating future studies investigating interoception's potential as a biomarker for diagnosis, monitoring, and treatment.

In this chapter, we have highlighted multiple links between disrupted interoception and body image both in physical illness and psychopathology. However, the precise directional nature of this relationship remains unclear and understudied (for discussion, see Verdonk and Khalsa 2025). Longitudinal and developmental studies examining how interoception emerges and evolves across the lifespan could provide important insights into the causal mechanisms underlying disrupted interoception, its contribution to illness, and its relevance for treatment development. Investigating interoception in childhood, particularly in relation to the emergence of psychopathology, may help identify early biomarkers for mental health symptoms and related disorders. This is particularly important given that many psychiatric conditions can be traced to childhood, making this developmental period a promising target for early idenfication and intervention.

Interoceptive deficits may also precede or exacerbate body image concerns, particularly due to impaired integration of bodily signals within the insula. This opens possibilities for targeted neurological interventions in conditions where body image is distorted. For example, deep brain stimulation of the posterior insula has been proposed as a treatment for chronic pain (Bergeron et al. 2021). Although this approach has not yet been applied to phantom limb pain, deep brain stimulation in areas involved in affective pain processing, such as the thalamus and anterior cingulate cortex (ACC), has shown promise for pain relief (Nüssel et al. 2022). Another frontier area relates to understanding how the reconstitution of the body following transplantation impacts interoceptive representation and sensitivity. Studies have demonstrated intact cardiac interoception following heart transplant (Barsky et al. 1998; Salamone et al. 2020). However, the impacts of transplantation of other visceral organs, such as orthotopic liver or kidney transplantation, or even appendicular transplant of hand (Philip et al. 2022) or face (Furr et al. 2007; Rifkin et al. 2018) on interoception and body image, may be considered low-hanging fruit awaiting investigation.

Given the frequent coactivation of the ACC and insula during subjective experiences, further research could help define effective treatments for phantom limb pain. Other therapeutic neuromodulation approaches targeting the anterior insula, such as neurofeedback, could be studied for enhancement of the integration of internal and external signals and alleviation of distorted body image. Neurofeedback approaches could conceivably be combined with psychotherapies to potentially augment therapist-patient communication and insight, although there are numerous hurdles to be overcome in developing an appropriate clinical framework for neurofeedback-assisted psychotherapy (Morrissey et al. 2024).

Traditional psychotherapies focusing on modulation of body signals, such as interoceptive exposure, are beginning to be adapted for use in treating eating disorders (Boswell et al. 2019; Plasencia et al. 2019). Given the aforementioned enhancement of interoceptive awareness and improved body image with floatation REST (Choquette et al. 2023; Feinstein et al. 2018a), the combination of this approach with interoceptive exposure techniques for eating disorders would be a logical extension. This approach, which we have called “float-assisted psychotherapy,” leverages the theoretical model described earlier (Fig. 1) in an effort to rebalance interoceptive and exteroceptive systems, thereby improving body image. A functional neuroimaging study of floating in healthy volunteers (Al Zoubi et al. 2021) observed reductions of resting state functional connectivity within and between parts of the default mode network, specifically the posterior cingulate and precuneus, and parts of the somatomotor network, which includes the somatosensory cortices extending into the posterior insula. While similar studies are needed in eating disorders, given their already reduced rsFC in regions involved in interoceptive/exteroceptive and self-referential processing (Angeletti et al. 2022), perhaps one therapeutic mechanism of float therapy is to “reset” neurocircuitry supporting maladaptive body image perception, which could explain the post-float reductions in body dissatisfaction.

As our understanding of the neurobiological processes linking brain and body deepens, novel interventions may emerge to target persistent symptoms such as body image disturbance. Realizing this potential will require integrative, cross-disciplinary research that bridges traditionally disconnected fields (Lane et al. 2009; Ritz et al. 2024). Although such an approach is inherently complex, it will be essential for translating insights from interoceptive neuroscience into meaningful clinical applications.

References

Abbate-Daga G, Delsedime N, Nicotra B, Giovannone C, Marzola E et al (2013) Psychosomatic syndromes and anorexia nervosa. BMC Psychiatry 13:1–11

Al Zoubi O, Misaki M, Bodurka J, Kuplicki R, Wohlrab C et al (2021) Taking the body off the mind: decreased functional connectivity between somatomotor and default-mode networks following floatation-REST. Hum Brain Mapp 42:3216–3227

American Psychological Association [APA] (2022) Diagnostic and statistical manual of mental disorder. American Psychological Association, Washington, DC

Angeletti L, Innocenti M, Felciai F, Ruggeri E, Cassioli E et al (2022) Anorexia nervosa as a disorder of the subcortical-cortical interoceptive-self. Eat Weight Disord 27:3063–3081

Avalos L, Tylka TL, Wood-Barcalow N (2005) The body appreciation scale: development and psychometric evaluation. Body Image 2:285–297

Barsky AJ, Ahern DK, Brener J, Surman OS, Ring C, Dec GW (1998) Palpitations and cardiac awareness after heart transplantation. Psychosom Med 60:557–562
Beilharz F, Castle DJ, Grace S, Rossell SL (2017) A systematic review of visual processing and associated treatments in body dysmorphic disorder. Acta Psychiatr Scand 136:16–36
Beisheim-Ryan EH, Hicks GE, Pohlig RT, Medina J, Sions JM (2023) Body image and perception among adults with and without phantom limb pain. PM R 15:278–290
Benarroch EE (1993) The central autonomic network: functional organization, dysfunction, and perspective. Mayo Clin Proc 68:988–1001
Bergeron D, Obaid S, Fournier-Gosselin MP, Bouthillier A, Nguyen DK (2021) Deep brain stimulation of the posterior insula in chronic pain: a theoretical framework. Brain Sci 11:639
Berner LA, Simmons AN, Wierenga CE, Bischoff-Grethe A, Paulus MP et al (2018) Altered interoceptive activation before, during, and after aversive breathing load in women remitted from anorexia nervosa. Psychol Med 48:142–154
Bluemel S, Menne D, Milos G, Goetze O, Fried M et al (2017) Relationship of body weight with gastrointestinal motor and sensory function: studies in anorexia nervosa and obesity. BMC Gastroenterol 17:4
Boswell JF, Anderson LM, Oswald JM, Reilly EE, Gorrell S, Anderson DA (2019) A preliminary naturalistic clinical case series study of the feasibility and impact of interoceptive exposure for eating disorders. Behav Res Ther 117:54–64
Boucard CC, Hernowo AT, Maguire RP, Jansonius NM, Roerdink JB et al (2009) Changes in cortical grey matter density associated with long-standing retinal visual field defects. Brain 132: 1898–1906
Bruch H (1962) Perceptual and conceptual disturbances in anorexia nervosa. Psychosom Med 24: 187–194
Byrne E, Gaffey J, Hayden L, Daly A, Gallagher P, Dunne S (2023) Body image and cancer-related lymphoedema: a systematic review. Psychooncology 32:1528–1538
Cannon WB (1929) Organization for physiological homeostasis. Physiol Rev 9:399–431
Cash T (2000) The multidimensional body-self relations questionnaire users' manual. Available from the author at www.body-images.com
Cash TF (2004) Body image: past, present, and future. Body Image 1:1–5
Choquette EM, Flux MC, Moseman SE, Chappelle S, Naegele J et al (2023) The impact of floatation therapy on body image and anxiety in anorexia nervosa: a randomised clinical efficacy trial. eClinicalMedicine. https://doi.org/10.1016/j.eclinm.2023.102173
Craig AD (2002) How do you feel? Interoception: the sense of the physiological condition of the body. Nat Rev Neurosci 3:655–666
Craig AD (2003) Interoception: the sense of the physiological condition of the body. Curr Opin Neurobiol 13:500–505
Craig AD (2008) Interoception and emotion: a neuroanatomical perspective. In: Handbook of emotions, 3rd edn. The Guilford Press, New York, pp 272–292
Craig AD (2009) How do you feel–now? The anterior insula and human awareness. Nat Rev Neurosci 10:59–70
Craig AD (2010) The sentient self. Brain Struct Funct 214:563–577
Craig AD, Reiman EM, Evans A, Bushnell MC (1996) Functional imaging of an illusion of pain. Nature 384:258–260
Craig AD, Chen K, Bandy D, Reiman EM (2000) Thermosensory activation of insular cortex. Nat Neurosci 3:184–190
Craig AD, Krout K, Andrew D (2001) Quantitative response characteristics of thermoreceptive and nociceptive lamina I spinothalamic neurons in the cat. J Neurophysiol 86:1459–1480
Damasio AR (1994) Descarte's error: emotion, reason, and the human brain. Avon, New York
Damasio AR (1999) The feeling of what happens: body and emotion in the making of consciousness. Harcourt Brace, New York
Damasio A (2000) The feeling of what Hppens: body and emotion in the making of consciousness. Harcourt, Inc, Orlando

De la Cruz F, Teed AR, Lapidus RC, Upshaw V, Schumann A et al (2022) Central autonomic network alterations in anorexia nervosa following peripheral adrenergic stimulation. Biol Psychiatry Cogn Neurosci Neuroimaging 8:720

Diaz-Fong JP, Feusner JD (2024) Visual perceptual processing abnormalities in body dysmorphic disorder. Curr Top Behav Neurosci

Dostrovsky JO, Craig AD (1996) Cooling-specific spinothalamic neurons in the monkey. J Neurophysiol 76:3656–3665

Feinstein JS, Khalsa SS, Yeh H-w, Wohlrab C, Simmons WK et al (2018a) Examining the short-term anxiolytic and antidepressant effect of floatation-REST. PLoS One 13:e0190292

Feinstein JS, Khalsa SS, Yeh H, Al Zoubi O, Arevian AC et al (2018b) The elicitation of relaxation and interoceptive awareness using floatation therapy in individuals with high anxiety sensitivity. Biol Psychiatry Cogn Neurosci Neuroimaging 3:555–562

Ferraro S, Klugah-Brown B, Tench CR, Bazinet V, Bore MC et al (2022) The central autonomic system revisited – convergent evidence for a regulatory role of the insular and midcingulate cortex from neuroimaging meta-analyses. Neurosci Biobehav Rev 142:104915

Feusner J, Deshpande R, Strober M (2017) A translational neuroscience approach to body image disturbance and its remediation in anorexia nervosa. Int J Eat Disord 50:1014–1017

Frederick DA, Tylka TL, Rodgers RF, Pennesi JL, Convertino L et al (2022) Pathways from sociocultural and objectification constructs to body satisfaction among women: the U.S. Body Project I. Body Image 41:195–208

Furr LA, Wiggins O, Cunningham M, Vasilic D, Brown CS et al (2007) Psychosocial implications of disfigurement and the future of human face transplantation. Plast Reconstr Surg 120:559–565

Gadsby S (2017) Distorted body representations in anorexia nervosa. Conscious Cogn 51:17–33

Garner DM (2004) Eating disorder inventory-3 (EDI-3). Professional manual. Psychological Assessment Resources, Odessa, p 1

Grogan S (2021) Body image: understanding body dissatisfaction in men. Routledge, Women and Children

Grunewald W, Fogelberg S, Ferguson W, Hines S, Fortenberry B, Smith AR (2023a) Longitudinal relationships between specific domains of interoception and muscle dysmorphia symptoms. Eat Behav 48:101686

Grunewald W, Gagliano E, Brown TA, Smith AR (2023b) Conformity to masculine norms, interoceptive dysfunction, and changes in muscle dysmorphia symptoms. Body Image 45: 343–354

Han ZS, Zhang ET, Craig AD (1998) Nociceptive and thermoreceptive lamina I neurons are anatomically distinct. Nat Neurosci 1:218–225

Herrador Colmenero L, Perez Marmol JM, Martí-García C, Querol Zaldivar MÁ, Tapia Haro RM et al (2018) Effectiveness of mirror therapy, motor imagery, and virtual feedback on phantom limb pain following amputation: a systematic review. Prosthetics Orthot Int 42:288

Jenkinson PM, Rossell SL (2024) Disturbed interoception in body dysmorphic disorder: a framework for future research. Aust N Z J Psychiatry 58:300–307

Keigan J, De Los SB, Gaither SE, Walker DC (2024) The relationship between racial/ethnic identification and body ideal internalization, hair satisfaction, and skin tone satisfaction in black and black/white biracial women. Body Image 50:101719

Kerr KL, Moseman SE, Avery JA, Bodurka J, Zucker NL, Simmons WK (2016) Altered insula activity during visceral Interoception in weight-restored patients with anorexia nervosa. Neuropsychopharmacology 41:521–528

Khalsa SS, Craske MG, Li W, Vangala S, Strober M, Feusner JD (2015) Altered interoceptive awareness in anorexia nervosa: effects of meal anticipation, consumption and bodily arousal. Int J Eat Disord 48:889–897

Khalsa SS, Adolphs R, Cameron OG, Critchley HD, Davenport PW et al (2018a) Interoception and mental health: a roadmap. Biol Psychiatry Cogn Neurosci Neuroimaging 3:501–513

Khalsa SS, Hassanpour MS, Strober M, Craske MG, Arevian AC, Feusner JD (2018b) Interoceptive anxiety and body representation in anorexia nervosa. Front Psych 9:444

Khalsa SS, Moseman SE, Yeh H-W, Upshaw V, Persac B et al (2020) Reduced environmental stimulation in anorexia nervosa: an early-phase clinical trial. Front Psychol 11:567499

Kikkert S, Johansen-Berg H, Tracey I, Makin TR (2018) Reaffirming the link between chronic phantom limb pain and maintained missing hand representation. Cortex 106:174–184

Kikkert S, Mezue M, O'Shea J, Henderson Slater D, Johansen-Berg H et al (2019a) Neural basis of induced phantom limb pain relief. Ann Neurol 85:59–73.

Kikkert S, Mezue M, O'Shea J, Henderson Slater D, Johansen-Berg H et al (2019b) Neural basis of induced phantom limb pain relief. Ann Neurol 85:59–73

Kołodziejczyk A, Pawłowski T (2019) Negative body image in breast cancer patients. Adv Clin Exp Med 28:1137–1142

Lane RD, Waldstein SR, Critchley HD, Derbyshire SW, Drossman DA et al (2009) The rebirth of neuroscience in psychosomatic medicine, Part II: clinical applications and implications for research. Psychosom Med 71:135–151

Lapidus RC, Puhl M, Kuplicki R, Stewart JL, Paulus MP et al (2020) Heightened affective response to perturbation of respiratory but not pain signals in eating, mood, and anxiety disorders. PLoS One 15:e0235346

Li W, Lai TM, Bohon C, Loo SK, McCurdy D et al (2015a) Anorexia nervosa and body dysmorphic disorder are associated with abnormalities in processing visual information. Psychol Med 45: 2111–2122

Li W, Lai TM, Loo SK, Strober M, Mohammad-Rezazadeh I et al (2015b) Aberrant early visual neural activity and brain-behavior relationships in anorexia nervosa and body dysmorphic disorder. Front Hum Neurosci 9:301

Luo L, Rodriguez E, Jerbi K, Lachaux JP, Martinerie J et al (2010) Ten years of nature reviews neuroscience: insights from the highly cited. Nat Rev Neurosci 11:718–726

MacLachlan M, McDonald D, Waloch J (2004) Mirror treatment of lower limb phantom pain: a case study. Disabil Rehabil 26:901–904

Madsen SK, Bohon C, Feusner JD (2013) Visual processing in anorexia nervosa and body dysmorphic disorder: similarities, differences, and future research directions. J Psychiatr Res 47:1483–1491

Makin TR, Scholz J, Filippini N, Henderson Slater D, Tracey I, Johansen-Berg H (2013) Phantom pain is associated with preserved structure and function in the former hand area. Nat Commun 4: 1570

McKinley NM, Hyde JS (1996) The objectified body consciousness scale: development and validation. Psychol Women Q 20:181–215

Morrissey G, Tsuchiyagaito A, Takahashi T, McMillin J, Aupperle RL, Misaki M, Khalsa SS (2024) Could neurofeedback improve therapist-patient communication? Considering the potential for neuroscience informed examinations of the psychotherapeutic relationship. Neurosci Biobehav Rev 161:105680

Nüssel M, Zhao Y, Knorr C, Regensburger M, Stadlbauer A et al (2022) Deep brain stimulation, stereotactic radiosurgery and high-intensity focused ultrasound targeting the limbic pain matrix: a comprehensive review. Pain Ther 11:459–476

Ordaz DL, Schaefer LM, Choquette E, Schueler J, Wallace L, Thompson JK (2018) Thinness pressures in ethnically diverse college women in the United States. Body Image 24:1–4

Pearce JM (2004) Richard Morton: origins of anorexia nervosa. Eur Neurol 52:191–192

Philip BA, Valyear KF, Cirstea CM, Baune NA, Kaufman C, Frey SH (2022) Changes in primary somatosensory cortex following allogeneic hand transplantation or autogenic hand replantation. Front Neuroimaging 1:919694

Plasencia M, Sysko R, Fink K, Hildebrandt T (2019) Applying the disgust conditioning model of food avoidance: a case study of acceptance-based interoceptive exposure. Int J Eat Disord 52: 473–477

Preißler S, Feiler J, Dietrich C, Hofmann GO, Miltner WHR, Weiss T (2012) Gray matter changes following limb amputation with high and low intensities of phantom limb pain. Cereb Cortex 23:1038–1048

Preißler S, Feiler J, Dietrich C, Hofmann GO, Miltner WHR, Weiss T (2013) Gray matter changes following limb amputation with high and low intensities of phantom limb pain. Cereb Cortex 23:1038–1048

Ralph-Nearman C, Arevian AC, Puhl M, Kumar R, Villaroman D et al (2019) A novel Mobile tool (Somatomap) to assess body image perception pilot tested with fashion models and nonmodels: cross-sectional study. JMIR Ment Health 6:e14115

Ralph-Nearman C, Arevian AC, Moseman S, Sinik M, Chappelle S et al (2021) Visual mapping of body image disturbance in anorexia nervosa reveals objective markers of illness severity. Sci Rep 11:12262

Ramachandran VS, Rogers-Ramachandran D (1996) Synaesthesia in phantom limbs induced with mirrors. Proc Biol Sci 263:377–386

Ramachandran VS, Rogers-Ramachandran D, Cobb S (1995) Touching the phantom limb. Nature 377:489–490

Ramsay DS, Woods SC (2014) Clarifying the roles of homeostasis and allostasis in physiological regulation. Psychol Rev 121:225–247

Rifkin WJ, Kantar RS, Ali-Khan S, Plana NM, Diaz-Siso JR et al (2018) Facial disfigurement and identity: a review of the literature and implications for facial transplantation. AMA J Ethics 20: 309–323

Ritz T, Schulz A, Khalsa S (2024) The golden age of integrative neuroscience? The brain joins the body in the latest renaissance of interoception research. Biol Psychol 192:108851

Riva G, Gaudio S (2012) Allocentric lock in anorexia nervosa: new evidences from neuroimaging studies. Med Hypotheses 79:113–117

Salamone PC, Sedeño L, Legaz A, Bekinschtein T, Martorell M et al (2020) Dynamic neurocognitive changes in interoception after heart transplant. Brain Commun 2:fcaa095

Skarderud F (2007) Eating one's words, part I: 'concretised metaphors' and reflective function in anorexia nervosa-an interview study. Eur Eat Disord Rev 15:163–174

Stanghellini G, Ballerini M, Mancini M (2019) The optical-coenaesthetic disproportion hypothesis of feeding and eating disorders in the light of neuroscience. Front Psychiatry 10:630

Stunkard AJ, Sørensen T, Schulsinger F (1983) Use of the Danish adoption register for the study of obesity and thinness. Res Publ Assoc Res Nerv Ment Dis 60:115–120

Swami V, Salem N, Furnham A, Tovée MJ (2008) Initial examination of the validity and reliability of the female photographic figure rating scale for body image assessment. Personal Individ Differ 44:1752–1761

Swami V, Frederick DA, Aavik T, Alcalay L, Allik J et al (2010) The attractive female body weight and female body dissatisfaction in 26 countries across 10 world regions: results of the international body project I. Personal Soc Psychol Bull 36:309–325

Swami V, Stieger S, Harris AS, Nader IW, Pietschnig J et al (2012) Further investigation of the validity and reliability of the photographic figure rating scale for body image assessment. J Pers Assess 94:404–409

Travagli RA, Anselmi L (2016) Vagal neurocircuitry and its influence on gastric motility. Nat Rev Gastroenterol Hepatol 13:389–401

Tylka TL, Wood-Barcalow NL (2015) The body appreciation Scale-2: item refinement and psychometric evaluation. Body Image 12:53–67

Vandereycken W, van Deth R (1989) Who was the first to describe anorexia nervosa: Gull or Lasègue? Psychol Med 19:837–845

Vani MF, Lucibello KM, Trinh L, Santa Mina D, Sabiston CM (2021) Body image among adolescents and young adults diagnosed with cancer: a scoping review. Psychooncology 30: 1278–1293

Verdonk C, Khalsa SS (2025) Interoception and mental health. In: Murhpy J, Brewer R (eds) Interoception: a comprehensive guide. Springer Nature

Westlund KN, Craig AD (1996) Association of spinal lamina I projections with brainstem catecholamine neurons in the monkey. Exp Brain Res 110:151–162

Zanchi D, Depoorter A, Egloff L, Haller S, Mählmann L et al (2017) The impact of gut hormones on the neural circuit of appetite and satiety: a systematic review. Neurosci Biobehav Rev 80:457–475

The Intersection of Interoception and Anticipation Related to Pain

Irina A. Strigo and Alan N. Simmons

Contents

Abstract This chapter focuses on the interconnectedness of bodily awareness, emotion, and time perception, focusing on the roles of interoception and anticipation, pioneered by A.D. (Bud) Craig and advanced by contemporary research. Pain, a pivotal aspect of human experience, serves as a conduit for understanding our relationship with the world. Anticipation of pain, vital for survival, influences subjective pain experiences and is modulated by factors such as physiological reactivity and contextual cues. Emotional states significantly shape pain perception, with chronic pain conditions and affective disorders characterized by dysregulated pain modulation mechanisms. We discuss a state space model for pain, wherein pain functions as a latent construct shaped by both anticipatory and contemporaneous

I. A. Strigo (✉)
Veterans Affairs & University of California, San Francisco, San Francisco, CA, USA
e-mail: irina.strigo@ucsf.edu

A. N. Simmons (✉)
Veterans Affairs & University of California, San Diego, San Diego, CA, USA
e-mail: ansimmons@health.ucsd.edu

Curr Topics Behav Neurosci (2025) 70: 127–142
https://doi.org/10.1007/7854_2025_588
Published Online: 14 May 2025

factors. Understanding these mechanisms is critical for informing clinical interventions aimed at pain management and highlights the intersection of interoception, emotion, and pain.

Keywords Anticipation · Chronic pain · Emotional allodynia · Emotions · Insula · Interoception · Pain · Predictive coding · State-space model

1 Introduction

The bodily self, emotion, and subjective time are fundamentally linked through the processes of interoception and anticipation. Appreciation of this linkage was pioneered by Bud Craig (2002) and has been forwarded by countless other researchers over the last few decades. It was no coincidence that much of the original mapping of these pathways was from the field of pain research, as pain is fundamental to the way that we understand and behave in relationship to the world around us (de C Williams 2023). To best understand, treat, and predict pain, the brain formulates a predictive model. The premise of most pain as predictive signal model is based on continuously updating pertinent information (Friston 2003). When sensory evidence fails to confirm the prediction, either the sensory evidence is ignored and, in turn, biases perception, or the brain learns from the error and updates subsequent predictions and evaluations to keep bias in check (Feldman and Friston 2010; Rigoli et al. 2016). In this chapter, we will review the current work in understanding the role of anticipation in pain through an interoceptive lens.

2 Pain as a Modifiable Construct

The continual formulation of predictions about future pain is implicit to survival and influences the pain we experience (Fields 2018; Keltner et al. 2006; Koyama et al. 2005), particularly when the experienced pain is difficult to quantify (Strigo et al. 2024c). Modification of expectation can lead to significant reduction (Tracey 2010) or amplification (Bingel et al. 2011; Schmid et al. 2013, 2015) of subjective pain. Thus, anticipation of pain is vital in understanding, and clinically addressing, subjective pain (Fields 2018). One of the key aspects that can strongly influence pain is our emotional state, or psychiatric status, when experiencing pain (Aupperle et al. 2012; Gorka et al. 2017; Grillon et al. 2009; Simmons et al. 2013; Strigo et al. 2008a).

Pain is not solely a physical sensation. According to the International Association for Study of Pain, pain is "an unpleasant sensory and emotional experience associated with, or resembling that associated with, actual or potential tissue damage" (Raja et al. 2020), (i.e., warning that damage has or could happen or a warning signal of estimated damage). The goal of the pain signal produced in the brain is to motivate action (Craig 2003b). The physical signal is insufficient to properly determine the correct amount of pain to motivate the correct response as the needed compendium of information may be delayed by seconds, minutes, days, or more. Thus we produce a predictive model (Friston 2003), not dissimilar from the "interoceptive moment" model by Bud Craig (2015). This predictive model depends on (1) nerve signals, such as from Aδ and C fibers (Craig 2003c), (2) physiological reactivity (Cowen et al. 2015; Koenig et al. 2014), and context from the (3) anticipated pain onset, (4) contemporaneous pain contextual cues, and (5) expectations of pain offset (Fields 2018). In this way chronic pain can be seen as an interaction between predicted and contemporaneous information. Related concept is that of "prediction error," i.e., the mismatch between prediction and afferent information, triggering the behavioral component of pain, the drive/motivation to do something about it (Hoskin and Talmi 2023). This echoes the theory that chronic pain experience may not be directly linked solely to nociceptive pain (Fields 2018), rather, it is a state that is influenced by the brain mechanisms of anticipation and expectation, both the anticipation of what pain will be like (onset anticipation) and the expectation that pain will end (offset anticipation).

2.1 The Role of Anticipation in Pain

The construct of anticipation can be categorized into two main types: (1) acute or imminent threat, where pain occurs immediately following a cue and (2) prolonged impending threat, where the likelihood of experiencing pain increases over an extended period, often with unpredictable timing. These categories exist along a continuum of the perceived imminence and as expressions of phases of dealing with the perceived threat (Abend 2023; Fanselow 2022). There is strong evidence of increased brain activation during anticipation of heat pain versus warm stimuli (Atlas et al. 2010; Ploghaus et al. 1999; Strigo et al. 2008a, 2010a). Similar networks have been shown when the prolonged, instead of acute, threat of pain is experienced (Abend et al. 2022). In a recent review, Pavy and colleagues suggest that the predictability of pain does not significantly alter pain perception (Pavy et al. 2024); however, expectations of pain intensity and basal affective state can moderate these effects. In a sample of 147 chronic pain-free subjects, we saw that large individual differences in the brain of positive and negative pain predictors during unpredictable cues exist even when pain ratings remained similar between subgroups (Strigo et al. 2024c). A key modulator of the pain experience and the experienced anticipation can be contemporaneous information that can lead to biasing (Atlas and Wager 2012; Geuter et al. 2013).

2.2 *The Role of Emotion in Pain*

One critical aspect of the pain experience is the interaction with affective state (Lumley et al. 2011). Clinically, mental and physical health are closely linked, and mental health problems and physical health conditions often occur together (Bondesson et al. 2018). Experimentally, this relationship is evidenced by noted differences in pain and pain anticipation across a variety of disorders including depression (Bar et al. 2007; Strigo et al. 2008a, 2013b), posttraumatic stress (Geuze et al. 2007; Moeller-Bertram et al. 2012; 2014; Strigo et al. 2010b), eating disorders (Strigo et al. 2013a), and traumatic brain injury (Strigo et al. 2014, 2018). The pain experience itself represents a sophisticated integration of sensory-discriminative components (intensity and location) and affective components (unpleasantness and suffering), deeply rooted in interoceptive processing (Khera and Rangasamy 2021). While experimental pain stimuli typically create balanced responses between these dimensions in healthy subjects, certain types of sensations like visceral or ischemic stimuli tend to weigh more heavily on the emotional dimension, causing greater suffering even at lower, non-painful intensity levels (Rainville et al. 1992; Strigo et al. 2002). The affective component of sensory processing can be experimentally modulated – changing one's emotional state affects pain ratings to experimental stimuli, particularly in their affective dimension (Lehoux and Abbott 2011; Loggia et al. 2008; Villemure and Bushnell 2002; Zelman et al. 1991). Specifically, negative emotional states like sadness increase the distress associated with the same level of sensory input, an effect that was initially overlooked as most studies only assessed unpleasantness when stimuli were reported as painful. This heightened affective response to sensory stimulation is particularly pronounced in affective disorders, a phenomenon we have termed emotional allodynia (Strigo et al. 2008b) (see below).

This emotional-interoceptive relationship is processed through an interconnected network of brain regions including the insula and anterior cingulate cortex – areas crucial for integrating bodily signals with emotional states (Craig 2002, 2015; Critchley and Patchitt 2024) (see chapters "Insula and Interoceptive Pathways" and "Emotions and Physiological Feelings" in this volume). When individuals experience negative emotions, these same neural circuits show heightened activation, which can prime the brain for enhanced pain perception through altered interoceptive sensitivity. This creates a complex feedback loop where negative emotions can amplify pain perception through enhanced body awareness, and pain can further intensify negative emotions, ultimately affecting how internal bodily signals are processed and interpreted. The role of interoception in this process is particularly evident in how anticipation and emotional context can significantly modify the pain experience even before a stimulus is applied, highlighting how our brain's anticipation of bodily signals is heavily influenced by emotional state.

2.3 Emotional Allodynia

Emotional allodynia, first described by Strigo et al. (2008b) in depression, refers to a phenomenon where individuals experience non-painful stimuli as unpleasant. In this research, depressed individuals rated warm, non-painful temperature stimuli as highly unpleasant, with their unpleasantness ratings matching those that healthy individuals gave to suprathreshold and clearly painful stimuli (Strigo et al. 2008b). The heightened sensitivity to unpleasantness was profound – depressed participants began reporting unpleasantness at temperatures well below both healthy controls' thresholds and their own pain thresholds. These findings are in direct agreement with the literature showing that sensory pain is rated as more distressing by depressed participants (Bar et al. 2010). Importantly, this altered pain perception proved to be a stable characteristic of depression, persisting regardless of attentional states or acute mood changes (Strigo et al. 2008a; Ushinsky et al. 2013), suggesting that emotional allodynia is not simply a reflection of lower baseline affect or increased self-focus found in depression.

A comprehensive meta-analysis by (Thompson et al. 2016) examined pain thresholds across various experimental methods in depression. While results were inconsistent across different types of sensory testing (including electrical, heat, cold, pressure, ischemic), there was one notable exception: all five studies using ischemic stimulation showed consistent findings – individuals with depression demonstrated increased pain sensitivity, reporting pain at significantly lower levels of ischemic stimulation compared to healthy controls ($g = -0.81$; 95% CI, -1.39 to -0.24; $p = 0.006$). This is particularly noteworthy as ischemic sensations are often experienced as highly unpleasant even if they are not painful (Rainville et al. 1992). Using linear mixed effects modeling of BOLD responses to non-painful warm stimuli applied to the left forearm, our neuroimaging studies revealed a distinct hemispheric lateralization in the insula (see Fig. 1). Depression severity showed a strong positive association with increased right insula activation, while emotional allodynia was negatively associated with decreased left insula activation (total of ~16 ratings) (ICC3, $k = 0.865$, 95% CI: 0.736–0.931, 36 subjects, and 2 raters/measurements; ICC type as referenced by (Shrout and Fleiss 1979).

This lateralization aligns with Craig's (2005) model and is in line with neuroimaging evidence (Strigo and Craig 2016) showing that the left anterior insula is associated predominantly with parasympathetic activity and positive emotions, whereas the right insula is associated with sympathetic activity and negative emotions. The neural findings within the insula were further supported by heart rate variability (HRV) data, which revealed a significant negative correlation between emotional allodynia and RMSSD (Root Mean Square of the Successive Differences; $r = -0.4$, $p < 0.05$), which reflects vagally mediated changes in autonomic tone. This sympathovagal imbalance, characterized by low parasympathetic and increased sympathetic tone, has been previously associated with decreased pain tolerance in individuals with somatoform disorder and comorbid major depression (Pollatos et al.

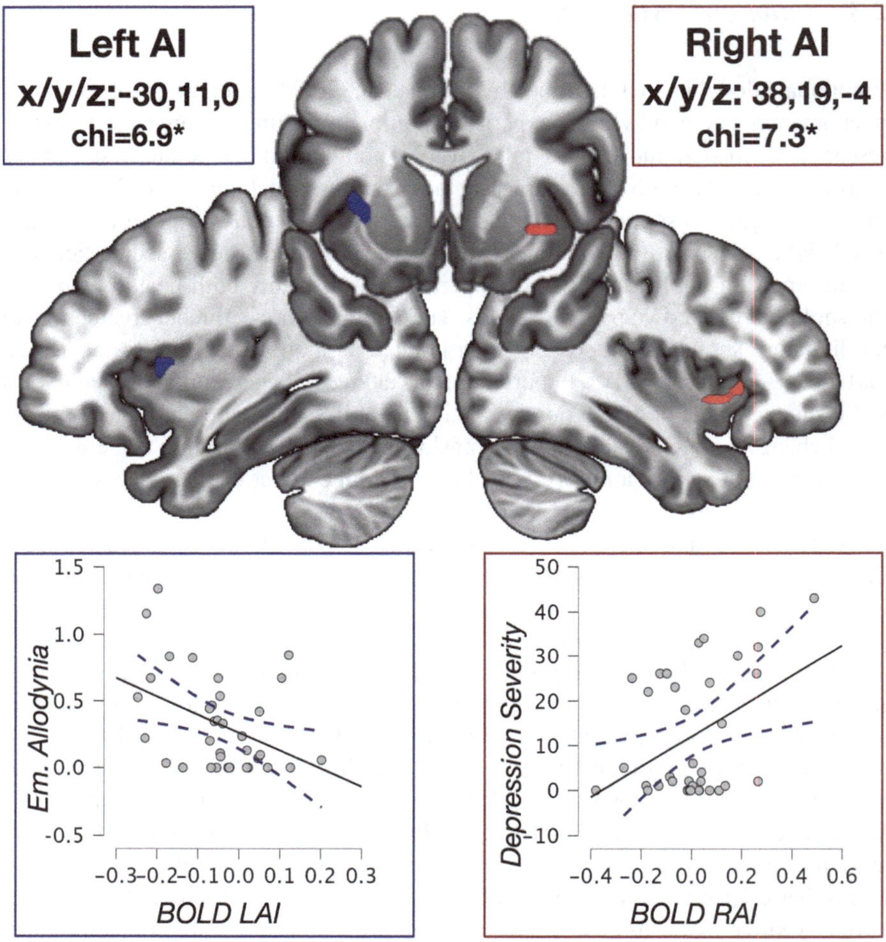

Fig. 1 Dissociable anterior insula responses to emotional allodynia and depression in non-painful warmth processing. Linear mixed effects model – non-painful warmth-related BOLD response (anterior insula mask) was regressed against warmth affective bias ratings (emotional allodynia) and depression symptoms severity, as measured by Beck Depression Inventory-2. Activation within the anterior insula showed that emotional allodynia when controlling for depression severity showed association predominantly with the left anterior insula (blue, inverse relationship), while depression severity, when controlling for emotional allodynia showed association predominantly with the right anterior insula (red, positive correlation)

2011). These findings align with prior research showing lower HRV association with somatic depressive symptoms (Bosch et al. 2009; de Jonge et al. 2007).

The relationship between emotional allodynia and insula function is further supported by (Mutschler et al. 2012), who found evidence of functional reorganization in the insula, with emotion-related responses shifting to areas normally involved in pain processing in healthy individuals. Similarly DeVille et al. (2018) linked depression to altered interoceptive processing and insula pathology, showing how disrupted body signal processing may contribute to emotional allodynia. These findings have influenced both theoretical models of interoception, pain and emotion, and practical approaches to treatment. Garland et al. (2013) integrated this into understanding of chronic pain, showing how negative emotions prime threat processing in affective brain circuits, increasing the likelihood of interpreting bodily sensations as painful. This helps explain the strong epidemiological association between depression and painful conditions (Patten et al. 2015), how depression can increase pain sensitivity following sad mood induction (Lehoux and Abbott 2011), and poor outcomes in chronic pain conditions (Carroll et al. 2016; Giesecke et al. 2005).

Recent research has expanded our understanding of emotional allodynia's broader implications. Raison et al. (2014) connected it to thermoregulatory dysfunction in depression, suggesting altered temperature processing may play a role. Newer treatment approaches have incorporated these insights with developing mindfulness-based pain treatments for chronic pain management (Garland et al. 2023) that also include the "savoring" of positive emotions and positive physical sensations thereby potentially disrupting altered pain appraisals characteristics of emotional allodynia. Tyler et al. (2024) found that emotional allodynia may influence how depressed individuals respond to thermal interventions, though noting some complexity in the relationship.

3 A Neural Model of Pain

The neural model of pain (Fig. 2a, see chapters "The Life and Legacy of Arthur Bud Craig: A Pioneer in Interoception", "Insula and Interoceptive Pathways", and "Emo tions and Physiological Feelings" in this volume for further details) operates through a sophisticated network of interconnected pathways. The primary interoceptive pathway begins with small-diameter (A-delta and C) sensory fibers, which detect thermal, mechanical, and chemical conditions throughout body tissues. These signals are initially processed in lamina I of the spinal cord, then conveyed through a specific thalamic relay to reach the posterior insula, which functions as the primary interoceptive cortex (Craig 2002, 2003a, c; Craig and Blomqvist 2002). From here, information is progressively processed along a posterior-to-anterior gradient within the insula. The middle insula serves as an integration hub, combining interoceptive information with hedonic, exteroceptive, and homeostatic motor signals. The anterior insula then incorporates top-down appraisal from the frontal cortex, functioning as the center for subjective awareness (Craig 2009b, 2015; Craig et al. 2000). Running parallel to this main pathway, a secondary projection from lamina I reaches the cingulate cortex, which operates as the limbic motor cortex and is particularly crucial for generating motivational responses and maintaining homeostatic control (Strigo and Craig 2016). This integrated system becomes further refined through the reward/motivational mesolimbic and interoceptive circuits, where positive or negative attributions of pain are coded (Navratilova and Porreca 2014; Strigo et al.

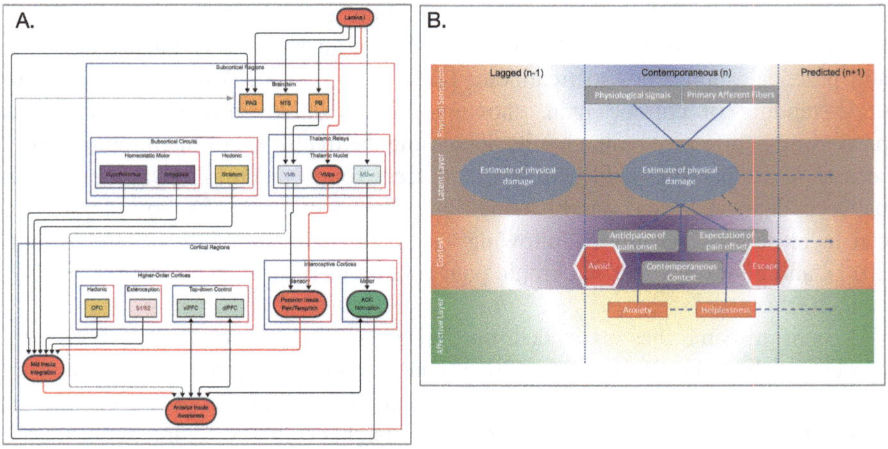

Fig. 2 Neural models of interoception and pain. (**a**) *Schematic of the neural circuitry underlying interoception and pain.* Based on A.D. (Bud) Craig's neuroanatomical work. Spinal lamina I projections (orange) relay through thalamus to cortical targets: interoceptive sensory (posterior insula) and interoceptive motor (anterior cingulate) regions. Interoceptive information flows from posterior insula to middle insula (where it integrates with subcortical inputs) and then to anterior insula (where it integrates with higher-order cortical regions). Known projections are shown with solid arrows. Note: While parabrachial (PB) projections to thalamus (VMb) do not directly project to posterior insula, they likely contribute to anterior insula activity via pathways not yet fully characterized (see chapter "Insula and Interoceptive Pathways"). Recent work also reveals feedback loops from anterior insula to brainstem nuclei. (**b**) *State-space model of interoception and pain.* This dynamic model illustrates the contemporaneous mechanisms of pain and interoception, color-coded to correspond with neural systems in panel A. Feed-forward effects to predicted states occur within and across layers (dashed arrows). For example, persistent pain perpetuates affective distress. Additionally, avoidance behaviors prevent learning opportunities, reducing modulation of anxiety and helplessness, thus maintaining symptoms. These processes unfold in the context of ongoing physical sensations. *Abbreviations*: *PAG* periaqueductal gray, *NTS* nucleus of solitary tract, *PB* parabrachial nucleus, *VMb* basal ventromedial thalamus, *VMpo* posterior ventromedial thalamus, *MDvc* mediodorsal thalamus, *OFC* orbitofrontal cortex, *S1/S2* primary/secondary somatosensory cortices, *vl/dl PFC* ventrolateral/dorsolateral prefrontal cortex, *ACC* anterior cingulate cortex

2024c). The anterior insula plays a key role in aversive expectation of impending pain (Atlas et al. 2010; Craig 2009b; Kong et al. 2008; Strigo et al. 2024c), with notable asymmetry – the right anterior insula specifically mediating negative valence emotions (Brooks et al. 2005; Craig 2005, 2009b; Duerden et al. 2013; Strigo et al. 2013a). In chronic pain conditions, such as back pain, this system can become dysregulated: the reward/motivation circuit shows diminished response (Baliki et al. 2010), potentially impairing pain-escape motivation pain (Goedhoop et al. 2022; Xie et al. 2014), while the interoceptive pathway becomes sensitized, particularly when painful stimuli occur at the site of endogenous pain (Strigo et al. 2021). This comprehensive network forms the neurobiological basis for how we experience, interpret, and respond to pain signals.

4 A State Space Model of Pain

We have proposed a state space model (Fig. 2b) for pain and posttraumatic stress (Strigo et al. 2024a). Similar computational models for pain (Chen and Wang 2023; Poublan-Couzardot and Talmi 2024) have been suggested that propose that pain functions as a latent construct in a continuously updating and adjusting dynamic model. Thus, in this way pain is a latent variable constructed by lagged and contemporaneous features, such as information from the nervous system and the contextual situation (Craig 2015; Pace-Schott et al. 2019). This latent variable is modeled based on the available training data and is dependent on the inputs to this model continuing to operate within the expected range. The latent construct of subjective pain is dependent on expected onset/offset and contemporaneous signals continuing to operate within expected ranges. Thus, emotional trauma or damage to a part of the body can cause signals to operate outside this range causing failure of the model. If recalibration of the pain model is not possible, due to continued avoidance of pain and/or escape from pain, new training and recalibration do not occur and the failed model of pain continues in a chronic state. In the case of chronic lower back pain or peripheral neuropathy, the context of pain is specific to the learned prediction that the sensation in the affected body part (e.g., back) or limb (e.g., feet) is indicative of damage or suffering (Baliki et al. 2010; Strigo et al. 2021). Misconceptions on the contemporaneous pain can be substantially modified by anticipation of context and expectations of offset, but, unlike these features, we proposed that the contemporaneous assessment is less directly linked to an affective condition than that of predictive evaluations (Strigo et al. 2008a, b). Thus, these components of the pain model (i.e., onset and offset anticipation) line up explicitly with different neuroanatomical substrates outlined above (Sect. 3), such that the "salient self" experienced in the anterior insula functions as the projected future self that is based on the integration of the prior moment, expectations, and contemporaneous inputs that update this model. When body signals operate outside our prior experience, the model's predictions become unreliable, leading to prediction errors that manifest in the context of tissue damage and/or emotional trauma. These prediction errors are then incorporated into the state-space model, affecting how future pain is processed. Emotional allodynia described above (Sect. 2.3) elegantly fits into the state space model of pain by modulation of the processes supporting the anticipation of onset (Strigo et al. 2008a) and offset (Thones and Oberfeld 2015). While emotional state directly affects regulatory pathways of pain (Fields 2018), the separation of contemporaneous and affective components does provide leverage into addressing the additive effect of emotional allodynia in pain management.

5 Further Considerations

5.1 Interoception and Additive Models

The model of pain and anticipation presented in this chapter is primarily an interoceptive model (Craig 2002, 2009b, 2015). There are other models for the perception of pain that are complimentary to this model. Those include models that are based on animal literature that are more amygdala centric (Thompson and Neugebauer 2017), models that focus more of the contributions to the perception of pain that come through the sensory motor cortex (Gombaut and Holmes 2022), models that focus on the modulatory effects of the nucleus accumbens (Altier and Stewart 1999), and the default mode network, respectively. The advantage of viewing pain through an interoceptive perspective is that, by understanding the integrative function of the insula (particularly its mid and anterior regions) (Barrett and Simmons 2015), we can conceptualize pain and its anticipation. This approach allows us to consider the complex experience of pain as a hidden variable within a state space model, which can then be used to make clinically relevant predictions.

5.2 Time

Our understanding of when pain will occur shapes how we prepare for and cope with it, showing the fundamental connection between time perception and pain (Fanselow 2022). This interaction of pain anticipation and time is most likely founded on the use of interoception as the underlying construct that supports both processes. Several chapters in this book elegantly present how time and interoception (Wittmann and Droit-Volet 2024) and pain and interoception (Mehling 2024) have key commonalities. Bud Craig's seminal work (Craig 2009a) on this matter was key to better understanding how the insula plays a critical role in the intersection of these functions. As described above, the biological goal of pain as a predictive latent variable make the substrates that support this function well tuned for utilization of available internal body signals for mapping specific phases of experience (Wittmann et al. 2010).

5.3 Clinical Implications

Understanding pain anticipation has important implications for how we treat people with pain disorders, especially considering mental comorbidities. The latent construct of pain is influenced by anticipation of onset of pain and the expectation of the offset of pain, or pain relief. These factors can be influenced by affective and contemporaneous factors that can help modify the pain experience. An important aspect of keeping the predicted current pain state within normal bounds is to maintain recursive feedback to the model. Thus, approaches that (1) minimize avoidance, distraction, and/or misinterpretations of potential pain signals, such as interoceptive attention training (Mehling et al. 2024; Strigo et al. 2024b), (2) minimize overinterpretation of contemporaneous information, such as mindfulness (Mehling et al. 2025), or (3) minimize escape behavior or exaggeration of the duration of pain, such as exposure therapy (Cayoun et al. 2020) can be important parts of the clinical approaches. Similarly, neurostimulation or pharmacological interventions help disrupt existent predictive weighting and allow for the formation of predictive pain states that can delay or postpone chronification.

6 Conclusion

A complete interoceptive model of pain anticipation requires appreciation of a multitude of internal and external signals, utilizing prior experience and future predictions (Strigo et al. 2024a). Much of this understanding can be extended to a broad domain of experience beyond pain. Experiences of emotional states and emotional pain may also be viewed within a state space model. However, as pain is critical to survival, it may be the core pathway that other systems have coopted to deal with the complexities of being human in our ever more abstracted and complicated experience. The model of interoception forwarded by Bud Craig's work provides the brain architecture that would logically manifest as a substratum that elegantly combines the complexities of prediction of the homeostatic state and the interoceptive self in the interoceptive moment.

Competing Interests The authors report no competing interests.

Funding This work was supported by in part by the US Department of Veterans Affairs I01-CX-000816(IAS), I01-CX- 001652(IAS), I01-BX-005918 (ANS), and Painless Research Foundation. Dr. Strigo is a recipient of a Research Career Scientist Award (IK6CX002926) from the Department of Veterans Affairs.

References

Abend R (2023) Understanding anxiety symptoms as aberrant defensive responding along the threat imminence continuum. Neurosci Biobehav Rev 152:105305

Abend R, Ruiz SG, Bajaj MA, Harrewijn A, Linke JO, Atlas LY, Winkler AM, Pine DS (2022) Threat imminence reveals links among unfolding of anticipatory physiological response, cortical-subcortical intrinsic functional connectivity, and anxiety. Neurobiol Stress 16:100428

Altier N, Stewart J (1999) The role of dopamine in the nucleus accumbens in analgesia. Life Sci 65: 2269–2287

Atlas LY, Wager TD (2012) How expectations shape pain. Neurosci Lett 520:140–148

Atlas LY, Bolger N, Lindquist MA, Wager TD (2010) Brain mediators of predictive cue effects on perceived pain. J Neurosci 30:12964–12977

Aupperle RL, Allard CB, Grimes EM, Simmons AN, Flagan T, Behrooznia M, Cissell SH, Twamley EW, Thorp SR, Norman SB, Paulus MP, Stein MB (2012) Dorsolateral prefrontal cortex activation during emotional anticipation and neuropsychological performance in posttraumatic stress disorder. Arch Gen Psychiatry 69:360

Baliki MN, Geha PY, Fields HL, Apkarian AV (2010) Predicting value of pain and analgesia: nucleus accumbens response to noxious stimuli changes in the presence of chronic pain. Neuron 66:149–160

Bar KJ, Wagner G, Koschke M, Boettger S, Boettger MK, Schlosser R, Sauer H (2007) Increased prefrontal activation during pain perception in major depression. Biol Psychiatry 62:1281

Bar KJ, Terhaar J, Boettger MK, Boettger S, Berger S, Weiss T (2010) Pseudohypoalgesia on the skin: a novel view on the paradox of pain perception in depression. J Clin Psychopharmacol 31:103–107

Barrett LF, Simmons WK (2015) Interoceptive predictions in the brain. Nat Rev Neurosci 16:419–429

Bingel U, Wanigasekera V, Wiech K, Ni Mhuircheartaigh R, Lee MC, Ploner M, Tracey I (2011) The effect of treatment expectation on drug efficacy: imaging the analgesic benefit of the opioid remifentanil. Sci Transl Med 3:70ra14

Bondesson E, Larrosa Pardo F, Stigmar K, Ringqvist A, Petersson IF, Joud A, Schelin MEC (2018) Comorbidity between pain and mental illness – evidence of a bidirectional relationship. Eur J Pain 22:1304–1311

Bosch NM, Riese H, Dietrich A, Ormel J, Verhulst FC, Oldehinkel AJ (2009) Preadolescents' somatic and cognitive-affective depressive symptoms are differentially related to cardiac autonomic function and cortisol: the TRAILS study. Psychosom Med 71:944–950

Brooks JC, Zambreanu L, Godinez A, Craig AD, Tracey I (2005) Somatotopic organisation of the human insula to painful heat studied with high resolution functional imaging. NeuroImage 27: 201–209

Carroll CP, Lanzkron S, Haywood C Jr, Kiley K, Pejsa M, Moscou-Jackson G, Haythornthwaite JA, Campbell CM (2016) Chronic opioid therapy and central sensitization in sickle cell disease. Am J Prev Med 51:S69–S77

Cayoun B, Simmons A, Shires A (2020) Immediate and lasting chronic pain reduction following a brief self-implemented mindfulness-based interoceptive exposure task: a pilot study. Mindfulness 11:112–124

Chen ZS, Wang J (2023) Pain, from perception to action: a computational perspective. iScience 26: 105707

Cowen R, Stasiowska MK, Laycock H, Bantel C (2015) Assessing pain objectively: the use of physiological markers. Anaesthesia 70:828–847

Craig AD (2002) How do you feel? interoception: the sense of the physiological condition of the body. Nat Rev Neurosci 3:655–666

Craig AD (2003a) Interoception: the sense of the physiological condition of the body. Curr Opin Neurobiol 13:500–505

Craig AD (2003b) A new view of pain as a homeostatic emotion. Trends Neurosci 26:303–307

Craig AD (2003c) Pain mechanisms: labeled lines versus convergence in central processing. Annu Rev Neurosci 26:1–30

Craig AD (2005) Forebrain emotional asymmetry: a neuroanatomical basis? Trends Cogn Sci 9: 566–571

Craig AD (2009a) Emotional moments across time: a possible neural basis for time perception in the anterior insula. Philos Trans R Soc Lond Ser B Biol Sci 364:1933–1942

Craig AD (2009b) How do you feel – now? The anterior insula and human awareness. Nat Rev Neurosci 10:59–70

Craig AD (2015) How do you feel?: an interoceptive moment with your neurobiological self. Princeton University Press

Craig AD, Blomqvist A (2002) Is there a specific lamina I spinothalamocortical pathway for pain and temperature sensations in primates? J Pain 3:95–101

Craig AD, Chen K, Bandy D, Reiman EM (2000) Thermosensory activation of insular cortex. Nat Neurosci 3:184–190

Critchley HD, Patchitt J (2024) Interoception, insula, and autonomic integration: relevance to the expression and treatment of psychiatric symptoms. Curr Top Behav Neurosci

de C Williams AC (2023) Pain: behavioural expression and response in an evolutionary framework. Evol Med Public Health 11:429–437

de Jonge P, Mangano D, Whooley MA (2007) Differential association of cognitive and somatic depressive symptoms with heart rate variability in patients with stable coronary heart disease: findings from the heart and soul study. Psychosom Med 69:735–739

DeVille DC, Kerr KL, Avery JA, Burrows K, Bodurka J, Feinstein JS, Khalsa SS, Paulus MP, Simmons WK (2018) The neural bases of interoceptive encoding and recall in healthy adults and adults with depression. Biol Psychiatry Cogn Neurosci Neuroimaging 3:546–554

Duerden EG, Arsalidou M, Lee M, Taylor MJ (2013) Lateralization of affective processing in the insula. NeuroImage 78:159–175

Fanselow MS (2022) Negative valence systems: sustained threat and the predatory imminence continuum. Emerg Top Life Sci 6:467–477

Feldman H, Friston KJ (2010) Attention, uncertainty, and free-energy. Front Hum Neurosci 4:215

Fields HL (2018) How expectations influence pain. Pain 159(Suppl 1):S3–S10

Friston K (2003) Learning and inference in the brain. Neural Netw 16:1325–1352

Garland EL, Froeliger B, Zeidan F, Partin K, Howard MO (2013) The downward spiral of chronic pain, prescription opioid misuse, and addiction: cognitive, affective, and neuropsychopharmacologic pathways. Neurosci Biobehav Rev 37:2597–2607

Garland EL, Roberts RL, Hanley AW, Zeidan F, Keefe FJ (2023) The mindful reappraisal of pain scale (MRPS): validation of a new measure of psychological mechanisms of mindfulness-based analgesia. Mindfulness (NY) 14:192–204

Geuter S, Eippert F, Hindi Attar C, Buchel C (2013) Cortical and subcortical responses to high and low effective placebo treatments. NeuroImage 67:227–236

Geuze E, Westenberg HG, Jochims A, de Kloet CS, Bohus M, Vermetten E, Schmahl C (2007) Altered pain processing in veterans with posttraumatic stress disorder. Arch Gen Psychiatry 64: 76–85

Giesecke T, Gracely RH, Williams DA, Geisser ME, Petzke FW, Clauw DJ (2005) The relationship between depression, clinical pain, and experimental pain in a chronic pain cohort. Arthritis Rheum 52:1577–1584

Goedhoop JN, van den Boom BJG, Robke R, Veen F, Fellinger L, van Elzelingen W, Arbab T, Willuhn I (2022) Nucleus accumbens dopamine tracks aversive stimulus duration and prediction but not value or prediction error. elife 11

Gombaut C, Holmes SA (2022) Sensorimotor integration and pain perception: mechanisms integrating nociceptive processing. A systematic review and ALE-meta analysis. Front Integr Neurosci 16:931292

Gorka SM, Lieberman L, Shankman SA, Phan KL (2017) Startle potentiation to uncertain threat as a psychophysiological indicator of fear-based psychopathology: an examination across multiple internalizing disorders. J Abnorm Psychol 126:8–18

Grillon C, Pine DS, Lissek S, Rabin S, Bonne O, Vythilingam M (2009) Increased anxiety during anticipation of unpredictable aversive stimuli in posttraumatic stress disorder but not in generalized anxiety disorder. Biol Psychiatry 66:47–53

Hoskin R, Talmi D (2023) Adaptive coding of pain prediction error in the anterior insula. Eur J Pain 27:766–778

Keltner JR, Furst A, Fan C, Redfern R, Inglis B, Fields HL (2006) Isolating the modulatory effect of expectation on pain transmission: a functional magnetic resonance imaging study. J Neurosci 26:4437–4443

Khera T, Rangasamy V (2021) Cognition and pain: a review. Front Psychol 12:673962

Koenig J, Jarczok MN, Ellis RJ, Hillecke TK, Thayer JF (2014) Heart rate variability and experimentally induced pain in healthy adults: a systematic review. Eur J Pain 18:301–314

Kong J, Gollub RL, Polich G, Kirsch I, Laviolette P, Vangel M, Rosen B, Kaptchuk TJ (2008) A functional magnetic resonance imaging study on the neural mechanisms of hyperalgesic nocebo effect. J Neurosci 28:13354–13362

Koyama T, McHaffie JG, Laurienti PJ, Coghill RC (2005) The subjective experience of pain: where expectations become reality. Proc Natl Acad Sci USA 102:12950–12955

Lehoux CP, Abbott FV (2011) Pain, sensory function, and neurogenic inflammatory response in young women with low mood. J Psychosom Res 70:241–249

Loggia ML, Mogil JS, Bushnell MC (2008) Experimentally induced mood changes preferentially affect pain unpleasantness. J Pain 9:784–791

Lumley MA, Cohen JL, Borszcz GS, Cano A, Radcliffe AM, Porter LS, Schubiner H, Keefe FJ (2011) Pain and emotion: a biopsychosocial review of recent research. J Clin Psychol 67:942–968

Mehling WE (2024) How Bud Craig's insights reshape the research on pain and mind-body therapies. Curr Top Behav Neurosci

Mehling WE, Strigo IA, Goldman V, Hartogensis W, Adler SR, Lotz J, Hecht FM (2024) Mind your pain: a single-arm feasibility study to assess a smartphone-based interoceptive attention training for patients with chronic low back pain. PLoS One 19:e0307690

Mehling WE, Brintz CE, Hartogensis W, Wolf C, Rogers K, Adler SR, Strigo IA, Hecht FM (2025) Development and initial validation of mindfulness-based pain reduction (MBPR) in patients with chronic low back pain. J Pain Res 18:785

Moeller-Bertram T, Keltner J, Strigo IA (2012) Pain and post traumatic stress disorder, Äì review of clinical and experimental evidence. Neuropharmacology 62:586–597

Moeller-Bertram T, Strigo IA, Simmons AN, Schilling JM, Patel P, Baker DG (2014) Evidence for acute central sensitization to prolonged experimental pain in posttraumatic stress disorder. Pain Med 15:762–771

Mutschler I, Ball T, Wankerl J, Strigo IA (2012) Pain and emotion in the insular cortex: evidence for functional reorganization in major depression. Neurosci Lett 520:204–209

Navratilova E, Porreca F (2014) Reward and motivation in pain and pain relief. Nat Neurosci 17: 1304–1312

Pace-Schott EF, Amole MC, Aue T, Balconi M, Bylsma LM, Critchley H, Demaree HA, Friedman BH, Gooding AEK, Gosseries O, Jovanovic T, Kirby LAJ, Kozlowska K, Laureys S, Lowe L, Magee K, Marin MF, Merner AR, Robinson JL, Smith RC, Spangler DP, Van Overveld M, VanElzakker MB (2019) Physiological feelings. Neurosci Biobehav Rev 103:267–304

Patten SB, Wilkes TC, Williams JV, Lavorato DH, El-Guebaly N, Schopflocher D, Wild C, Colman I, Bulloch AG (2015) Retrospective and prospectively assessed childhood adversity in association with major depression, alcohol consumption and painful conditions. Epidemiol Psychiatr Sci 24:158–165

Pavy F, Zaman J, Van den Noortgate W, Scarpa A, von Leupoldt A, Torta DM (2024) The effect of unpredictability on the perception of pain: a systematic review and meta-analysis. Pain 165: 1702

Ploghaus A, Tracey I, Gati JS, Clare S, Menon RS, Matthews PM, Rawlins JN (1999) Dissociating pain from its anticipation in the human brain. Science 284:1979–1981

Pollatos O, Dietel A, Herbert BM, Wankner S, Wachsmuth C, Henningsen P, Sack M (2011) Blunted autonomic reactivity and increased pain tolerance in somatoform patients. Pain 152: 2157–2164

Poublan-Couzardot A, Talmi D (2024) Pain perception as hierarchical Bayesian inference: a test case for the theory of constructed emotion. Ann N Y Acad Sci 1536:42–59

Rainville P, Feine JS, Bushnell MC, Duncan GH (1992) A psychophysical comparison of sensory and affective responses to four modalities of experimental pain. Somatosens Mot Res 9:265–277

Raison CL, Hale MW, Williams LE, Wager TD, Lowry CA (2014) Somatic influences on subjective well-being and affective disorders: the convergence of thermosensory and central serotonergic systems. Front Psychol 5:1580

Raja SN, Carr DB, Cohen M, Finnerup NB, Flor H, Gibson S, Keefe FJ, Mogil JS, Ringkamp M, Sluka KA, Song X-J, Stevens B, Sullivan MD, Tutelman PR, Ushida T, Vader K (2020) The revised International Association for the Study of Pain definition of pain: concepts, challenges, and compromises. Pain 161:1976–1982

Rigoli F, Friston KJ, Dolan RJ (2016) Neural processes mediating contextual influences on human choice behaviour. Nat Commun 7:12416

Schmid J, Theysohn N, Gass F, Benson S, Gramsch C, Forsting M, Gizewski ER, Elsenbruch S (2013) Neural mechanisms mediating positive and negative treatment expectations in visceral pain: a functional magnetic resonance imaging study on placebo and nocebo effects in healthy volunteers. Pain 154:2372–2380

Schmid J, Bingel U, Ritter C, Benson S, Schedlowski M, Gramsch C, Forsting M, Elsenbruch S (2015) Neural underpinnings of nocebo hyperalgesia in visceral pain: a fMRI study in healthy volunteers. NeuroImage 120:114–122

Shrout PE, Fleiss JL (1979) Intraclass correlations: uses in assessing rater reliability. Psychol Bull 86:420

Simmons AN, Norman SB, Spadoni AD, Strigo IA (2013) Neurosubstrates of remission following prolonged exposure therapy in veterans with posttraumatic stress disorder. Psychother Psychosom 82:382–389

Strigo IA, Craig AD (2016) Interoception, homeostatic emotions and sympathovagal balance. Philos Trans R Soc Lond Ser B Biol Sci 371:20160010

Strigo IA, Bushnell MC, Boivin M, Duncan GH (2002) Psychophysical analysis of visceral and cutaneous pain in human subjects. Pain 97:235–246

Strigo IA, Simmons AN, Matthews SC, Craig AD, Paulus MP (2008a) Association of major depressive disorder with altered functional brain response during anticipation and processing of heat pain. Arch Gen Psychiatry 65:1275–1284

Strigo IA, Simmons AN, Matthews SC, Craig AD, Paulus MP (2008b) Increased affective bias revealed using experimental graded heat stimuli in young depressed adults: evidence of "emotional allodynia". Psychosom Med 70:338–344

Strigo IA, Matthews SC, Simmons AN (2010a) Right anterior insula hypoactivity during anticipation of homeostatic shifts in major depressive disorder. Psychosom Med 72:316–323

Strigo IA, Simmons AN, Matthews SC, Grimes EM, Allard CB, Reinhardt LE, Paulus MP, Stein MB (2010b) Neural correlates of altered pain response in women with posttraumatic stress disorder from intimate partner violence. Biol Psychiatry 68:442–450

Strigo IA, Matthews SC, Simmons AN (2013a) Decreased frontal regulation during pain anticipation in unmedicated subjects with major depressive disorder. Transl Psychiatry 3:e239

Strigo IA, Matthews SC, Simmons AN, Oberndorfer T, Klabunde M, Reinhardt LE, Kaye WH (2013b) Altered insula activation during pain anticipation in individuals recovered from anorexia nervosa: evidence of interoceptive dysregulation. Int J Eat Disord 46:23–33

Strigo IA, Spadoni AD, Lohr J, Simmons AN (2014) Too hard to control: compromised pain anticipation and modulation in mild traumatic brain injury. Transl Psychiatry 4:e340

Strigo IA, Spadoni AD, Inslicht SS, Simmons AN (2018) Repeated exposure to experimental pain differentiates combat traumatic brain injury with and without post-traumatic stress disorder. J Neurotrauma 35:297–307

Strigo IA, Keltner JR, Ellis RJ, Simmons AN (2021) Association of painful human immunodeficiency virus distal sensory polyneuropathy with aberrant expectation of pain relief: functional magnetic resonance imaging evidence. Brain Commun 3:fcab260

Strigo IA, Craig ADB, Simmons AN (2024a) Expectation of pain and relief: a dynamical model of the neural basis for pain-trauma co-morbidity. Neurosci Biobehav Rev 163:105750

Strigo IA, Guerra SG, Torrisi S, Murphy E, Toor T, Goldman V, Alter BJ, Vu AT, Hecht R, Lotz J, Simmons AN, Mehling WE (2024b) Enhancing chronic low back pain management: an initial neuroimaging study of a mobile interoceptive attention training. Front Pain Res (Lausanne) 5: 1408027

Strigo IA, Kadlec M, Mitchell JM, Simmons AN (2024c) Identification of group differences in predictive anticipatory biasing of pain during uncertainty: preparing for the worst but hoping for the best. Pain 165:1735–1747

Thompson JM, Neugebauer V (2017) Amygdala plasticity and pain. Pain Res Manag 2017: 8296501

Thompson T, Correll CU, Gallop K, Vancampfort D, Stubbs B (2016) Is pain perception altered in people with depression? A systematic review and meta-analysis of experimental pain research. J Pain 17:1257–1272

Thones S, Oberfeld D (2015) Time perception in depression: a meta-analysis. J Affect Disord 175: 359–372

Tracey I (2010) Getting the pain you expect: mechanisms of placebo, nocebo and reappraisal effects in humans. Nat Med 16:1277–1283

Tyler MP, Wright BJ, Raison CL, Lowry CA, Evans L, Hale MW (2024) Greater severity of depressive symptoms is associated with changes to perceived sweating, preferred ambient temperature, and warmth-seeking behavior. Temperature (Austin) 11:266–279

Ushinsky A, Reinhardt LE, Simmons AN, Strigo IA (2013) Further evidence of emotional allodynia in unmedicated young adults with major depressive disorder. PLoS One 8:e80507

Villemure C, Bushnell MC (2002) Cognitive modulation of pain: how do attention and emotion influence pain processing? Pain 95:195–199

Wittmann M, Droit-Volet S (2024) Subjective time in ordinary and non-ordinary states of consciousness: how interoceptive feelings inform us about the passage of time. Curr Top Behav Neurosci

Wittmann M, Simmons AN, Aron JL, Paulus MP (2010) Accumulation of neural activity in the posterior insula encodes the passage of time. Neuropsychologia 48:3110–3120

Xie JY, Qu C, Patwardhan A, Ossipov MH, Navratilova E, Becerra L, Borsook D, Porreca F (2014) Activation of mesocorticolimbic reward circuits for assessment of relief of ongoing pain: a potential biomarker of efficacy. Pain 155:1659–1666

Zelman DC, Howland EW, Nichols SN, Cleeland CS (1991) The effects of induced mood on laboratory pain. Pain 46:105–111

How Bud Craig's Insights Reshape the Research on Pain and Mind–Body Therapies

Wolf E. Mehling

Contents

Abstract With his elegant studies, Bud Craig determined the structural neural basis for interoception and critically expanded our conceptual understanding of it. Importantly, he placed pain in the framework of interoception and redefined pain as a homeostatic emotion. Craig understood emotions and pain as experiences based on inferential brain processes within the theoretical model of prediction processing. This chapter aims to give a brief overview of relevant research. Mind–body therapies, such as meditation, mindfulness, yoga, Tai Chi, and others, are included as first-line non-pharmacological approaches in clinical guidelines for the management of chronic pain. Craig's groundbreaking work provided the background for our contemporary understanding of mind–body therapies and for the key role that interoceptive processes play in these therapies as they apply to a wide range of clinical conditions, including pain. This chapter reviews the tremendous influence that Craig's work had on the current state of research on mind–body

W. E. Mehling (✉)
Department of Family and Community Medicine, Osher Center for Integrative Health, University of California, San Francisco, San Francisco, CA, USA
e-mail: Wolf.Mehling@ucsf.edu

Curr Topics Behav Neurosci (2025) 70: 143–158
https://doi.org/10.1007/7854_2024_521
Published Online: 23 October 2024

therapies for managing chronic pain and how it led to new directions for cutting-edge clinical and neuroscientific research.

Keywords Interoception · Homeostatic emotion · MAIA · Meditation · Mindfulness · Mind–body therapies

1 Introduction

A simple quick PubMed search for articles authored or co-authored by Bud Craig with "pain" in the abstract shows 36 publications starting in 1981. Of these, by far, the two most quoted articles are from 2003, one summarizing his experimental findings of a newly discovered neurological pathway for pain (Craig 2003b) and the other on the implications of these findings for the general concept of pain "as a homeostatic emotion" (Craig 2003a). These publications contributed and dramatically expanded our understanding of pain. Prior to Craig, the neurology of pain was conventionally viewed as a pattern of convergent activity within the somatosensory system, in which pain is represented similarly to the exteroceptive sense of touch. Craig added to our knowledge by describing phylogenetically novel primate lamina I trigeminal- and spinothalamic projections that are of key importance for the sensory and motivational aspects of pain, temperature, itch, muscle ache, sensual touch, and other interoceptive feelings from the body. This placed pain into a revised and expanded conceptualization of interoception (Craig 2002), the perception of feelings from our bodies that relate to our state of well-being, our energy and stress levels, and our mood and disposition. According to Craig, beginning with the initiation of the primate lineage and the enormous encephalization in primate brains, we humans are now able to consciously experience bodily feelings generated through an extension of the ancient homeostatic sensorimotor hierarchy from the spinal and brainstem levels to thalamocortical forebrain circuits. This potentially frees us from the constraints engendered by instinctual hypothalamic fixed action patterns. Craig based this broadened conceptualization of interoception on his functional anatomical work, which revealed and described in detail an afferent neural system in primates – including humans – that represents all aspects of the physiological condition of the physical body. He argued that this system might represent "the material me" and provide a foundation for subjective feelings, emotions, self-awareness [page 225 in (Craig 2015a)], and a specific form of self-reference (Farb et al. 2007), to be discussed below.

2 Craig's Discovery and the Evolutionary Perspective on Pain as a Homeostatic Emotion

Craig identified the primary nociceptive sensory cortex in the dorsal posterior insula as part of the interoceptive cortex. This area receives the homeostatic afferent input from all tissues of the entire body, which connects forward to the mid and anterior insula cortex with its specialized cytoarchitecture and connectivity, which in turn allows us humans – and a few other species – to become subjectively aware of this afferent input. Importantly for our topic, Craig's discovery suggests a radically new perspective on pain as a homeostatic emotion, as a warning message triggered by an adverse condition in the body, transmitted to and processed along the spinal cord, brainstem, and brain. It comprises both a felt sensation and a behavioral drive, motivating a response to a physiological condition in the body that the autonomic (subconscious) homeostatic system cannot rectify by itself. The "intention" is to bring the body back to homeostasis, that is to maintain its functional integrity (pages 42–43 in (Craig 2015b)). Most interoceptive alterations can initiate autonomous responses that do not need to enter awareness. But pain *demands* attention (Eccleston and Crombez 1999), interrupts daily functioning, and motivates an active response. Craig and his team provided the neurological correlates to the complex anatomical and physiological details of this integrated response system (Craig 2003a). Importantly, although animals perceive pain and mount reflexive as well as learned responses to pain, the way humans experience pain is subjective and unique due to the evolution of the anterior insular cortex and its neural connections to the forebrain, which provide the neural base for the phenomenal awareness of all feelings (incl. pain), the "sentient self" (Craig 2009a, b, 2010). Neural connections to the anterior cingulate cortex (ACC) can guide adaptive behaviors (Strigo and Craig 2016). Being aware of, sensitive to, and able to understand one's own feelings – and the feelings of others – may have provided an evolutionary advantage. However, this highly complex integration of bodily and emotional signals with multiple other regions in the primate brain comes with a trade-off, an increased human capacity for pain-related suffering (Chapman and Gavrin 1999).

Homeostasis is the maintenance of the physiological condition of the body, which is constantly exposed to change, disruptions, and stress, by bringing it back to relatively stable bounds. Allostasis refers to self-regulation that acts in advance of expected change, disruption, and stress. The relatively stable bounds are what we perceive as health, self-regulated in anticipation of disruptions by allostasis or, once disruption has occurred, by homeostatic emotions inducing an adaptive behavioral response. The afferent pathways in this bidirectional or circulatory reflexive system were originally outlined by Craig. Pointing out the neurological parallels with emotions, Craig firmly rooted our current understanding of pain in the neurology of interoception and the homeostatic behavioral regulatory system.

3 Integrating Predictive Coding with Interoception in Understanding Pain

Later on, Craig connected the neurology of interoception with the predictive coding (also referred to as predictive processing) model (included in (Craig 2009a) and with more detail in (Craig 2010)) that emphasizes the bi-directionality of interoception and the top-down brain activities that process, filter, and modulate interoceptive afferents *before* we become aware of them. This is a function of the salience network with its hub in the anterior insula (Craig 2015b). Essentially, the predictive processing model of interoceptive inference (Seth 2013) posits that we never immediately perceive nociceptive afferents as pain sensations as these are transmitted bottom-up from the body's periphery to the – presumed central – brain. Rather, the human brain generates top-down predictions – or priors – for all bottom-up afferent input about the environment and our body's physiological state. Top-down neural activity is computed and compared in the insula and ACC with the afferent bottom-up neural activity. Another familiar term for these predictions is expectations, which may match or not match afferent input, be erroneous, or be subjected to surprise and the need for updating. There is abundant research on how expectations – with their personal, historical, social, and cultural biases – shape our perception of pain in its intensity and valence (Bushnell et al. 2013; Fields 2018; Koyama et al. 2005). This predictive process, comparing afferent nociceptive input to priors, generates a synthesized inference that consequently may become conscious as our subjective experience. How we experience afferents – including nociceptive pain – is highly dependent on the relative weight we give, in this comparison, to our priors versus the afferent sensory input (Hauck et al. 2015). And this process in itself, in turn, depends on the willingness and ability to learn and update our priors in the situational context of our culture and umwelt (Farb et al. 2015).

This is beautifully and – as to be expected from Lorimer Moseley – rather dramatically demonstrated in his popular video presentation https://www.youtube.com/watch?v=Sjes9CuN6Ko. In this, Lorimer, the world-renowned expert in pain education, shares two incidents, separated by a few years, on his way by foot to a refreshing morning plunge in a nearby lake in the Australian jungle. He explains how both incidents are processed in his brain: the first incident, sensing "something" at his lower leg just above the lateral ankle, was dismissed as likely irrelevant, not expecting any potential danger in this sunny and warm vacation setting, before he woke up in intensive care recovering from a highly venomous snake bite. There was not much pain with the bite. The second incident was a scratch from a small twig on the same path, but it was processed in his brain together with the memory of his prior incident as potentially fatal. The rather harmless scratch was subjectively experienced as intense pain. Predictive processing and the brain's inference amplified the peripheral afferent signal into a very dramatic warning Lorimer was unable to ignore.

The predictive processing model can also explain how pain can persist and become chronic after the initial injury that created the nociceptive input has resolved, or how chronic pain can develop even without any such afferent input. According to this model, *chronic* pain may be understood as the inference from dysfunctional interoceptive predictions being processed with or without nociceptive input. Adhering to this predictive processing model, several theoretical variations have been proposed in the past decade that aimed at a better understanding of chronic pain. A few examples within this theoretical framework are Barrett and Simmons's Embodied Predictive Interoception Coding (Barrett 2017), Di Lernia et al.'s Residual Interoceptive Ghosts across Time (Di Lernia et al. 2016a), the embodied predictive processing (EPP) theory by Kiverstein et al. (2022), the transdiagnostic "Better-Safe-Than-Sorry" processing strategy by Van den Bergh et al. (2017, 2021), and the computational model of interoception by Stephan and Petzschner (Stefanics et al. 2018). All of these theoretical models refer back to Craig, his broad definition of interoception, and its neurological underpinnings that he discovered and taught us.

The view of pain as an interoceptive inference and as a homeostatic emotion has been taken up by psychologists in the – currently dominating – conceptualization of emotions as being entirely constructed (Barrett 2006, 2017). However, the question may remain open whether the currently widely held constructivist understanding of emotions does full justice to the experience of pain. Pain as a homeostatic emotion may potentially include – at its very core – an evolutionarily derived innate element onto which humans superimpose cortically constructed elements (Panksepp 2007). Maybe the nature of emotions and – with these – of pain is not either/or but both/and (Gilam et al. 2020).

Based on his anatomical findings for interoceptive afferents, Craig also includes the afferent pathways of C-fiber-mediated pleasure sensations from touch in interoception. Our skin is at the border between the inside and the outside of the body. Affective touch elicits interoceptive feelings that follow the homeostatic pathway (Bjornsdotter et al. 2010; Craig 2010; Crucianelli and Ehrsson 2023). However, it is noteworthy that Oliver Desmedt and colleagues have made a fine distinction by considering touch as "the successive contribution of exteroception (i.e., the processing of external cutaneous stimulations) and interoception (i.e., the processing of internal consequences induced by external stimulations)" (Desmedt et al. 2023). Regarding pain, all mothers know, and research confirms, that pleasant affective touch – using interoceptive pathways – has analgesic effects (Di Lernia et al. 2020; Reddan et al. 2020). I will come back to pleasure and its relation to pain.

4 Evolving Operationalizations of Interoception and Their Relationship to Pain

Since Craig, the multidimensional concept of interoception has developed and undergone a variety of operationalizations for research. As described in more detail in other chapters of this book, until recently the most common one was Garfinkel's three-dimensional model, which has subsequently been cautiously broadened (Farb et al. 2015; Khalsa et al. 2018) and, more recently, again been widened and modified by Suksasilp and Garfinkel (2022). Pain is “intended” to enter consciousness as part of *conscious* interoception, an element of interoception that enters awareness and is accessible to self-report. Here, the term “interoceptive awareness” has frequently been used but subsequently had a rather confusing circuitous journey within the broader conceptualization of interoception (Ceunen et al. 2016) and – since Garfinkel – has commonly been substituted by the term interoceptive sensibility. However, a seemingly stubborn use of the older term and definition of interoceptive awareness, that part of interoception that is accessible to consciousness and self-report – as used in the Multidimensional Assessment of Interoceptive Awareness (MAIA) (Mehling et al. 2022), appears to survive all label changes (Gnall et al. 2024).

If, following Craig, we want to understand pain as a form of interoception, how are measures of interoception related to common measures of pain intensity, either by self-report or semi-objectively by quantitative sensory testing (QST)? Di Lernia and colleagues conducted a systematic review of interoception and pain (Di Lernia et al. 2016b) essentially concluding that “*chronic pain subjects might present low interoceptive accuracy and, allegedly, that interoceptive accuracy negatively correlates with symptom severity*.” However, data were inconclusive for interoceptive sensibility and interoceptive awareness (operationalized as confidence in one’s interoceptive sensibility), which are commonly assessed using a variety of self-report measures that, unfortunately, are based on potentially contradictory operationalizations of interoceptive sensibility, an ambiguous multidimensional construct (Mehling 2016).

Another systematic review and meta-analysis by Horsburgh and colleagues was recently published that summarized the research on interoception and chronic pain (Horsburgh et al. 2024). Its authors concluded that “*an individual’s ability to accurately sense internal signals is decreased in those with chronic pain, despite them reporting being more aware of internal sensations.*” However, the authors did not differentiate among the various included self-report measures of interoceptive sensibility. The references included numerous studies in the Forest plot (Fig. 4) (Horsburgh et al. 2024) that strongly drove the latter conclusion of chronic pain being associated with higher interoceptive sensibility (e.g., Dick et al. 2002; Geisser et al. 2003; Martínez et al. 2018; Rost et al. 2017; Herman 2022). However, these studies on the right side of the plot measured interoceptive sensibility by hypervigilance-related self-report instruments, which all essentially might – and should – be viewed as proxy measures for anxiety: Body Perception Questionnaire, Modified Somatic Perception Questionnaire, and Body Vigilance Scale. The review’s authors failed to differentiate between beneficial and maladaptive styles of interoceptive sensibility, namely between anxiety-driven and mindful interoceptive sensibility, a key distinction for clinicians (Mehling 2016). Not clearly differentiating between these contradictory facets of interoceptive sensibility has already led to confusing results in other fields before (Trevisan et al. 2021).

5 Integrating Interoception into Non-pharmacological Approaches for Chronic Pain Management

For the role of understanding pain as a part of interoception in the treatment or management of chronic pain, let us take a look at chronic low back pain (cLBP), the number 1 reason for disability worldwide (Collaborators 2024). National guidelines recommend non-pharmacological approaches as first-line therapies (Qaseem et al. 2017). Among these are yoga and mindfulness-based interventions. These and numerous others (see below) so-called mind–body practices (Wahbeh et al. 2008) have long claimed to enhance body awareness, a construct closely related to interoceptive sensibility or awareness (Mehling et al. 2009), and have shown benefits for patients with pain (C.J. Price and Mehling 2016). In the past, researchers have struggled with the ambiguous concept of bodily awareness, as clinicians were used to view it as a maladaptive concept associated with anxiety, hypochondriasis, and somatization (Mehling 2016). Craig's definition of interoception provided clarification with a more scientific framework that was taken up and embraced by researchers of modern psychotherapeutic methods and integrative medicine approaches, particularly for what has become known as mindfulness and/or mind–body approaches. The enhancement of a more mindful style of body awareness was seen as potentially underlying the beneficial effects of practices from integrative health, such as mindfulness, meditation, breath therapy, Feldenkrais method, yoga, Tai Chi, Alexander technique, Eutony, Somatic Experiencing and others, often summarized under the label of mind–body therapies or somatic approaches to mental health (Farb et al. 2015; Mehling 2016). It should be noted that practitioners of these approaches may prefer the term "self-awareness" (Mehling et al. 2011), in order to overcome the conceptual split of body and mind, which goes along with Craig's word choice in the "sentient self" (Craig 2010). Craig provided a neurological basis for the discourse and research on body and self-awareness. Although there are controversies about whether this is reductionistic and biased by being "cortex-centered" (as the obliteration of the bilateral insula cortices does not abolish self-awareness) (Philippi et al. 2012), research on mindfulness and other mind–body approaches has largely embraced Craig's concept of interoception (Farb et al. 2015). At times to the very surprise of these practitioners (Mehling 2001), these approaches have shown that they are able to modify the perception of pain (Villemure et al. 2013) and benefit patients with chronic pain.

Craig himself contributed to few studies regarding the relevance of his findings on interoception for mind–body approaches, mindfulness, meditation, or yoga. An experimental study showed a reduction in trait negative affect from slow-breathing yoga during thermal pain stimuli in fibromyalgia patients (Zautra et al. 2010). The authors concluded that slow breathing as a means for natural parasympathetic stimulation may facilitate emotional regulation and the maintenance of homeostasis under the challenging conditions of acute pain induction consistent with the model of pain as a homeostatic emotion (Zautra et al. 2010). Craig also studied brain metabolic changes with electric vagus nerve stimulation for patients with depression

(Conway et al. 2013). But Craig's work is used as a key reference in the now-established understanding that mindfulness provides some of its benefits through training mindful interoceptive attention to unexpected or unwelcomed bodily sensations, be it from symptoms of depression or pain (Farb et al. 2015; Farb and Segal 2024; Gawande et al. 2019; Hölzel et al. 2011; Schuman-Olivier et al. 2020). A handbook on assessments for mindfulness research includes a chapter for a common self-report measure of interoception (Mehling et al. 2022). These approaches have gained popularity as non-pharmacological therapies for chronic pain and entered national guidelines for, e.g., the treatment of chronic low back pain. Mindfulness-based interventions have been shown to reduce cLBP (Cherkin et al. 2016), reduce behavioral factors implicated in pain persistence (Cramer et al. 2012; Turner et al. 2016), and may even be cost-effective (Cherkin and Herman 2018). Similarly, yoga is helpful with cLBP and cost-effective (Chang et al. 2016; Groessl et al. 2020). Furthermore, a cross-sectional study of experienced yoga practitioners and healthy controls demonstrated a higher threshold for experimental pain with yoga that was correlated with the neural density in the insula cortex, the hub for interoception. Asked how these experienced yoga practitioners coped with their experimental pain (submerging a hand in icy water), they reported using a mindful, accepting, present-moment attention style, maintaining the focus toward the pain sensation, whereas controls used distraction (Villemure et al. 2013).

6 Assessing Interoception

The mind–body interventions mentioned above appear to have in common that their practitioners claim to improve their clients' body and/or self-awareness and thereby deliver benefits of well-being (Mehling et al. 2011). To overcome the difficulties in investigating this claim (Mehling et al. 2009), my colleagues and I developed the MAIA (Mehling et al. 2012) and its updated version MAIA-2 (Mehling et al. 2018). The MAIA has been translated into numerous other languages (Mehling 2012) and – despite its shortcomings (Vig et al. 2022) – is currently still the most used self-report measure to assess interoception in its elements that are accessible to conscious awareness (Desmedt et al. 2021). This measure attempts to distinguish between anxiety-driven and mindful attention styles of interoceptive body awareness (Mehling 2016) and includes regulatory aspects, which are important in clinically distinguishing the ways in which patients experience pain.

Since Craig, research studies on mind–body practices for chronic pain that included at least self-report measures for interoception have dramatically increased (Voss et al. 2023). A meta-analysis of mind–body interventions for changes in interoception in patients with chronic pain conditions was conducted by Gnall et al. (2024). Although the authors found a lack of studies that used objective measures for interoception, all but one study used the self-report MAIA (Mehling et al. 2012). In a total of 11 studies, it revealed significant improvements in total interoceptive awareness (Becker's $d = 1.168$, $p < 0.01$) and improvements in seven of eight scales of the MAIA ($ds = 0.28$ to 0.81). Furthermore, the included interventions were associated with reductions in pain intensity ($d = -1.46$, $p = 0.01$) and pain interference ($d = -1.07, p < 0.001$).

7 Interoceptive Exposure: An Innovative Approach to Chronic Pain Management

Based on Craig's work and studies that found interoception to be a key factor in emotion regulation (C. J. Price and Hooven 2018; Zamariola et al. 2019), interoceptive exposure has been used in the psychotherapeutic treatment of anxiety disorders (Sabourin et al. 2015). Whether interoceptive exposure can be helpful with chronic pain is an old unresolved question within the field of the psychology of chronic pain, that goes beyond the concept of pain as a homeostatic emotion and asks: could paying close attention to the phenomenological details of the experience of pain be an alternative approach to chronic pain management compared with the typical coping style of distraction and ignoring? (Boersma et al. 2004) That latter habitual coping style is prevalent (Johnson 2005; Mehling et al. 2013) but of questionable benefit in chronic pain (Goubert et al. 2004) (review in Rad and Wippert (2024)). In the language of the predictive processing model, paying closer attention to bottom-up afferent sensations would be understood as giving the interoceptive afference more precision weight and thereby potentially updating priors and top-down predictions. Mindfulness-based approaches to emotions entrain acceptance and acknowledgment of unpleasant feelings including pain, staying with these, not attempting to change one‘s perception but to decenter and detach from pain (de Jong et al. 2016), letting go of pain-related stories of the past, worries about the future, and reactive responses, to disidentify with pain, and focus on the whole body rather than on its painful parts. These approaches practice the distinction between in-the-moment sensing of one's feelings and sensations on one hand, and the narration and thoughts about these sensations on the other hand. This distinction between narrative, thinking self-reference, and sensing one's bodily sensations, a sensation-based self-reference, is, in general, less available to untrained individuals, for which thinking and sensing are inseparably intertwined. In a groundbreaking study, Norman Farb has shown that this distinction is based on separate neurological pathways and can be learned (Farb and Segal 2024; Farb et al. 2007).

In an editorial for a journal edition on the psychological processes within mindfulness approaches for pain, Marc Williams, a co-developer of Mindfulness-Based Cognitive Therapy (MBCT), differentiated between cognitive attention and present-moment sensory attention to one‘s unpleasant experience and raised the question of whether the former cognitive attention style is a form of distracting in itself as well (Williams 2010). Focusing on the phenomenal details of unpleasant sensations in a curious, detached, present-moment attention style has been labeled mindfulness-based interoceptive exposure, an emerging therapeutic approach generally applied with MBCT for symptoms of depression. Early research suggested benefits with pain (Cayoun et al. 2017). Practices such as the body scan enhance interoceptive awareness and practice "sense foraging" (Segal 2022). Similarly, a mindfulness-based approach to chronic pain and opiate dependency, Mindfulness-Oriented Recovery Enhancement (MORE), applies a mindful interoceptive exposure task to both unpleasant and pleasant sensations, resulting in reappraising pain and savoring pleasure (Garland 2020; Garland et al. 2024). Combining the training of interoceptive body and emotional awareness with affective C-fiber mediated touch (Bjornsdotter et al. 2009) is another approach applied in Mindful Awareness in Body-Oriented Therapy (MABT) that has shown improvements for patients with chronic pain (C. J. Price et al. 2007; C.J. Price and Mehling 2016).

Based on Cayoun's Mindfulness-based Interoceptive Exposure Task (MIET), which had shown preliminary evidence for sustained pain relief (Cayoun et al. 2017), we developed a phone app that asked patients with cLBP to purposefully attend to the very characteristics of their pain sensations, such as feeling tone (sharp vs dull), three-dimensional shape, temperature, motion, and density, guided by a recorded voice for about a minute at two or more times per day over 8 weeks (Mehling et al. 2024). Using the PEG scale (Krebs et al. 2009), the majority (21 of 29) of participants reported a more than 30% improvement in pain intensity and pain interference, which is generally considered clinically important. The participants noted that this approach was challenging to adapt, as it was contrary to their usual habits of distraction and ignoring. Not everybody was able to change their habit and learn this new approach of a curious sensory attention focus on pain. Responders to the intervention showed pre-post fMRI changes in experimental heat pain-related insula activation with increased activation within the contralateral posterior insula, Craig's interoceptive sensory cortex, and bilateral ventral anterior insula in responders (Z. S. Chen 2023). This finding is particularly important, as prior research had revealed that the transition from acute to chronic LBP was associated with a reduction in neural density in the insula cortex (Hashmi et al. 2013). One could speculate that the reduction in insula cortex density with chronic pain was due to habitually avoiding and distracting from the pain experience. In the setting of orthopedic care with surgery and physical therapy, it had been shown that recovery from chronic pain was associated with partial recovery from that neural loss (Ceko et al. 2015). In qualitative exit interviews at the end of the interoceptive attention training, patients were surprised about the remarkable variety of pain descriptions for their chronic pain they had not previously noticed regarding the shapes of pain (American footballs of various opacity and density, stars, bricks, pool balls . . .), colors (orange, black, blue, white, red . . .), temperature (hot cold), dead or alive, and other elements.

In another study, we aimed at improving and adapting the Mindfulness-Based Stress Reduction (MBSR) program specifically for patients with cLBP by incorporating current pain neurophysiology education, training interoceptive body awareness through pain-adapted yoga exercises, sensory awareness of breath movements in the pain region, and practicing curious and self-compassionate mindful interoceptive exposure to the details of their pain experience. Patient-oriented outcome measures, exit interviews, and brain images are still in the analysis phase, but preliminary findings show substantial improvements in pain interference, showing a stronger impact when compared to the interoceptive attention task.

8 Conclusions

Since Craig, pain is strongly associated with – and incorporated into – interoception as a conscious perception of sensations from the entire body, traditionally conceptualized as bodily awareness. Craig's definition and neuro-anatomical outline of interoception provided a framework in which mind–body interventions and contemplative practices and their effect on pain and other clinical conditions are increasingly being studied (Gnall et al. 2024). Craig also embraced the theoretical predictive processing model of interoception, which has added a theoretical understanding of the psychological processes active in mind–body approaches including mindfulness. Craig has mentored younger researchers following in his footsteps, who are editing this volume and expanding on his research. Studying interoception has become a high-priority topic for the National Center of Complementary and Integrative Health in the USA at the National Institute of Health (W. G. Chen et al. 2021). Craig's research is the basis for interoceptive exposure training in anxiety disorders. One of the newest applications of his work is the currently ongoing investigation into mindful attention to pain and mindful interoceptive exposure (Mehling et al. 2024). Craig established the structural anatomical foundation for the key role that interoceptive processes play in mind–body therapies.

References

Barrett LF (2006) Solving the emotion paradox: categorization and the experience of emotion. Pers Soc Psychol Rev 10(1):20–46. https://doi.org/10.1207/s15327957pspr1001_2

Barrett LF (2017) How emotions are made. The secret life of the brain

Bjornsdotter M, Loken L, Olausson H, Vallbo A, Wessberg J (2009) Somatotopic organization of gentle touch processing in the posterior insular cortex. J Neurosci 29(29):9314–9320. 29/29/9314 [pii]. https://doi.org/10.1523/JNEUROSCI.0400-09.2009

Bjornsdotter M, Morrison I, Olausson H (2010) Feeling good: on the role of C fiber mediated touch in interoception. Exp Brain Res 207(3–4):149–155. https://doi.org/10.1007/s00221-010-2408-y

Boersma K, Linton S, Overmeer T, Jansson M, Vlaeyen J, de Jong J (2004) Lowering fear-avoidance and enhancing function through exposure in vivo. A multiple baseline study across six patients with back pain. Pain 108(1–2):8–16. https://doi.org/10.1016/j.pain.2003.03.001

Bushnell MC, Ceko M, Low LA (2013) Cognitive and emotional control of pain and its disruption in chronic pain. Nat Rev Neurosci 14(7):502–511. https://doi.org/10.1038/nrn3516

Cayoun B, Simmons A, Shires A (2017) Immediate and lasting chronic pain reduction following a brief self-implemented mindfulness-based interoceptive exposure task: a pilot study. Mindfulness. https://doi.org/10.1007/s12671-017-0823-x

Ceko M, Shir Y, Ouellet JA, Ware MA, Stone LS, Seminowicz DA (2015) Partial recovery of abnormal insula and dorsolateral prefrontal connectivity to cognitive networks in chronic low back pain after treatment. Hum Brain Mapp 36(6):2075–2092. https://doi.org/10.1002/hbm.22757

Ceunen E, Vlaeyen JW, Van Diest I (2016) On the origin of Interoception. Front Psychol 7:743. https://doi.org/10.3389/fpsyg.2016.00743

Chang DG, Holt JA, Sklar M, Groessl EJ (2016) Yoga as a treatment for chronic low back pain: a systematic review of the literature. J Orthop Rheumatol 3(1):1–8. Retrieved from https://www.ncbi.nlm.nih.gov/pubmed/27231715

Chapman CR, Gavrin J (1999) Suffering: the contributions of persistent pain. Lancet 353(9171): 2233–2237. https://doi.org/10.1016/S0140-6736(99)01308-2

Chen ZS (2023) Hierarchical predictive coding in distributed pain circuits. Front Neural Circuits 17: 1073537. https://doi.org/10.3389/fncir.2023.1073537

Chen WG, Schloesser D, Arensdorf AM, Simmons JM, Cui C, Valentino R et al (2021) The emerging science of interoception: sensing, integrating, interpreting, and regulating signals within the self. Trends Neurosci 44(1):3–16. https://doi.org/10.1016/j.tins.2020.10.007

Cherkin DC, Herman PM (2018) Cognitive and mind-body therapies for chronic low back pain and neck pain: effectiveness and value. JAMA Intern Med 178(4):556–557. https://doi.org/10.1001/jamainternmed.2018.0113

Cherkin DC, Sherman KJ, Balderson BH, Cook AJ, Anderson ML, Hawkes RJ et al (2016) Effect of mindfulness-based stress reduction vs cognitive behavioral therapy or usual care on back pain and functional limitations in adults with chronic low back pain: a randomized clinical trial. JAMA 315(12):1240–1249. https://doi.org/10.1001/jama.2016.2323

Conway CR, Chibnall JT, Gebara MA, Price JL, Snyder AZ, Mintun MA et al (2013) Association of cerebral metabolic activity changes with vagus nerve stimulation antidepressant response in treatment-resistant depression. Brain Stimul 6(5):788–797. https://doi.org/10.1016/j.brs.2012.11.006

Craig AD (2002) How do you feel? Interoception: the sense of the physiological condition of the body. Nat Rev Neurosci 3(8):655–666. https://doi.org/10.1038/nrn894

Craig AD (2003a) A new view of pain as a homeostatic emotion. Trends Neurosci 26(6):303–307. Retrieved from http://www.ncbi.nlm.nih.gov/pubmed/12798599

Craig AD (2003b) Pain mechanisms: labeled lines versus convergence in central processing. Annu Rev Neurosci 26:1–30. https://doi.org/10.1146/annurev.neuro.26.041002.131022

Craig AD (2009a) How do you feel – now? The anterior insula and human awareness. Nat Rev Neurosci 10(1):59–70. Retrieved from http://www.ncbi.nlm.nih.gov/entrez/query.fcgi?cmd=Retrieve&db=PubMed&dopt=Citation&list_uids=19096369

Craig AD (2009b) A rat is not a monkey is not a human: comment on Mogil (Nature Rev. Neurosci. 10, 283–294 (2009)). Nat Rev Neurosci 10(6):466. https://doi.org/10.1038/nrn2606-c1

Craig AD (2010) The sentient self. Brain Struct Funct 214(5–6):563–577. https://doi.org/10.1007/s00429-010-0248-y

Craig AD (2015a) How do you feel?: An interoceptive moment with your neurobiological self princeton. Princeton University Press

Craig AD (2015b) How do you feel?: An interoceptive moment with your neurobiological self. Princeton University Press, Princeton

Cramer H, Haller H, Lauche R, Dobos G (2012) Mindfulness-based stress reduction for low back pain. A systematic review. BMC Complement Altern Med 12:162. https://doi.org/10.1186/1472-6882-12-162

Crucianelli L, Ehrsson HH (2023) The role of the skin in interoception: a neglected organ? Perspect Psychol Sci 18(1):224–238. https://doi.org/10.1177/17456916221094509

de Jong M, Lazar SW, Hug K, Mehling WE, Holzel BK, Sack AT et al (2016) Effects of mindfulness-based cognitive therapy on body awareness in patients with chronic pain and comorbid depression. Front Psychol 7:967. https://doi.org/10.3389/fpsyg.2016.00967

Desmedt O, Heeren A, Corneille O, Luminet O (2021) What do measures of self-report interoception measure? Insights from a systematic review, latent factor analysis, and network approach. Retrieved from https://doi.org/10.31234/osf.io/8mpz9

Desmedt O, Luminet O, Maurage P, Corneille O (2023) Discrepancies in the definition and measurement of human interoception: a comprehensive discussion and suggested ways forward. Perspect Psychol Sci:17456916231191537. https://doi.org/10.1177/17456916231191537

Di Lernia D, Serino S, Cipresso P, Riva G (2016a) Ghosts in the machine. Interoceptive modeling for chronic pain treatment. Front Neurosci 10:314. https://doi.org/10.3389/fnins.2016.00314

Di Lernia D, Serino S, Riva G (2016b) Pain in the body. Altered interoception in chronic pain conditions: a systematic review. Neurosci Biobehav Rev 71:328–341. https://doi.org/10.1016/j.neubiorev.2016.09.015

Di Lernia D, Lacerenza M, Ainley V, Riva G (2020) Altered Interoceptive perception and the effects of interoceptive analgesia in musculoskeletal, primary, and neuropathic chronic pain conditions. J Pers Med 10(4). https://doi.org/10.3390/jpm10040201

Dick B, Eccleston C, Crombez G (2002) Attentional functioning in fibromyalgia, rheumatoid arthritis, and musculoskeletal pain patients. Arthritis Rheum 47(6):639–644. https://doi.org/10.1002/art.10800

Eccleston C, Crombez G (1999) Pain demands attention: a cognitive-affective model of the interruptive function of pain. Psychol Bull 125(3):356–366. Retrieved from http://www.ncbi.nlm.nih.gov/entrez/query.fcgi?cmd=Retrieve&db=PubMed&dopt=Citation&list_uids=10349356

Farb NA, Segal ZV (2024) The senses: a pathway to well-being. Retrieved from https://www.mindandlife.org/insight/the-senses-a-pathway-to-well-being/

Farb NA, Segal ZV, Mayberg H, Bean J, McKeon D, Fatima Z, Anderson AK (2007) Attending to the present: mindfulness meditation reveals distinct neural modes of self-reference. Soc Cogn Affect Neurosci 2(4):313–322. https://doi.org/10.1093/scan/nsm030

Farb NA, Daubenmier J, Price CJ, Gard T, Kerr C, Dunn BD et al (2015) Interoception, contemplative practice, and health. Front Psychol 6:763. https://doi.org/10.3389/fpsyg.2015.00763

Fields HL (2018) How expectations influence pain. Pain 159(Suppl 1):S3–S10. https://doi.org/10.1097/j.pain.0000000000001272

Garland EL (2020) Psychosocial intervention and the reward system in pain and opioid misuse: new opportunities and directions. Pain 161(12):2659–2666. https://doi.org/10.1097/j.pain.0000000000001988

Garland EL, Nakamura Y, Bryan CJ, Hanley AW, Parisi A, Froeliger B et al (2024) Mindfulness-oriented recovery enhancement for veterans and military personnel on long-term opioid therapy for chronic pain: a randomized clinical trial. Am J Psychiatry 181(2):125–134. https://doi.org/10.1176/appi.ajp.20230272

Gawande R, To MN, Pine E, Griswold T, Creedon TB, Brunel A et al (2019) Mindfulness training enhances self-regulation and facilitates health behavior change for primary care patients: a randomized controlled trial. J Gen Intern Med 34(2):293–302. https://doi.org/10.1007/s11606-018-4739-5

GBD 2021 Forecasting Collaborators (2024) Burden of disease scenarios for 204 countries and territories, 2022–2050: a forecasting analysis for the Global Burden of Disease Study 2021. Lancet 403(10440):2204–2256. https://doi.org/10.1016/S0140-6736(24)00685-8

Geisser ME, Casey KL, Brucksch CB, Ribbens CM, Appleton BB, Crofford LJ (2003) Perception of noxious and innocuous heat stimulation among healthy women and women with fibromyalgia: association with mood, somatic focus, and catastrophizing. Pain 102(3):243–250. https://doi.org/10.1016/S0304-3959(02)00417-7

Gilam G, Gross JJ, Wager TD, Keefe FJ, Mackey SC (2020) What is the relationship between pain and emotion? Bridging constructs and communities. Neuron 107(1):17–21. https://doi.org/10.1016/j.neuron.2020.05.024

Gnall KE, Sinnott SM, Laumann LE, Park CL, David A, Emrich M (2024) Changes in interoception in mind-body therapies for chronic pain: a systematic review and meta-analysis. Int J Behav Med. https://doi.org/10.1007/s12529-023-10249-z

Goubert L, Crombez G, Eccleston C, Devulder J (2004) Distraction from chronic pain during a pain-inducing activity is associated with greater post-activity pain. Pain 110(1–2):220–227. https://doi.org/10.1016/j.pain.2004.03.034. S0304395904001769 [pii]

Groessl EJ, Liu L, Richard EL, Tally SR (2020) Cost-effectiveness of Yoga for chronic low back pain in veterans. Med Care 58(Suppl 2 9S):S142–S148. https://doi.org/10.1097/MLR.0000000000001356

Hashmi, J. A., Baliki, M. N., Huang, L., Baria, A. T., Torbey, S., Hermann, K. M.,. .. Apkarian, A. V. (2013). Shape shifting pain: chronification of back pain shifts brain representation from nociceptive to emotional circuits. Brain, 136(Pt 9), 2751-2768. doi:https://doi.org/10.1093/brain/awt211.

Hauck M, Domnick C, Lorenz J, Gerloff C, Engel AK (2015) Top-down and bottom-up modulation of pain-induced oscillations. Front Hum Neurosci 9:375. https://doi.org/10.3389/fnhum.2015.00375

Herman AM, T.R. (2022) Stanton delay and effort-based discounting, and the role of bodily awareness, in people experiencing long-term pain: a cross-sectional study. J Pain 23(3):487–500. https://doi.org/10.1016/j.jpain.2021.10.001

Hölzel BK, Lazar SW, Gard T, Schuman-Olivier Z, Vago DR, Ott U (2011) How does mindfulness meditation work? Proposing mechanisms of action from a conceptual and neural perspective. Perspect Psychol Sci 6(6):537–559

Horsburgh A, Summers SJ, Lewis A, Keegan RJ, Flood A (2024) The relationship between pain and interoception: a systematic review and meta-analysis. J Pain. https://doi.org/10.1016/j.jpain.2024.01.341

Johnson MH (2005) How does distraction work in the management of pain? Curr Pain Headache Rep 9(2):90–95. Retrieved from http://www.ncbi.nlm.nih.gov/pubmed/15745617

Khalsa SS, Adolphs R, Cameron OG, Critchley HD, Davenport PW, Feinstein JS et al (2018) Interoception and mental health: a roadmap. Biol Psychiatry Cogn Neurosci Neuroimaging 3(6):501–513. https://doi.org/10.1016/j.bpsc.2017.12.004

Kiverstein J, Kirchhoff MD, Thacker M (2022) An embodied predictive processing theory of pain experience. Rev Philos Psychol 2022(13):973–998. https://doi.org/10.1007/s13164-022-00616-2

Koyama T, McHaffie JG, Laurienti PJ, Coghill RC (2005) The subjective experience of pain: where expectations become reality. Proc Natl Acad Sci U S A 102(36):12950–12955. 0408576102 [pii]. https://doi.org/10.1073/pnas.0408576102

Krebs, E. E., Lorenz, K. A., Bair, M. J., Damush, T. M., Wu, J., Sutherland, J. M.,. .. Kroenke, K. (2009). Development and initial validation of the PEG, a three-item scale assessing pain intensity and interference. J Gen Intern Med, 24(6), 733–738. doi:https://doi.org/10.1007/s11606-009-0981-1.

Martínez E, Aira Z, Buesa I, Aizpurua I, Rada D, Azkue JJ (2018) Embodied pain in fibromyalgia: disturbed somatorepresentations and increased plasticity of the body schema. PLoS One 13(4):e0194534. https://doi.org/10.1371/journal.pone.0194534

Mehling WE (2001) The experience of breath as a therapeutic intervention – psychosomatic forms of breath therapy. A descriptive study about the actual situation of breath therapy in Germany, its relation to medicine, and its application in patients with back pain. Forsch Komplementarmed Klass Naturheilkd 8(6):359–367. Retrieved from http://www.ncbi.nlm.nih.gov/pubmed/11799304

Mehling WE (2012) The MAIA website: multidimensional assessment of interoceptive awareness. Retrieved from https://osher.ucsf.edu/research/maia

Mehling WE (2016) Differentiating attention styles and regulatory aspects of self-reported interoceptive sensibility. Philos Trans R Soc Lond B Biol Sci 371(1708). https://doi.org/10.1098/rstb.2016.0013

Mehling WE, Gopisetty V, Daubenmier J, Price CJ, Hecht FM, Stewart A (2009) Body awareness: construct and self-report measures. PLoS One 4(5):e5614. https://doi.org/10.1371/journal.pone.0005614

Mehling WE, Wrubel J, Daubenmier JJ, Price CJ, Kerr CE, Silow T et al (2011) Body awareness: a phenomenological inquiry into the common ground of mind-body therapies. Philos Ethics Humanit Med 6:6. https://doi.org/10.1186/1747-5341-6-6

Mehling WE, Price C, Daubenmier JJ, Acree M, Bartmess E, Stewart A (2012) The multidimensional assessment of interoceptive awareness (MAIA). PLoS One 7(11):e48230. https://doi.org/10.1371/journal.pone.0048230

Mehling WE, Daubenmier J, Price CJ, Acree M, Bartmess E, Stewart AL (2013) Self-reported interoceptive awareness in primary care patients with past or current low back pain. J Pain Res 6: 403–418. https://doi.org/10.2147/JPR.S42418

Mehling WE, Acree M, Stewart A, Silas J, Jones A (2018) The multidimensional assessment of interoceptive awareness, version 2 (MAIA-2). PLoS One 13(12):e0208034. https://doi.org/10.1371/journal.pone.0208034

Mehling WE, Todd J, Schuman-Olivier Z (2022) Multidisciplinary assessment of interoceptive awareness, version 2 (MAIA-2). In: Medvedev ON, Krägeloh CU, Siegert RJ, Singh NN (eds) Handbook of assessment in mindfulness research. Springer, Cham

Mehling WE, Strigo IA, Goldman V, Hartogensis W, Adler SR, Lotz J, Hecht FM (2024) Mind your pain: assessment of a smartphone-based interoceptive attention training for patients with chronic low back pain. In press

Panksepp J (2007) Neurologizing the psychology of affects: how appraisal-based constructivism and basic emotion theory can coexist. Perspect Psychol Sci 2(3):281–296. https://doi.org/10.1111/j.1745-6916.2007.00045.x

Philippi, C. L., Feinstein, J. S., Khalsa, S. S., Damasio, A., Tranel, D., Landini, G.,. .. Rudrauf, D. (2012). Preserved self-awareness following extensive bilateral brain damage to the insula, anterior cingulate, and medial prefrontal cortices. PLoS One, 7(8), e38413. doi:https://doi.org/10.1371/journal.pone.0038413.

Price CJ, Hooven C (2018) Interoceptive awareness skills for emotion regulation: theory and approach of mindful awareness in body-oriented therapy (MABT). Front Psychol 9:798. https://doi.org/10.3389/fpsyg.2018.00798

Price CJ, Mehling W (2016) Body awareness and pain. In: Thompson D, Brooks M (eds) Integrative pain management. Handspring Publishing. ISBN13: 978-1-909141-26-1

Price CJ, McBride B, Hyerle L, Kivlahan DR (2007) Mindful awareness in body-oriented therapy for female veterans with post-traumatic stress disorder taking prescription analgesics for chronic pain: a feasibility study. Altern Ther Health Med 13(6):32–40. Retrieved from http://www.ncbi.nlm.nih.gov/pubmed/17985809

Qaseem A, Wilt TJ, McLean RM, Forciea MA, Clinical Guidelines Committee of the American College of Physicians (2017) Noninvasive treatments for acute, subacute, and chronic low back pain: a clinical practice guideline from the American College of Physicians. Ann Intern Med. https://doi.org/10.7326/M16-2367

Rad AA, Wippert P-M (2024) Insights into pain distraction and the impact of pain catastrophizing on pain perception during different types of distraction tasks. Front Pain Res 5(2024). https://doi.org/10.3389/fpain.2024.1266974

Reddan MC, Young H, Falkner J, Lopez-Sola M, Wager TD (2020) Touch and social support influence interpersonal synchrony and pain. Soc Cogn Affect Neurosci 15(10):1064–1075. https://doi.org/10.1093/scan/nsaa048

Rost S, Van Ryckeghem DML, Schulz A, Crombez G, C. (2017) Vögele Generalized hypervigilance in fibromyalgia: normal interoceptive accuracy, but reduced self-regulatory capacity. J Psychosom Res 93:48–54. https://doi.org/10.1016/j.jpsychores.2016.12.003

Sabourin BC, Stewart SH, Watt MC, Krigolson OE (2015) Running as interoceptive exposure for decreasing anxiety sensitivity: replication and extension. Cogn Behav Ther 44(4):264–274. https://doi.org/10.1080/16506073.2015.1015163

Schuman-Olivier Z, Trombka M, Lovas DA, Brewer JA, Vago DR, Gawande R et al (2020) Mindfulness and behavior change. Harv Rev Psychiatry 28(6):371–394. https://doi.org/10.1097/HRP.0000000000000277

Segal Z (2022) Advances in evidence-based mindfulness programs. Online presentation for Mindful Society Global Institute. https://www.mindfulinstitute.org: Sept 28, Stream On Demand
Seth AK (2013) Interoceptive inference, emotion, and the embodied self. Trends Cogn Sci 17(11): 565–573. https://doi.org/10.1016/j.tics.2013.09.007
Stefanics G, Heinzle J, Horvath AA, Stephan KE (2018) Visual mismatch and predictive coding: a computational single-trial ERP study. J Neurosci 38(16):4020–4030. https://doi.org/10.1523/JNEUROSCI.3365-17.2018
Strigo IA, Craig AD (2016) Interoception, homeostatic emotions and sympathovagal balance. Philos Trans R Soc Lond B Biol Sci 371(1708). https://doi.org/10.1098/rstb.2016.0010
Suksasilp C, Garfinkel SN (2022) Towards a comprehensive assessment of interoception in a multidimensional framework. Biol Psychol 168(108):262. https://doi.org/10.1016/j.biopsycho.2022.108262
Trevisan DA, Mehling WE, McPartland JC (2021) Adaptive and maladaptive bodily awareness: distinguishing interoceptive sensibility and interoceptive attention from anxiety-induced somatization in autism and alexithymia. Autism Res 14(2):240–247. https://doi.org/10.1002/aur.2458
Turner JA, Anderson ML, Balderson BH, Cook AJ, Sherman KJ, Cherkin DC (2016) Mindfulness-based stress reduction and cognitive behavioral therapy for chronic low back pain: similar effects on mindfulness, catastrophizing, self-efficacy, and acceptance in a randomized controlled trial. Pain 157(11):2434–2444. https://doi.org/10.1097/j.pain.0000000000000635
Van den Bergh O, Witthoft M, Petersen S, Brown RJ (2017) Symptoms and the body: taking the inferential leap. Neurosci Biobehav Rev 74(Pt A):185–203. https://doi.org/10.1016/j.neubiorev.2017.01.015
Van den Bergh O, Brosschot J, Critchley H, Thayer JF, Ottaviani C (2021) Better safe than sorry: a common signature of general vulnerability for psychopathology. Perspect Psychol Sci 16(2): 225–246. https://doi.org/10.1177/1745691620950690
Vig L, Koteles F, Ferentzi E (2022) Questionnaires of interoception do not assess the same construct. PLoS One 17(8):e0273299. https://doi.org/10.1371/journal.pone.0273299
Villemure C, Ceko M, Cotton VA, Bushnell MC (2013) Insular cortex mediates increased pain tolerance in yoga practitioners. Cereb Cortex. https://doi.org/10.1093/cercor/bht124
Voss S, Boachie DA, Nieves N, Gothe NP (2023) Mind-body practices, interoception and pain: a scoping review of behavioral and neural correlates. Ann Med 55(2):2275661. https://doi.org/10.1080/07853890.2023.2275661
Wahbeh H, Elsas SM, Oken BS (2008) Mind-body interventions: applications in neurology. Neurology 70(24):2321–2328. https://doi.org/10.1212/01.wnl.0000314667.16386.5e
Williams JM (2010) Mindfulness and psychological process. Emotion 10(1):1–7. https://doi.org/10.1037/a0018360
Zamariola G, Frost N, Van Oost A, Corneille O, Luminet O (2019) Relationship between interoception and emotion regulation: new evidence from mixed methods. J Affect Disord 246:480–485. https://doi.org/10.1016/j.jad.2018.12.101
Zautra AJ, Fasman R, Davis MC, Craig ADB (2010) The effects of slow breathing on affective responses to pain stimuli: an experimental study. Pain 149(1):12–18. https://doi.org/10.1016/j.pain.2009.10.001

Interoceptive Processing in Substance Use Disorders (SUDs): From the Neuroanatomy to Insights from Computational Models and Predictive Coding Frameworks

Martin P. Paulus

Contents

Abstract Substance use disorders (SUDs) represent complex public health challenges characterized by a blend of genetic, cognitive, environmental, and psychosocial factors. This chapter explores the critical role of interoceptive processing – the

M. P. Paulus (✉)
Laureate Institute for Brain Research, Tulsa, OK, USA

College of Health Sciences, Community Medicine, University of Tulsa, Tulsa, OK, USA

Department of Psychiatry, University of California, San Diego, La Jolla, CA, USA
e-mail: mpaulus@laureateinstitute.org

Curr Topics Behav Neurosci (2025) 70: 159–178
https://doi.org/10.1007/7854_2024_519
Published Online: 26 November 2024

internal sensing of physiological states – in the neurobiology and treatment of SUDs. Interoceptive dysfunctions are highlighted as central to the craving, emotional regulation, and decision-making processes that underpin addictive behaviors. The importance of the insula in sustaining drug use, particularly nicotine, underscores a broader involvement of interoceptive pathways in SUDs. Altered interoceptive processing is evident across various SUDs, where individuals demonstrate both a heightened sensitivity to drug-related cues and a diminished ability to process aversive stimuli, suggesting substantial neurobiological underpinnings that complicate treatment outcomes. Moreover, we delve into the theoretical and computational approach to understanding interoceptive processing in SUDs. This perspective utilizes a predictive coding framework, positing that the brain continuously generates and updates predictions about internal states based on sensory inputs. In SUDs, disruptions in this predictive mechanism can lead to inaccuracies in interoceptive perception, contributing significantly to the compulsive nature of drug-seeking behaviors and the challenges associated with treatment. We explore how computational models, such as Bayesian inference, provide insights into the interplay between expected and received interoceptive signals, highlighting the role of hyper-precise prior beliefs in the persistence of craving and impulsivity. This theoretical approach not only deepens our understanding of the neural and cognitive bases of addiction but also suggests novel intervention strategies. By recalibrating interoceptive predictions through targeted therapies, such as neuromodulation and mindfulness training, we can potentially restore the interoceptive accuracy, thereby offering new avenues for effective treatment of SUDs.

1 Substance Use Disorders

Substance use disorders in general (Diseases and Injuries 2024) and opioid use disorder (OUD (Kosten and Petrakis 2021)) and methamphetamine use disorder (MUD (Paulus and Stewart 2020)) in particular represent significant public health challenges globally. These disorders, though pharmacologically distinct, share complex interplays of genetic predispositions, cognitive impacts, environmental factors, and psychosocial behaviors. For example, studies have reported a high prevalence of non-opioid substance use disorders among individuals with OUD, with cocaine use disorder being particularly common (Santo et al. 2024). Similarly, co-drug use is also common among individuals with MUD, who show a high use of cannabis (Daldegan-Bueno et al. 2022). The emergence of these disorders is complex and includes both genetic (Chawar et al. 2021; Falconnier et al. 2023; Guerin et al. 2021) and environmental factors (Deol et al. 2023; Lodge et al. 2024) such as early childhood aversive experiences or sexual preference behaviors. Although treatment options are substantially better for OUD (Sugarman et al. 2022) relative to MUD (Minozzi et al. 2024), treatment outcomes are still modest. Thus, these disorders pose significant challenges for public health and a better understanding of their underlying neurobiology is sorely needed to develop novel, evidence-based interventions.

Interoception, the process by which the nervous system senses, interprets, and integrates signals originating from within the body (Craig 2002, 2009), has been considered an important construct to understand the behavior and the neural underpinnings and SUDs. The relationship between interoception and addiction is complex and will be elaborated below but it involves aspects of craving (Gray and Critchley 2007), emotional regulation (Wisniewski et al. 2021), and decision-making (Paulus 2007), all of which are central to the development and persistence of addictive behaviors. Moreover, interventions that manipulate interoceptive pathways, such as mindfulness and other neuromodulation techniques, have shown promise in improving symptoms and possibly altering the underlying pathophysiology of SUDs(Nayok et al. 2023; Parlier-Ahmad et al. 2023). Finally, the discovery that the insula might be critically important in sustaining continued use of nicotine (Naqvi et al. 2007) has provided further evidence that interoceptive pathways might be critically important for substance use disorder.

2 Core Findings and Neural Mechanisms

Studies have shown that individuals with SUDs and other mental health conditions exhibit significant alterations in their interoceptive processing capabilities. For instance, we have recently demonstrated a transdiagnostic failure among patients to adjust beliefs about the reliability of interoceptive signals (Lavalley et al. 2023). This inability to accurately perceive changes in one's bodily state suggests a fundamental disruption in the interoceptive precision measures that could have broad implications for treatment outcomes and the identification of novel therapeutic targets aimed at improving visceral regulation. Specifically, our group (Smith et al. 2021) has shown a perceptual insensitivity to the modulation of interoceptive signals across several psychiatric disorders. This finding suggests that interoceptive deficits manifest most prominently during states of homeostatic perturbation, a condition common in various forms of psychopathology, including addiction. The altered neural processing of interoceptive signals, particularly in conditions such as stimulant use disorder (Stewart et al. 2020) further underscores the significant impact of interoceptive dysfunctions on the neural underpinnings of addiction. These findings extend to other substances as well. For example, adolescents with cannabis/alcohol use disorder also demonstrate an altered response to interoceptive perturbations (May et al. 2020) highlighting the critical role of developmental factors in the emergence of interoceptive dysfunctions. This alteration in interoceptive processing is not merely limited to the perception of internal signals but extends to hypersensitivity to aversive stimuli (Berk et al. 2015) and exaggerated neural processing of pleasant stimuli (Stewart et al. 2015), revealing a complex interplay between interoception and the valence of experienced stimuli. Taken together, the sensing of the internal body state is fundamentally affected in individuals with SUD and may occur even before the disorder develops. However, the precise interplay between pre-existing interoceptive deficits and insults via various environmental factors, which together result in altered body-relevant processing is still unclear.

2.1 Altered Interoceptive Processing in Adolescent Substance Users

We have conducted a series of investigations with adolescents with substance use disorders and shown that these individuals show distinct alterations in their interoceptive processing (Berk et al. 2015; May et al. 2020; Migliorini et al. 2013). These studies collectively reveal that adolescent substance users show an altered response to interoceptive perturbations, including both aversive and pleasant interoceptive stimuli. Such alterations suggest that substance use during adolescence may disrupt the normal development of interoceptive awareness, potentially contributing to the maintenance of substance use behaviors and increasing the risk of developing SUDs. For example, adolescents with cannabis or alcohol use disorder demonstrate an altered response to interoceptive perturbations (May et al. 2020), which suggest that these dysfunctions are notable early during the development of substance use disorder. These individuals also show hypersensitivity to aversive interoceptive stimuli (May et al. 2020) and exaggerated neural processing of pleasant interoceptive stimuli (Stewart et al. 2015). Together, these findings further underscore the complex relationship between interoception and substance use during adolescence.

The identification of altered interoceptive processing across a spectrum of disorders offers a unique lens through which the mechanisms of addiction and mental health conditions can be understood. The insular cortex (Augustine 1985), a key neural region implicated in these studies, serves as a potential focal point for therapeutic intervention. For example, mindfulness-based approaches and intense physical exercise (Paulus et al. 2013) have been proposed to modulate interoceptive function, suggesting a pathway to enhance or restore disrupted interoceptive processing. Moreover, the potential predictive value of interoceptive precision measures (Lavalley et al. 2023) opens up new avenues for personalized treatment strategies that could significantly improve clinical outcomes. By targeting the specific interoceptive dysfunctions characteristic of each disorder, interventions can be tailored to address the root causes of these alterations, potentially leading to more effective and sustainable treatment outcomes. The examination of altered interoception across disorders sheds light on the fundamental disruptions in how individuals perceive and interpret bodily signals. Understanding these alterations not only provides a deeper insight into the neural basis of addiction and related mental health conditions but also offers promising directions for the development of novel therapeutic interventions aimed at rectifying these dysfunctions. Future research will undoubtedly expand on these findings, further elucidating the complex relationship between interoception, addiction, and mental health.

2.2 *Neural Mechanisms and Insular Cortex Involvement in Altered Interoception*

The insular cortex, a critical region for integrating interoceptive signals, plays a pivotal role in our understanding of how the brain processes signals from within the body (Strigo and Craig 2016; Torrealba et al. 2017). This understanding is particularly relevant in the context of addiction and mental health disorders, where altered interoceptive processing has been linked to various neural mechanisms and insular cortex dysfunctions. Here, we describe some of the neural underpinnings of altered interoception, focusing on the involvement of the insular cortex and its implications for addiction and psychiatric conditions.

2.3 *Abnormal Functional Connectivity and Insular Hypo-Activity*

There is some evidence abnormal functional connectivity in the Default Mode Network (DMN) across Substance Use Disorders (SUDs) (Zhang and Volkow 2019), which is closely connected to the insula (Sridharan et al. 2008), emphasizing the impact of these alterations on self-awareness, emotional regulation, and craving. This suggests a foundational disruption in how individuals with SUDs process and integrate interoceptive signals, with the insular cortex playing a crucial role in this dysregulation. We (Paulus and Stewart 2014) have reported hypo-activity in the insular cortex during cognitive control processes in addicted individuals. This finding indicates a decreased ability to integrate interoceptive signals properly, potentially contributing to the compulsive nature of drug-seeking behavior and the difficulty in exerting control over substance use. Others (Zhang et al. 2016) have reported decreased functional connectivity between insula-based networks and motor/executive control regions in individuals with Internet Gaming Disorder, suggesting that such disruptions may extend beyond substance addiction to other forms of compulsive behaviors and addictions.

2.4 *Insular Cortex: A Hub for Interoceptive Processing*

The insular cortex has a critical role in representing the interoceptive effects of drug taking in general and in processing the subjective experience of drug effects in particular (Naqvi and Bechara 2010). When considering the molecular and circuitry mechanisms in SUD, the insula (Lovelock et al. 2021; Stewart et al. 2020) is important for modulating the perception of internal bodily states and its connection to drug-taking behaviors. The dorsal dysgranular insula has been identified as a critical region for attending to bodily sensations in SUD, suggesting that disruptions in this area may contribute to the altered interoceptive awareness observed in individuals with SUD. This region receives intense interoceptive afferents via the vagal nerve (Ibanez et al. 2021), which provides a direct link to the autonomic nervous system, which can alter interoceptive signals, influencing an individual's susceptibility to and experience of substances. The notion that connectivity between insula and other body-relevant processing areas in the brain is altered in SUD is consistent with findings of altered functional connectivity of the insular cortex in individuals with cocaine use disorder (Cisler et al. 2013), which – in addition to physiological processing areas – also affects cognitive control and possibly contribute to the persistence of addictive behaviors. Moreover, in rodent models (Daniel et al. 2017) of addictive behaviors insular dysfunction can lead to impaired decision-making, a common feature in addiction and related disorders. These insights into the neural mechanisms and insular cortex involvement offer a deeper understanding of the pathophysiology of addiction, which may be a treatment target for these disorders. For example, brain stimulation techniques modulating the neural circuits involved in interoceptive processing (Ibrahim et al. 2019) could offer new approaches for treating substance use disorders. Moreover, insular hypo-activity and abnormal insular functional connectivity could be potential biomarkers for addiction and compulsive disorders, which may ultimately lead to prognostic tools that can guide personalized treatment approaches.

2.5 *Aberrant Network Connectivity in SUD*

Some have suggested that individuals with SUD have abnormal functional connectivity within the Default Mode Network (DMN) (Zhang and Volkow 2019), which impacts crucial aspects of self-awareness, emotional regulation, and craving, suggesting a foundational disruption in the integration of interoceptive signals. In addition, investigators have shown that acute administration of alcohol affects the interoceptive network (Lovelock et al. 2021), which alters the individual's ability to utilize interoceptive afferents to guide motivated behavior. The identification of specific network dysfunctions provides a roadmap for developing interventions that target these neural circuits. For instance, interventions that aim to recalibrate the functional connectivity of the DMN could potentially enhance interoceptive awareness and improve self-regulation in individuals with SUD. This could involve neuromodulation techniques, psychotherapeutic interventions designed to enhance mindfulness and bodily awareness, or physical exercises that impact brain network connectivity. Moreover, DMN Resting-State Functional Connectivity (RSFC) has been proposed as a potential biomarker (Zhang and Volkow 2019) for clinical outcomes in addiction, which might eventually lead to prediction of treatment responsiveness. Future research in this area should focus on expanding our understanding of how different brain networks interact to mediate interoceptive processing and its dysregulation in addiction. Longitudinal studies examining changes in network connectivity over the course of treatment could provide valuable insights into the neural mechanisms underlying recovery and relapse. Additionally, exploring the genetic and environmental factors that influence network connectivity could enhance our understanding of the vulnerability to SUD and identify potential preventive measures. Network-based frameworks represent a useful approach for dissecting the complex neural underpinnings of interoceptive dysfunctions in addiction. By focusing on the connectivity patterns and interactions between different brain networks, this approach offers new perspectives on the pathophysiology of SUD and paves the way for the development of targeted, effective interventions.

2.6 Predictive Value of Interoceptive Measures

Recently, our group (Lavalley et al. 2023) has proposed a computational approach that underscores the potential of interoceptive precision measures to predict treatment outcomes across various disorders. This research demonstrates that a transdiagnostic failure among patients to adjust beliefs about the reliability of interoceptive signals reliably and robustly differentiates those individuals who have substance use disorders from those that do not. Specifically, we propose that an individual's ability to perceive bodily states accurately is important in guiding motivated behavior. The focus on interoceptive processing has also emphasized the possibility of using mindfulness-based approaches or intense physical exercise (Paulus et al. 2013) to adjust motivated behavior in individuals with substance use disorder. These interventions aim to enhance the integration of interoceptive signals, potentially restoring disrupted processing pathways. Mindfulness practices, for instance, could increase interoceptive awareness and accuracy, enabling individuals to better navigate emotional and physiological states. On the other hand, intense physical exercise could recalibrate the interoceptive system, improving overall mental health and resilience. In addition, others (Ibrahim et al. 2019) have proposed the use of brain stimulation techniques, including transcranial magnetic stimulation (TMS), transcranial direct current stimulation (tDCS), and deep brain stimulation (DBS), as a way of modulating the insula directly. These techniques offer a direct means of modulating brain regions implicated in interoceptive dysfunctions, presenting a novel avenue for intervention that could complement traditional therapeutic approaches.

2.7 Interoception and Psychopathology

Interoceptive dysfunctions have emerged as a significant factor in the psychopathology of addiction, affecting individuals' ability to accurately perceive and respond to internal bodily cues. This altered interoceptive awareness can lead to a disconnection from physiological states that signal distress, craving, or satiety, potentially contributing to the maladaptive coping mechanisms seen in SUD (Nayok et al. 2023). For instance, the inability to accurately perceive internal signals may drive individuals toward substance use as a misguided attempt to regulate their emotional and physiological state (Leganes-Fonteneau et al. 2019). Many substances have profound effects on interoceptive processing, e.g. alcohol generates the sensation of light-headedness that has been tight to interoceptive processing dysfunctions (Leganes-Fonteneau et al. 2019). The exploration of interoception within the aetiopathogenesis of addiction provides valuable insights into the origins and development of addictive behaviors. By examining how individuals with SUD process interoceptive signals differently, researchers can identify potential pathways through which addiction emerges and progresses (Stewart et al. 2020). Some have reported dysregulated neural systems underlying interoceptive processing in cocaine used disorder which may contribute to the development and maintenance of addiction (de la Fuente et al. 2019).

3 Conceptual and Theoretical Insights

The integration of interoception and alliesthesia (Cabanac 1971) with theoretical approaches to SUD (Paulus et al. 2009) proposes that altered internal bodily states change the hedonic properties of drugs, potentially contributing to the development and maintenance of addiction. This conceptual framework posits that the rewarding or aversive experiences associated with drug use are not static but are influenced by the individual's current interoceptive state. Such an approach emphasizes the dynamic interplay between the body's internal signals and the external cues associated with drug use, providing a nuanced understanding of how bodily perceptions influence addictive behaviors. Moreover, the review of the role of interoception in addiction (Paulus and Stewart 2014) suggests that dysfunctions in processing bodily signals contribute significantly to drug use behavior. This perspective underscores the importance of the salience network (Fedota and Stein 2015) as exemplified by the insular cortex and related neural circuits in mediating the interoceptive experience of drug effects, further supporting the notion that interoception is a key component of the addiction process. This approach bridges the gap between the subjective experience of drug effects and the neurobiological mechanisms underlying addiction. By acknowledging the role of interoception in the embodiment of drug experiences, these frameworks offer a comprehensive approach that considers both the psychological and physiological dimensions of addiction. Moreover, understanding the dynamic nature of drug reward and aversion through the lens of alliesthesia provides a framework acknowledging that the interoceptive contribution to motivated behavior is highly dependent on the individual's current state, i.e. the individual's specific interoceptive profile which directly influences drug use behavior. Taken together, we had proposed that altered body prediction errors (Kruschwitz et al. 2019), i.e. the difference between the expected body state and the current body state fundamentally modulated motivated behaviors in individuals with SUD. This difference, however, is shaped by the combination of pre-existing interoceptive processing differences and the profound effect that the repeated exposure to drugs with strong interoceptive effects has on the individual.

3.1 *Bayesian Framework for Craving and Impulsivity in Addiction*

The application of a Bayesian framework to understand craving and impulsivity in addiction offers a novel approach to delineate the complex interplay between cognitive processes and behavioral outcomes in substance use disorders (SUD). This section examines how Bayesian models provide a structured perspective on the mechanisms underlying craving and impulsivity, drawing on key research findings to illuminate the role of predictive coding and belief updating in addiction. Through this lens, we explore the potential for Bayesian frameworks to enhance our understanding of addiction and inform the development of targeted interventions.

3.2 *Understanding Craving and Impulsivity Through Bayesian Inference*

A recent review emphasized the Bayesian computation framework (Kulkarni et al. 2023) to better understand the dysfunctions observed in substance users. This account begins by acknowledging that craving is a core feature of substance use disorders and is increasingly recognized as a significant component of behavioral addictions like gambling disorder. These authors propose that an interoceptive inference framework, building upon the Bayesian brain hypothesis, provides a comprehensive understanding of the subjective experience of craving. In this view, craving arises from a mismatch between the brain's prior beliefs about the physiological states associated with obtaining an addictive substance or performing an action, and the sensory evidence of its completion. These authors hypothesize that in behavioral addictions, prior beliefs about the desirability of the action become hyper-precise, leading to intense and persistent craving. Importantly, this computational formalization of craving has significant therapeutic implications. Current addiction treatments often focus on reducing overt behaviors, but this model highlights the importance of addressing the subjective experience of craving itself. Interventions that successfully reduce the precision of prior beliefs may effectively alleviate craving.

Our group has developed in parallel a computational approach to better delineate the interoceptive dysfunctions in substance use and integrate them with cognitive and affective processes. The use of computational models, such as those based on predictive coding or reinforcement learning, offers novel insights into how individuals with SUD might predict, experience, and learn from internal bodily signals in the context of substance use (Harle et al. 2013). For instance, predictive coding models suggest that SUD is characterized by a disbalance in the prediction of interoceptive signals and the actual sensory input, leading to a heightened or distorted sense of internal states (Schwartenbeck et al. 2015). This disbalance may drive substance-seeking behavior as an attempt to correct perceived interoceptive errors. Similarly, reinforcement learning models highlight how the valuation of drug-related cues can be modulated by interoceptive signals, shaping decision-making processes that favor substance use. Neuroimaging studies, particularly those employing functional magnetic resonance imaging (fMRI), provide concrete evidence of the neural underpinnings of altered interoceptive processing in SUD. These studies frequently report alterations in the activity of key brain regions involved in interoception, such as the insula (Harle et al. 2015a) and anterior cingulate cortex (Harle et al. 2014), among individuals with SUD. Such alterations are associated with increased craving and substance use, suggesting a neural pathway through which interoceptive processing may contribute to SUD. Moreover, the integration of computational models with neuroimaging data allows for a more precise understanding of the neural computations underlying interoceptive processing in SUD, offering a bridge between theoretical models and empirical evidence. For example, we have shown that individuals with methamphetamine use disorder are less sensitive to surprising task events, both across trials and upon making commission errors, which may help explain why these individuals may not engage in switching strategy when the environment changes (Harle et al. 2016). In general, we have found that substance users may have impaired abilities to accurately model and update their predictions regarding the probabilistic demands for inhibition, leading to suboptimal behavioral adjustment (Harle et al. 2014), which can predict relapse in methamphetamine users (Harle et al. 2019). Interestingly, in those at risk for stimulant use disorder, we find exaggerated neural responses to Bayesian prediction errors are associated with future problem use, underscore the role of maladaptive belief updating (Harle et al. 2015a). Finally, we found that SUD individuals fail to appropriately integrate information about future consequences into current decision-making processes, leading to a bias toward immediate rewards despite long-term costs (Harle et al. 2015b). Integrating these findings through a predictive coding lens suggests that SUDs may involve fundamental disruptions in how predictions are generated and updated based on sensory inputs and outcomes. These disruptions manifest as altered decision-making, impaired inhibitory control, and a heightened risk of substance dependence and relapse. By elucidating the mechanisms through which altered interoception contributes to substance use and identifying potential neural targets for intervention, this line of research holds significant potential for improving the prevention and treatment of SUD. Further interdisciplinary studies combining computational, neuroimaging, and clinical methodologies are essential to fully realize this potential and translate these insights into effective interventions for individuals with SUD.

3.3 Animal Models

Recent studies have elucidated the critical role of the anterior insular cortex (aIC) in mediating compulsive drug-taking behaviors (Chen et al. 2022), particularly in the context of cocaine use. Specifically, the activity of glutamatergic neurons within the aIC and their interactions with projections from the orbitofrontal cortex (OFC) plays a crucial role in the emergence of compulsivity. In this study about 30% of rats in continued cocaine use despite adverse consequences, mirroring the compulsivity observed in human addiction, and underscoring the aIC's pivotal role in this process. The aIC, through its hyperactivity, appears to distort the incentive value of drug-related stimuli and the anticipation of reward, while simultaneously minimizing the perceived risk of negative outcomes. This imbalance fosters a state of compulsive drug seeking by enhancing reward salience and diminishing punishment sensitivity. The involvement of the OFC-aIC circuit highlights a crucial regulatory mechanism, updating the relative value outcomes of reward and punishment, thus promoting compulsive behavior. Moreover, the identified imbalance in corticostriatal circuits, favoring the OFC-aIC "go" circuit over inhibitory control pathways, suggests a neural basis for the impaired decision-making observed in addiction. These authors suggest that inhibition of aIC glutamatergic neurons not only presents a potential therapeutic avenue by preventing conditioned aversion and reducing cocaine reward but also suggests a nuanced understanding of the aIC's role in encoding both aversion and reward.

Others (Livneh et al. 2017) have proposed a model that delineates a homeostatic pathway that intricately governs responses to food cues through a sophisticated interplay of sensory information processing, motivational state biases, emotional valence attribution, interoceptive assessment, and subsequent behavioral responses. These authors emphasize the role of the thalamus and primary olfactory cortex processing sensory information related to food cues, integrating visual, visceral, gustatory, and olfactory inputs to create a comprehensive sensory profile. The motivational state, influenced by factors such as caloric deficiency, modulates this pathway through mechanisms like the agouti-related protein signaling to the paraventricular nucleus of the thalamus, thereby introducing a bias based on the individual's current energy needs. The pathway progresses as the basolateral amygdala assigns emotional valence to these cues, a critical step that determines the attractiveness or aversiveness of the food based on the integrated sensory and motivational information. This emotional valuation is then refined by the insular cortex, which assesses the interoceptive significance of the food cue, aligning the body's internal state with external stimuli to guide decision-making processes. Ultimately, this culminates in a behavioral response mediated by the nucleus accumbens and the central nucleus of the amygdala, directing actions toward or away from the food cue based on the evaluated need and value. When extrapolating this model to drug use, parallels emerge that underscore the similarity in processing between food and drug cues. Drug cues, like food cues, initiate their influence through sensory processing in the thalamus and sensory cortices, engaging the brain's reward circuitry and decision-making processes. The motivational state, potentially mirrored by craving or withdrawal in the context of drug use, biases the individual's perception and response to these drug cues, akin to the bias introduced by caloric deficiency for food cues. Emotional valence attributed to drug cues by the basolateral amygdala could reflect the complex dynamics of drug euphoria and the aversiveness of withdrawal symptoms, while the insular cortex's role in interoceptive processing becomes pivotal in navigating the bodily sensations associated with drug craving and withdrawal, influencing the decision to seek and consume the drug. Finally, the behavioral response to drug cues, mediated by similar neural substrates involved in food cue responses, highlights the reinforcing nature of drug use, driven by the nucleus accumbens and the central nucleus of the amygdala.

4 Integrating Computational and Animal Models

Integrating the neuroscience-based conceptualization of interoceptive processing with a computational approach, particularly in the context of substance use disorders (SUD), offers a multifaceted view of the dysfunctions that underpin these conditions (Fig. 1). This integrated perspective highlights several specific processes that are likely dysfunctional in individuals with SUD:

4.1 Disrupted Prediction and Updating of Interoceptive Signals

At the core of the computational approach is the notion of predictive coding, which posits that the brain continuously generates predictions about internal and external states and updates these predictions based on incoming sensory information. In SUD, there appears to be a fundamental disruption in this process concerning interoceptive signals. Individuals with SUD may have impaired abilities to accurately predict interoceptive outcomes associated with substance use or abstention, leading to a mismatch between expected and actual bodily states. This disbalance can manifest as a heightened or distorted sense of craving and discomfort, driving substance-seeking behavior to alleviate these perceived interoceptive errors.

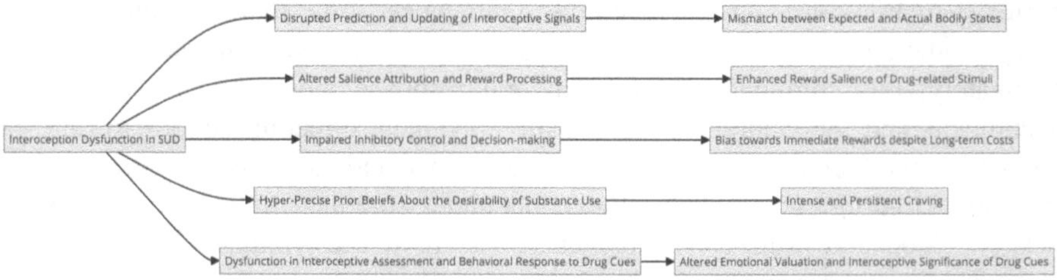

Fig. 1 Summarizing the proposed interoceptive dysfunctions in SUD

4.2 Altered Salience Attribution and Reward Processing

The anterior insular cortex (aIC) and its interactions with the orbitofrontal cortex (OFC) is crucial for updating the value outcomes of reward and punishment, promoting compulsive behavior. In SUD, the hyperactivity of this circuit can distort the incentive value of drug-related stimuli, enhancing reward salience and diminishing the perceived risk of negative outcomes. This imbalance fosters a state of compulsive drug seeking, where the immediate reward of substance use is overvalued in comparison to its long-term adverse consequences.

4.3 Impaired Inhibitory Control and Decision-Making

The identified imbalance in corticostriatal circuits, favoring the OFC-aIC "go" circuit over inhibitory control pathways, suggests a neural basis for the impaired decision-making observed in addiction. This dysfunction is characterized by a failure to integrate information about future consequences into current decision-making processes, leading to a bias toward immediate rewards despite long-term costs. Substance users may have impaired abilities to accurately model and update their predictions regarding the probabilistic demands for inhibition, leading to suboptimal behavioral adjustments and increased risk of relapse.

4.4 Hyper-Precise Prior Beliefs About the Desirability of Substance Use

In the context of Bayesian inference, craving and impulsivity in addiction can be understood as arising from hyper-precise prior beliefs about the desirability of substance use or the actions leading to it. This computational formalization suggests that individuals with SUD hold overly confident beliefs about the positive outcomes of substance use, leading to intense and persistent craving that is difficult to modulate or control. Addressing the subjective experience of craving by reducing the precision of these prior beliefs may offer a therapeutic avenue.

4.5 Dysfunction in Interoceptive Assessment and Behavioral Response to Drug Cues

The homeostatic pathway model, which governs responses to food cues, can be paralleled in drug use, highlighting similar processing dysfunctions for drug cues. The pathway involving sensory processing, motivational state biases, emotional valence attribution, interoceptive assessment, and subsequent behavioral responses is disrupted. This disruption leads to an altered emotional valuation and interoceptive significance of drug cues, culminating in maladaptive behaviors directed toward drug use.

Taken together, the integration of neuroscience-based and computational approaches elucidates a complex network of dysfunctions in SUD, encompassing altered interoceptive signal processing, biased reward processing, impaired decision-making, hyper-precise prior beliefs about substance use, and dysfunctional behavioral responses to drug cues. Addressing these dysfunctions requires a multifaceted intervention strategy that targets the underlying neural and computational mechanisms, potentially through a combination of pharmacological, behavioral, and neuromodulatory treatments.

The integrated process model that combines neuroscience-based conceptualizations of interoceptive processing with computational approaches offers a nuanced framework for understanding the dysfunctions in substance users at various stages of addiction. By focusing on the interplay between neural mechanisms and computational models of prediction and decision-making, we can identify specific dysfunctional processes across different contexts: individuals at risk for substance use disorder (SUD), those currently experiencing SUD, and individuals in abstinence.

5 Conclusion

Developing an integrated process model of interoception and SUD may provide a robust framework for understanding the intricate dynamics of substance use disorders across different stages. The model proposed here synergizes neuroscience-based conceptualizations of interoceptive processing with advanced computational approaches to prediction and decision-making, offering a sophisticated lens through which to examine the nuances of SUD in diverse contexts (Fig. 2).

For individuals at risk for SUD, the model hypothesizes that there are specific interoceptive vulnerabilities (e.g., reduced interoceptive precision or attenuated sensitivity) rooted in dysfunctional predictive coding in the presence of an exaggerated sensitivity to reward prediction errors. These vulnerabilities manifest as an increased anticipation of rewards, possibly driven by an overactive dopaminergic system, which predisposes these individuals to substance experimentation and potentially, addiction. This heightened reward sensitivity, coupled with impaired interoceptive awareness – where individuals have a diminished capacity to perceive internal bodily states – leads to an increased reliance on external cues, such as drug-related stimuli. Additionally, the faulty integration of interoceptive signals with cognitive and emotional processes impairs risk assessment and decision-making, further heightening the risk of initial substance use and the development of SUD.

During the active substance use disorder state, the model proposes that interoceptive dysfunctions are closely linked to compulsive drug-seeking behaviors, hyperactivation of the anterior insular cortex (aIC), and significant impairments in

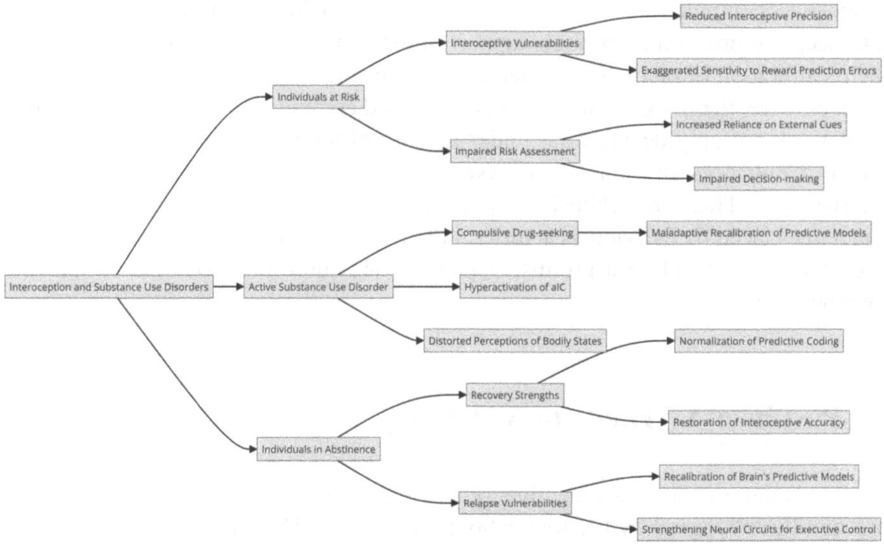

Fig. 2 State-dependent interoceptive process dysfunction in SUD

decision-making processes. These elements underscore a maladaptive recalibration of the brain's predictive models, where drugs are perceived as overwhelmingly valuable despite their detrimental consequences. The aIC's exaggerated focus on drug-related cues distorts perceptions of bodily states, propelling individuals toward substance use as a misguided corrective for perceived interoceptive discrepancies. This phase of SUD demonstrates a profound disruption in the computational balance between reward anticipation and punishment avoidance, particularly within key neural regions like the orbitofrontal cortex and the aIC.

For individuals in abstinence, the model focuses on identifying recovery strengths and relapse vulnerabilities. Central to this phase is the normalization of predictive coding and the restoration of interoceptive accuracy. The recalibration of the brain's predictive models to assess rewards and risks accurately, devoid of substance influence, is vital. Moreover, enhancing the ability to perceive and interpret bodily signals can reduce dependency on external cues, fostering emotional and physiological regulation. Strengthening neural circuits involved in executive control, particularly within the prefrontal cortex, is crucial for overriding compulsive behaviors and prioritizing long-term wellbeing.

Across these varied contexts, the integrated process model underscores the critical need for targeted interventions tailored to each stage of SUD. Preventive strategies for at-risk individuals might focus on bolstering interoceptive awareness and refining decision-making processes through cognitive-behavioral strategies and mindfulness practices. Treatment approaches for those actively struggling with SUD could aim to disrupt the cycle of compulsive drug seeking by rectifying distorted predictive coding and maladaptive learning processes, incorporating pharmacological treatments, cognitive behavioral therapy, and neurofeedback. For those in abstinence, continued support is necessary to reinforce healthy decision-making and interoceptive processing through support groups and therapies designed to sustain the recalibration of predictive models and enhance resilience against relapse.

In essence, the integration of neuroscience and computational models within this framework not only deepens our understanding of the multifaceted nature of SUD but also enhances the clinical approach toward personalized and stage-specific interventions. This comprehensive approach is instrumental in addressing the unique challenges faced by individuals at different points in their journey with substance use and recovery, providing a promising path toward more effective and sustainable treatment outcomes.

6 Post-hoc: A Tribute to A.D. Craig

The essential role of interoceptive processing in understanding SUD has been profoundly influenced by the foundational work of A. D. "Bud" Craig. His contributions have shaped the theoretical underpinnings of our current research approach, particularly in how we consider the insula's role in behavior. Firstly, Craig underscored the insula as a critical component of a regulatory homeostatic system, an insight he first proposed in 1996 (Craig 1996). This foundational idea emphasizes the insula's role in maintaining physiological equilibrium within the body, a concept that has significantly informed our understanding of its functions in emotional and substance-related contexts. Secondly, he posited that the insular cortex is integral to a network of structures involved in temporal processing. This perspective was further supported by empirical research from our group, notably by Wittmann et al. (2010), which elegantly explored how time perception is linked to interoceptive processes, underscoring the temporal dimension of interoceptive awareness. Thirdly, in his 2009 work (Craig 2009), Craig advanced the concept of the insula constructing a "global moment" that defines what he termed the "present state" of an individual. This idea of a synthesized moment in time is crucial for understanding how individuals integrate various physiological and environmental stimuli to form a coherent sense of the now, influencing their responses to substance-related cues. Lastly, Craig's analysis of the somatotopic organization of the left and right insula highlighted how different aspects of interoceptive processing are represented within these structures (Craig 2010). He viewed complex mental states as dynamic representations of activation levels and connectivity within this somatotopic map, providing a neuroanatomical basis for the diverse manifestations of SUD. Building on Craig's theoretical framework, our research continues to explore these complex mental states through sophisticated computational models and advanced neuroimaging techniques. By grounding our work in detailed neuroanatomical analyses, we aim to parse the dysfunctions characteristic of SUD more precisely. This approach not only honors Craig's legacy but also directs us toward the significant work still necessary to unravel the complex interplay between brain structure, function, and substance use pathology.

References

Augustine JR (1985) The insular lobe in primates including humans. Neurol Res 7(1): 2–10. PM:2860583 (NOT IN FILE)

Berk L, Stewart JL, May AC, Wiers RW, Davenport PW, Paulus MP, Tapert SF (2015) Under pressure: adolescent substance users show exaggerated neural processing of aversive interoceptive stimuli. Addiction 110(12):2025–2036. https://doi.org/10.1111/add.13090

Cabanac M (1971) Physiological role of pleasure. Science 173(2):1103–1107. PM:5098954 (NOT IN FILE)

Chawar C, Hillmer A, Sanger S, D'Elia A, Panesar B, Guan L, Xie DX, Bansal N, Abdullah A, Kapczinski F, Pare G, Thabane L, Samaan Z (2021) A systematic review of GWAS identified SNPs associated with outcomes of medications for opioid use disorder. Addict Sci Clin Pract 16(1):70. https://doi.org/10.1186/s13722-021-00278-y

Chen Y, Wang G, Zhang W, Han Y, Zhang L, Xu H, Meng S, Lu L, Xue Y, Shi J (2022) An orbitofrontal cortex-anterior insular cortex circuit gates compulsive cocaine use. Sci Adv 8(51): eabq5745. https://doi.org/10.1126/sciadv.abq5745

Cisler JM, Elton A, Kennedy AP, Young J, Smitherman S, Andrew James G, Kilts CD (2013) Altered functional connectivity of the insular cortex across prefrontal networks in cocaine addiction. Psychiatry Res 213(1):39–46. https://doi.org/10.1016/j.pscychresns.2013.02.007

Craig AD (1996) An ascending general homeostatic afferent pathway originating in lamina I. Prog Brain Res 107:225–242. PM:8782522 (NOT IN FILE)

Craig AD (2002) How do you feel? Interoception: the sense of the physiological condition of the body. Nat Rev Neurosci 3(8):655–666. PM:12154366 (NOT IN FILE)

Craig AD (2009) How do you feel - now? The anterior insula and human awareness. Nat Rev Neurosci 10(1):59–70. (NOT IN FILE)

Craig AD (2010) The sentient self. Brain Struct Funct 214(5–6):563–577. https://doi.org/10.1007/s00429-010-0248-y

Daldegan-Bueno D, Maia LO, Glass M, Jutras-Aswad D, Fischer B (2022) Co-exposure of cannabinoids with amphetamines and biological, behavioural and health outcomes: a scoping review of animal and human studies. Psychopharmacology (Berl) 239(5):1211–1230. https://doi.org/10.1007/s00213-021-05960-2

Daniel ML, Cocker PJ, Lacoste J, Mar AC, Houeto JL, Belin-Rauscent A, Belin D (2017) The anterior insula bidirectionally modulates cost-benefit decision-making on a rodent gambling task. Eur J Neurosci 46(10):2620–2628. https://doi.org/10.1111/ejn.13689

de la Fuente A, Sedeno L, Vignaga SS, Ellmann C, Sonzogni S, Belluscio L, Garcia-Cordero I, Castagnaro E, Boano M, Cetkovich M, Torralva T, Canepa ET, Tagliazucchi E, Garcia AM, Ibanez A (2019) Multimodal neurocognitive markers of interoceptive tuning in smoked cocaine. Neuropsychopharmacology 44(8):1425–1434. https://doi.org/10.1038/s41386-019-0370-3

Deol E, Siddiqui Z, Paracha A, Mujovic H, Abro Z, Chang SR, Abid A, Tyler JH, Rehman H, Bera S, Khan Z, Vasireddy S, Chalichama J, Fidahussain A, Paracha M, Ahmed Z, Alyasiry N, Engelhardt M, Kadariya B et al (2023) Exploring the link between ACEs and opioid use: a systematic review. J Opioid Manag 19(4):343–364. https://doi.org/10.5055/jom.2023.0791

Diseases GBD, Injuries C (2024) Global incidence, prevalence, years lived with disability (YLDs), disability-adjusted life-years (DALYs), and healthy life expectancy (HALE) for 371 diseases and injuries in 204 countries and territories and 811 subnational locations, 1990-2021: a systematic analysis for the global burden of disease study 2021. Lancet. https://doi.org/10.1016/S0140-6736(24)00757-8

Falconnier C, Caparros-Roissard A, Decraene C, Lutz PE (2023) Functional genomic mechanisms of opioid action and opioid use disorder: a systematic review of animal models and human studies. Mol Psychiatry 28(11):4568–4584. https://doi.org/10.1038/s41380-023-02238-1

Fedota JR, Stein EA (2015) Resting-state functional connectivity and nicotine addiction: prospects for biomarker development. Ann N Y Acad Sci 1349(1):64–82. https://doi.org/10.1111/nyas.12882

Gray MA, Critchley HD (2007) Interoceptive basis to craving. Neuron 54(2): 183–186. PM:17442239 (NOT IN FILE)

Guerin AA, Nestler EJ, Berk M, Lawrence AJ, Rossell SL, Kim JH (2021) Genetics of methamphetamine use disorder: a systematic review and meta-analyses of gene association studies. Neurosci Biobehav Rev 120:48–74. https://doi.org/10.1016/j.neubiorev.2020.11.001

Harle KM, Shenoy P, Paulus MP (2013) The influence of emotions on cognitive control: feelings and beliefs-where do they meet? Front Hum Neurosci 7:508. https://doi.org/10.3389/fnhum.2013.00508

Harle KM, Shenoy P, Stewart JL, Tapert SF, Yu AJ, Paulus MP (2014) Altered neural processing of the need to stop in young adults at risk for stimulant dependence. J Neurosci 34(13):4567–4580. https://doi.org/10.1523/JNEUROSCI.2297-13.2014

Harle KM, Stewart JL, Zhang S, Tapert SF, Yu AJ, Paulus MP (2015a) Bayesian neural adjustment of inhibitory control predicts emergence of problem stimulant use. Brain 138(Pt 11):3413–3426. https://doi.org/10.1093/brain/awv246

Harle KM, Yu AJ, Paulus MP (2019) Bayesian computational markers of relapse in methamphetamine dependence. Neuroimage Clin 22:101794. https://doi.org/10.1016/j.nicl.2019.101794

Harle KM, Zhang S, Ma N, Yu AJ, Paulus MP (2016) Reduced neural recruitment for Bayesian adjustment of inhibitory control in methamphetamine dependence. Biol Psychiatry Cogn Neurosci Neuroimaging 1(5):448–459. https://doi.org/10.1016/j.bpsc.2016.06.008

Harle KM, Zhang S, Schiff M, Mackey S, Paulus MP, Yu AJ (2015b) Altered statistical learning and decision-making in methamphetamine dependence: evidence from a two-armed bandit task. Front Psychol 6:1910. https://doi.org/10.3389/fpsyg.2015.01910

Ibanez C, Vicencio S, Quintanilla ME, Maldonado P (2021) Interoception and alcohol addiction: Vagotomy induces long-lasting suppression of relapse-type behavior. Addict Biol 26(1): e12836. https://doi.org/10.1111/adb.12836

Ibrahim C, Rubin-Kahana DS, Pushparaj A, Musiol M, Blumberger DM, Daskalakis ZJ, Zangen A, Le Foll B (2019) The insula: a brain stimulation target for the treatment of addiction. Front Pharmacol 10:720. https://doi.org/10.3389/fphar.2019.00720

Kosten TR, Petrakis IL (2021) The hidden epidemic of opioid overdoses during the coronavirus disease 2019 pandemic. JAMA Psychiatry 78(6):585–586. https://doi.org/10.1001/jamapsychiatry.2020.4148

Kruschwitz JD, Kausch A, Brovkin A, Keshmirian A, Paulus MP, Goschke T, Walter H (2019) Self-control is linked to interoceptive inference: craving regulation and the prediction of aversive interoceptive states induced with inspiratory breathing load. Cognition 193:104028. https://doi.org/10.1016/j.cognition.2019.104028

Kulkarni KR, O'Brien M, Gu X (2023) Longing to act: Bayesian inference as a framework for craving in behavioral addiction. Addict Behav 144:107752. https://doi.org/10.1016/j.addbeh.2023.107752

Lavalley CA, Hakimi N, Taylor S, Kuplicki R, Forthman KL, Stewart JL, Paulus MP, Khalsa SS, Smith R (2023) Transdiagnostic failure to adapt interoceptive precision estimates across affective, substance use, and eating disorders: a replication study. medRxiv. https://doi.org/10.1101/2023.10.11.23296870

Leganes-Fonteneau M, Cheang Y, Lam Y, Garfinkel S, Duka T (2019) Interoceptive awareness is associated with acute alcohol-induced changes in subjective effects. Pharmacol Biochem Behav 181:69–76. https://doi.org/10.1016/j.pbb.2019.03.007

Livneh Y, Ramesh RN, Burgess CR, Levandowski KM, Madara JC, Fenselau H, Goldey GJ, Diaz VE, Jikomes N, Resch JM, Lowell BB, Andermann ML (2017) Homeostatic circuits selectively gate food cue responses in insular cortex. Nature 546(7660):611–616. https://doi.org/10.1038/nature22375

Lodge W 2nd, Kelly PJA, Napoleon S, Plezia S, Mimiaga MJ, Biello KB (2024) Prevalence of methamphetamine use among gay, bisexual and other men who have sex with men: a systematic review and meta-analysis. Int J Drug Policy 123:104271. https://doi.org/10.1016/j.drugpo.2023.104271

Lovelock DF, Tyler RE, Besheer J (2021) Interoception and alcohol: mechanisms, networks, and implications. Neuropharmacology 200:108807. https://doi.org/10.1016/j.neuropharm.2021.108807

May AC, Jacobus J, Stewart JL, Simmons AN, Paulus MP, Tapert SF (2020) Do adolescents use substances to relieve uncomfortable sensations? A preliminary examination of negative reinforcement among adolescent cannabis and alcohol users. Brain Sci 10(4). https://doi.org/10.3390/brainsci10040214

Migliorini R, Stewart JL, May AC, Tapert SF, Paulus MP (2013) What do you feel? Adolescent drug and alcohol users show altered brain response to pleasant interoceptive stimuli. Drug Alcohol Depend 133(2):661–668. https://doi.org/10.1016/j.drugalcdep.2013.08.015

Minozzi S, Saulle R, Amato L, Traccis F, Agabio R (2024) Psychosocial interventions for stimulant use disorder. Cochrane Database Syst Rev 2(2):CD011866. https://doi.org/10.1002/14651858.CD011866.pub3

Naqvi NH, Bechara A (2010) The insula and drug addiction: an interoceptive view of pleasure, urges, and decision-making. Brain Struct Funct 214(5–6):435–450. https://doi.org/10.1007/s00429-010-0268-7

Naqvi NH, Rudrauf D, Damasio H, Bechara A (2007) Damage to the insula disrupts addiction to cigarette smoking. Science 315(5811):531–534. PM:17255515 (NOT IN FILE)

Nayok SB, Sreeraj VS, Shivakumar V, Venkatasubramanian G (2023) A primer on Interoception and its importance in psychiatry. Clin Psychopharmacol Neurosci 21(2):252–261. https://doi.org/10.9758/cpn.2023.21.2.252

Parlier-Ahmad AB, Eglovitch M, Legge C, Keyser-Marcus LA, Bjork JM, Adams A, Ramey T, Moeller FG, Martin CE (2023) Development and clinical feasibility study of a brief version of an addiction-focused phenotyping battery in females receiving buprenorphine for opioid use disorder. Brain Behav 13(8):e3128. https://doi.org/10.1002/brb3.3128

Paulus MP (2007) Decision-making dysfunctions in psychiatry – altered homeostatic processing? Science 318(5850):602–606. https://doi.org/10.1126/science.1142997

Paulus MP, Stewart JL (2014) Interoception and drug addiction. Neuropharmacology 76 (Pt B):342–350. https://doi.org/10.1016/j.neuropharm.2013.07.002

Paulus MP, Stewart JL (2020) Neurobiology, clinical presentation, and treatment of methamphetamine use disorder: a review. JAMA Psychiatry. https://doi.org/10.1001/jamapsychiatry.2020.0246

Paulus MP, Stewart JL, Haase L (2013) Treatment approaches for interoceptive dysfunctions in drug addiction. Front Psych 4:137. https://doi.org/10.3389/fpsyt.2013.00137

Paulus MP, Tapert SF, Schulteis G (2009) The role of interoception and alliesthesia in addiction. Pharmacol Biochem Behav 94(1):1–7. https://doi.org/10.1016/j.pbb.2009.08.005

Santo T Jr, Gisev N, Campbell G, Colledge-Friday S, Wilson J, Tran LT, Lynch M, Martino-Burke D, Taylor S, Degenhardt L (2024) Prevalence of comorbid substance use disorders among people with opioid use disorder: a systematic review & meta-analysis. Int J Drug Policy 128:104434. https://doi.org/10.1016/j.drugpo.2024.104434

Schwartenbeck P, FitzGerald TH, Mathys C, Dolan R, Wurst F, Kronbichler M, Friston K (2015) Optimal inference with suboptimal models: addiction and active Bayesian inference. Med Hypotheses 84(2):109–117. https://doi.org/10.1016/j.mehy.2014.12.007

Smith R, Feinstein JS, Kuplicki R, Forthman KL, Stewart JL, Paulus MP, Tulsa I, Khalsa SS (2021) Perceptual insensitivity to the modulation of interoceptive signals in depression, anxiety, and substance use disorders. Sci Rep 11(1):2108. https://doi.org/10.1038/s41598-021-81307-3

Sridharan D, Levitin DJ, Menon V (2008) A critical role for the right fronto-insular cortex in switching between central-executive and default-mode networks. Proc Natl Acad Sci U S A 105(34):12569–12574. https://doi.org/10.1073/pnas.0800005105

Stewart JL, Khalsa SS, Kuplicki R, Puhl M, Investigators T, Paulus MP (2020) Interoceptive attention in opioid and stimulant use disorder. Addict Biol 25(6):e12831. https://doi.org/10.1111/adb.12831

Stewart JL, May AC, Tapert SF, Paulus MP (2015) Hyperactivation to pleasant interoceptive stimuli characterizes the transition to stimulant addiction. Drug Alcohol Depend 154:264–270. https://doi.org/10.1016/j.drugalcdep.2015.07.009

Strigo IA, Craig AD (2016) Interoception, homeostatic emotions and sympathovagal balance. Philos Trans R Soc Lond B Biol Sci 371(1708). https://doi.org/10.1098/rstb.2016.0010

Sugarman A, Vittitow A, Cheng A, Malone M, McDonald R, Pace N, Williams O, Tofighi B, McNeely J, Schatz D, Roberts T, Hey SP, Garrity K, Lindquist K, Lee JD (2022) Opioid use disorder treatments: an evidence map. Drug Alcohol Depend 241:109657. https://doi.org/10.1016/j.drugalcdep.2022.109657

Torrealba F, Madrid C, Contreras M, Gomez K (2017) Plasticity in the interoceptive system. Adv Exp Med Biol 1015:59–74. https://doi.org/10.1007/978-3-319-62817-2_4

Wisniewski P, Maurage P, Jakubczyk A, Trucco EM, Suszek H, Kopera M (2021) Alcohol use and interoception – a narrative review. Prog Neuropsychopharmacol Biol Psychiatry 111:110397. https://doi.org/10.1016/j.pnpbp.2021.110397

Wittmann M, van Wassenhove V, Craig AD, Paulus MP (2010) The neural substrates of subjective time dilation. Front Hum Neurosci 4:2. https://doi.org/10.3389/neuro.09.002.2010

Zhang R, Volkow ND (2019) Brain default-mode network dysfunction in addiction. Neuroimage 200:313–331. https://doi.org/10.1016/j.neuroimage.2019.06.036

Zhang Y, Mei W, Zhang JX, Wu Q, Zhang W (2016) Decreased functional connectivity of insula-based network in young adults with internet gaming disorder. Exp Brain Res 234(9):2553–2560. https://doi.org/10.1007/s00221-016-4659-8

Interoception in Adolescence: Impacts on Mental Health and Adaptive Functioning

April C. May and Susan Tapert

Contents

Abstract Interoception plays a critical role in emotion regulation and mental health during adolescence, a critical period marked by profound biological, cognitive, and emotional changes. Variability in interoceptive processing during adolescence is shaped by biological, hormonal, and psychosocial factors, with implications for both resilience and vulnerability to affective disorders such as anxiety, depression, disordered eating, and substance use. This chapter reviews the trajectory of interoceptive development in adolescence, emphasizing its role in shaping emotional and behavioral outcomes. Interventions aimed at improving interoceptive awareness, including mindfulness practices and body-awareness training, are also reviewed as promising strategies to support emotional resilience and mental well-being during this formative stage. Understanding the complexities of interoception in adolescence provides a foundation for advancing research and interventions that promote adaptive functioning and long-term mental health.

Keywords Adolescence · Development · Interoception · Interventions · Mental health

A. C. May (✉) and S. Tapert
Department of Psychiatry, University of California, San Diego, CA, USA
e-mail: acmay@ucsd.edu; sftapert@ucsd.edu

Curr Topics Behav Neurosci (2025) 70: 179–197
https://doi.org/10.1007/7854_2025_580
Published Online: 18 April 2025

1 Introduction

Adolescence, defined by the World Health Organization as the period of life from 10 to 19, is a developmental stage following the onset of puberty marked by significant biological, cognitive, and social shifts (Dehne and Riedner 2001). These changes coincide with documented increases in rates of psychopathology and risky behaviors (e.g., substance use, reckless driving). Interoception, or the sensing of internal bodily states, is a fundamental component of human perception and plays a critical role in the regulation of emotions, motivation, and behavior (Barrett and Simmons 2015; Craig 2002). This ability to perceive physiological signals, such as heart rate, hunger, and thirst, also significantly influences both physical and mental health and is critical for higher-order cognitive processes (Critchley et al. 2004). Research directly examining how interoceptive abilities fluctuate during adolescence or how these changes intersect with broader developmental processes is limited. However, indirect evidence for altered interoception during adolescence comes from research linking interoception with emotional regulation and decision-making processes, both of which are undergoing significant maturation during adolescence. Here we hypothesize that developmental fluctuations in interoceptive abilities during adolescence are reflective of developmental changes in the brain, and in turn contribute to the increased rates of psychopathology and risk-taking observed during this developmental stage via influence on emotion regulation and decision-making processes (see Fig. 1 for Conceptual Model).

Adolescents experience fluctuations in their ability to accurately perceive and regulate internal signals, in part due to ongoing brain development and underdeveloped neural integration between regions associated with interoception, such as the insula, anterior cingulate cortex, and frontal cortex (Blakemore and Mills 2014;

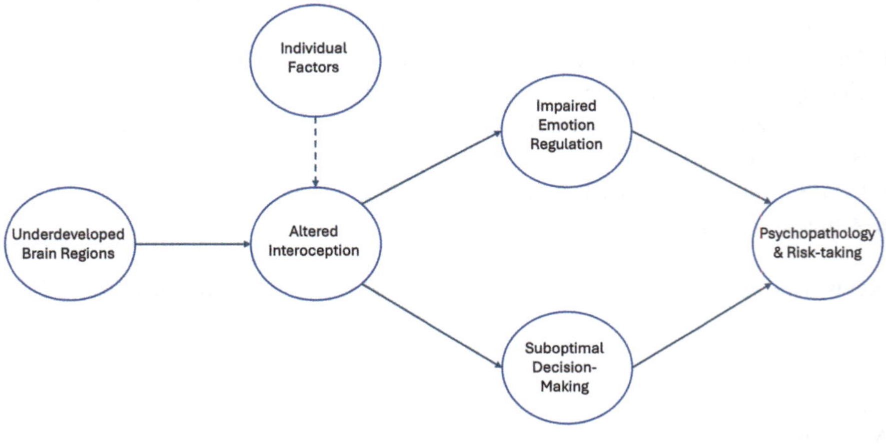

Fig. 1 Interconnected pathways linking adolescent brain development, interception, and behavioral outcomes. Proposed model illustrating how underdeveloped brain regions in adolescence contribute to altered interception, impaired emotion regulation, and suboptimal decision-making, which together drive psychopathology and risk-taking behaviors

Murphy et al. 2017). Adolescents' misinterpretation (interoceptive accuracy) or failure to recognize these internal bodily signals (interoceptive sensitivity) may result in disruptions to effective emotion regulation and poor decision-making, leaving some individuals particularly vulnerable to developing mental health issues and engaging in risky behaviors (Casey and Jones 2010; Casey et al. 2011). Over time, the maturation of these neural systems is critical for the ability to discern and respond appropriately to interoceptive signals, laying the groundwork for adaptive behavior, emotional resilience, and overall well-being (Khalsa et al. 2018; Murphy et al. 2017; Nord and Garfinkel 2022).

This chapter examines the developmental trajectory of interoceptive processing in adolescence, emphasizing its relationship with emotion regulation and decision-making, and potential influence on mental health and substance use in adolescence, reviewing empirical studies with adolescent samples when available. Additionally, we explore individual differences that may shape interoceptive awareness in adolescence and review current research on potential interventions to promote adaptive interoceptive processing. We argue the importance of understanding the intricacies of interoception during adolescence and the need for additional rigorous research so that we can better support young people in developing the resilience and emotional regulation skills necessary for well-being across the lifespan.

2 Developmental Trajectory of Interoception in Adolescence

During adolescence, substantial changes occur in brain regions critical for interoceptive processing, including the insular cortex, anterior cingulate cortex, frontal cortex, and somatosensory regions, which together support the monitoring, integration, and interpretation of bodily signals (Casey et al. 2008; Craig 2009). (See Chapter "Insula and Interoceptive Pathways" for a more in-depth discussion of the neural pathways underlying interoception.) These neurobiological changes lay the foundation for the adolescent's ability to detect and interpret internal signals, a skill essential for adaptive self-regulation and emotional awareness.

The insular cortex, a primary hub for interoceptive processing, is responsible for receiving and integrating signals from the body and converting them into conscious feelings of physiological states (Critchley et al. 2004). Neuroimaging studies indicate that the insula undergoes significant structural and functional changes throughout adolescence, with age-related increases in gray matter volume followed by functional changes that enhance specificity (Gennatas et al. 2017; Li et al. 2017; Smith et al. 2014). These changes are thought to improve the brain's efficiency in processing interoceptive information, allowing adolescents to gain better awareness of their bodily states. The maturation of the anterior insula, in particular, has been associated with enhanced interoceptive accuracy, enabling adolescents to more accurately perceive and interpret bodily cues, which is crucial for distinguishing between different emotional experiences (Kleckner et al. 2017). Concurrently, connectivity between the insula and prefrontal areas increases, reflecting a greater capacity to regulate interoceptive signals and utilize them in decision-making and self-regulation processes (Li et al. 2017).

The anterior cingulate cortex (ACC) also undergoes significant changes in adolescence, supporting the integration of interoceptive signals with affective and cognitive information (Steinberg 2005). Enhanced connectivity between the ACC, insula, and prefrontal cortex during this period allows for a more accurate appraisal of interoceptive cues and their emotional implications, aiding in the transition from reactive to more reflective forms of self-regulation (Kelly et al. 2009). This neurobiological maturation enables individuals to contextualize their physiological states within their social and environmental circumstances, enhancing their ability to manage emotions and behavior in complex social interactions (Nelson et al. 2005; Pfeifer and Blakemore 2012).

Only two known studies have directly examined neural activity elicited during interoceptive processing from a developmental perspective. First, within a sample of adolescents (ages 15–17; $n = 16$), young adults (ages 20–28; $n = 22$), and mature adults (ages 29–55; $n = 20$), neural activity during interoceptive processing of a pleasant soft touch was compared revealing enhanced bilateral insula activity in adolescent participants compared to the other age groups (May et al. 2014). More recently a pilot study demonstrated that activation within the ACC and orbital and prefrontal cortices during interoceptive processing (heartbeat detection task) was positively associated with age in a sample of 11 youth ages 6–17 (Klabunde et al. 2019). While various methodological limitations (e.g., small sample size, wide age range, cross-sectional design) limit the strength of these studies' conclusions, these findings provide preliminary evidence of functional changes in the interoceptive network with developmental age.

As the brain regions involved in interoceptive processing continue to mature, adolescents gain a more nuanced ability to interpret internal bodily cues. This maturation process not only supports adaptive behavior and social functioning but also prepares adolescents to navigate the emotional complexities of this developmental stage. By understanding the neural and developmental underpinnings of interoception, researchers and clinicians can better appreciate its impact on adolescent well-being.

3 Interoception and Mental Health Outcomes in Adolescence

The ability to regulate emotions is a complex, multi-faceted skill essential for adaptive functioning, particularly during adolescence (Casey et al. 2008; Steinberg 2005). Current research indicates that interoceptive sensitivity and accuracy play a critical role in one's ability to recognize, understand, and modulate emotions.

Improved interoceptive awareness has been associated with greater emotional insight and adaptive regulation strategies, while disrupted interoceptive processing can lead to challenges in managing emotions and is linked to affective disorders (Garfinkel et al. 2015; Paulus and Stein 2010).

One of the primary ways interoception supports emotion regulation is by facilitating emotional awareness, or the ability to identify and label emotional states based on bodily signals. Individuals who demonstrate higher interoceptive accuracy are better able to detect early physiological cues associated with emotional changes, which can act as signals for deploying regulatory strategies (Tsakiris and Critchley 2016). For example, the awareness of increased heart rate may signal the onset of anxiety or excitement, prompting individuals to engage in coping mechanisms such as cognitive reappraisal or relaxation techniques (Critchley and Garfinkel 2017). Such accurate interoceptive feedback is particularly important during adolescence, as young people begin to develop more sophisticated emotion regulation skills that rely on the ability to appraise and modify emotional experiences based on internal and external cues (Gross 2015; Shenhav et al. 2013). Interoceptive sensitivity can thus support a proactive approach to emotion regulation, allowing adolescents to manage emotional responses before they escalate.

Conversely, low interoceptive accuracy is often associated with emotional dysregulation, as individuals may struggle to distinguish between different emotional states or react disproportionately to minor physiological changes (Domschke et al. 2010; Pollatos et al. 2007). Adolescents with reduced interoceptive awareness may misinterpret benign bodily sensations as signs of negative emotional states, leading to heightened emotional reactivity or avoidance behaviors. The capacity to accurately interpret interoceptive signals is therefore crucial for distinguishing between varying emotional states and for adapting regulatory strategies accordingly.

Anxiety Disorders Anxiety disorders are among the most prevalent mental health conditions in adolescence (Hamm et al. 2014). Studies with adult samples indicate that anxiety disorders are closely linked to interoceptive processing, with disrupted interoceptive awareness often serving as both a symptom and contributing factor (Domschke et al. 2010). This suggests that adolescents with heightened interoceptive sensitivity, or a tendency to notice subtle physiological changes, may interpret benign bodily sensations as signs of impending danger, leading to heightened anxiety. This heightened sensitivity can create a feedback loop in which anxious interpretations of bodily cues amplify physiological arousal, reinforcing anxious thoughts and behaviors (Garfinkel et al. 2016).

Evidence to support this hypothesis comes from an examination of interoception in adolescents using resting state functional connectivity. This study revealed significant (1) hyperconnectivity between left insula and right amygdala compared to controls and (2) hypoconnectivity between prefrontal cortex and left amygdala in adolescents with anxiety disorders (mean age = 13.9; $n = 33$) compared to control participants (mean age = 14.6; $n = 23$; (Hamm et al. 2014)). Given the role of the amygdala in emotion processing and regulation, this finding suggests that adolescents with anxiety disorders may attribute an exaggerated and use this incorrect interpreted information when attempting to restore homeostasis via interoceptive processes. Similar alterations in connectivity between the insula and amygdala, suggesting dysfunction in the integration of interoceptive information, were reported in a sample of 15 adolescents with generalized anxiety disorder relative to 20 control participants (ages 12–17, mean age = 14.8; (Roy et al. 2013)). Small sample sizes limit the robustness of the conclusions drawn from these studies; nevertheless, these findings demonstrate an association between altered neural activity associated with interoceptive brain regions and anxiety disorders in adolescents.

These findings demonstrate a relationship between interoceptive processing and anxiety disorders in adolescence, suggesting that disruptions in the neural integration of interoceptive signals may contribute to heightened physiological arousal and maladaptive emotional regulation. The observed hyperconnectivity between the insula and amygdala, alongside hypoconnectivity between the prefrontal cortex and amygdala, supports the hypothesis that adolescents with anxiety disorders may struggle to accurately interpret bodily signals, leading to exaggerated threat appraisals and reinforcing anxiety symptoms.

Depression Research suggests that individuals with depression symptomatology often exhibit either hypo- or hyper-awareness of interoceptive signals, contributing to an overall sense of disconnection from or over-focus on their bodily states (Murphy et al. 2017; Paulus and Stein 2010). Hypo-awareness, or reduced sensitivity to bodily cues, has been linked to a blunted emotional experience, a common symptom of depression, which may make it difficult for individuals to engage with positive stimuli (Pollatos et al. 2007). On the other hand, hypersensitivity to negative physiological states, such as fatigue or bodily tension, can exacerbate ruminative thinking patterns, reinforcing the cycle of depressive symptoms (Dunn et al. 2010).

Research supporting the proposed connection between interoception and depressive symptomatology in adolescents primarily comes from self-report studies. For example, a sample of 37 youth (ages 8–13) demonstrated that self-reported interoceptive accuracy was significantly negatively associated with parent-reported youth internalizing symptoms ($\beta = -0.527$; (Brand et al. 2024)). Self-reported interoceptive deficits have also been found to predict suicidal ideation 6 months later in a sample of 436 adolescents (mean age = 13; (Perkins et al. 2021)). Together, these findings underscore the critical role of interoception in shaping the emotional and cognitive experiences associated with depression, particularly during adolescence.

The dual nature of interoceptive disruption, encompassing both hypo- and hyper-awareness, highlights the complex interplay between bodily awareness, emotional regulation, and symptomatology in depression. While much of the evidence relies on self-report measures, the observed associations between interoceptive deficits and internalizing symptoms, including suicidal ideation, point to interoception as a potential target for early intervention. Future research is needed to further elucidate the mechanisms underlying these relationships.

Disordered Eating Disordered eating behaviors, which often emerge during adolescence, have been consistently linked to disruptions in interoceptive processing, particularly in relation to hunger, fullness, and body image (Pollatos and Georgiou 2016). Impairments in interoceptive awareness, including sensations of hunger and satiety, are thought to contribute to disordered eating patterns (Khalsa et al. 2015). For instance, adults with anorexia nervosa often display diminished sensitivity to hunger cues, leading to restrictive eating behaviors, while those with bulimia nervosa may have trouble recognizing satiety cues, contributing to binge eating episodes (Herbert and Pollatos 2014; Pollatos et al. 2008). Heightened interoceptive sensitivity can also contribute to negative body awareness and distorted body image, which is particularly concerning during adolescence – a critical period for body image development. During this time, individuals may become overly attuned to subtle changes in their body size or shape, potentially leading to unhealthy weight control behaviors (Voelker et al. 2015; Zanetti et al. 2013).

In a sample of adolescents, insula activation during a decision-making task (Risky-Gains) negatively correlated with interoceptive sensitivity and positively correlated with external eating (i.e., eating in response to external food-related cues rather than internal signals of hunger) in adolescents with excess weight (body mass index >25). In contrast, healthy weight adolescents exhibited the opposite pattern, with positive correlations between insula activity and interoceptive sensitivity and negative correlation with external eating ($N = 54$, ages 12–18; (Mata et al. 2015)). Similarly, interoceptive dysfunction has been implicated in restrictive eating disorders among adolescents. A study of weight-restored female patients with anorexia nervosa (ages 13–24) found reduced middle insula activity during a gastric interoception attention task compared to healthy controls (partial $\eta^2 = 0.28$), suggesting persistent disruptions in interoceptive awareness even after weight restoration (Kerr et al. 2016). Supporting these findings, self-reported interoceptive deficits in a larger sample of 436 adolescents (mean age = 13) significantly predicted eating disorder symptoms 6 months later (Perkins et al. 2021). Variability in age range, sample size, and the interoceptive measures used limits the generalizability of these findings. Nevertheless, these results underscore the importance of considering interoceptive processes in understanding the development and maintenance of disordered eating behaviors during adolescence and highlight the need for additional research.

Interoception appears to play a complex and multifaceted role in adolescent mental health, acting as both a risk and protective factor depending on the individual's interoceptive accuracy and sensitivity. Disrupted interoception appears to be linked to common adolescent mental health disorders, including anxiety, depression, and disordered eating, where misinterpretations or avoidance of bodily cues can perpetuate maladaptive behaviors and emotional dysregulation. Limited existing research provides evidence of this association, yet further large-scale studies are crucial to directly examine the role of interoception in the development of psychopathology in adolescents.

4 Interoception and Adolescent Substance Use

Adolescence is characterized by heightened risk-taking and experimentation, including the initiation of substance use, which can have long-term implications for health and behavior. Adolescents often exhibit increased sensitivity to internal bodily signals, such as those related to emotional arousal, which may disrupt balanced decision-making by amplifying immediate reward-seeking over long-term considerations (Craig 2009; Paulus and Stewart 2014). While accurate interoceptive awareness can improve decision-making by enhancing the understanding of physiological markers of stress or reward, it follows that atypical interoceptive accuracy is associated with poor decision-making and heightened impulsivity (i.e., unplanned response to stimuli without consideration for the consequences), potentially due to over-reliance on interoceptive cues or misinterpretation of their salience (Murphy et al. 2017). In other words, if an individual is unable to detect, discriminate, and appropriately interpret the sensations they are experiencing (i.e., poor interoception), they may engage in impulsive actions as a maladaptive means of dealing with this misunderstood arousal. Adolescents may be particularly vulnerable to deficits in these processes due to ongoing maturation of interoceptive networks and the prefrontal cortex, which governs inhibitory control and executive functioning (Casey et al. 2008).

Research in adult samples indicates that individuals with low interoceptive accuracy, or difficulty interpreting bodily signals, are more likely to report using substances to regulate emotions, suggesting that substance use may serve as an external strategy to manage internal discomfort (Paulus and Stewart 2014). (See Chapter "Interoception, Emotion, Addiction, & Treatment Targets" for further review of interoception in substance use.) Individuals with higher interoceptive accuracy appear to be less impulsive and better able to modulate food-related cravings (Kruschwitz et al. 2019; Walter et al. 2020). Similarly, lower interoceptive accuracy was found to correlate with higher trait impulsivity levels, which is associated with an increased likelihood of experimenting with substances (Baiano et al. 2021; Herman et al. 2019). Although similar studies have not been conducted in adolescent samples, it follows that as brain regions underlying interoception develop during adolescence, these changes will coincide with fluctuation in interoceptive abilities, resulting in increased impulsivity and engagement in risky behaviors, such as substance use, as they seek external means to manage internal uncertainty (De Witte et al. 2016).

Research directly investigating the relationship between interoception and substance use specifically within adolescent samples is limited. One study found that adolescents (ages 15–17, $n = 18$) with alcohol or cannabis substance use disorder (SUD) exhibited increased activation, compared to control participants ($n = 15$), in regions of the insular cortex when experiencing an aversive interoceptive breathing load (Berk et al. 2015). In a related study, adolescents (ages 15–17) with SUD ($n = 15$) and controls ($n = 17$) differed in their response to positively valenced interoceptive stimulation with SUD participants showing reduced insula activation while experiencing a pleasant touch (*ps* =/< 0.02). Greater reduction in insular activity was also associated with more recent cannabis use (Migliorini et al. 2013). Together, these studies suggest that interoceptive processing in adolescents with SUD is altered compared to adolescents without problematic substance use. An additional study exploring neural connectivity in adolescents (average age = 17.6) with problematic substance use (alcohol, cannabis abuse/dependence) found that these individuals (n = 18) exhibited reduced functional connectivity within a subnetwork primarily composed of the insula (p = 0.037) compared to control participants (n = 96). Additionally, functional connectivity of this insula subnetwork was significantly negatively associated with frequency of alcohol, tobacco, and cannabis use (all *ps* </= 0.001; (Rakesh et al. 2021)).

These studies collectively suggest that adolescents who use drugs and alcohol exhibit neural alterations in interoceptive processing, particularly involving the insular cortex, and these alterations are associated with substance use characteristics (e.g., frequency, recency of use). These findings align with theories proposing that inaccurate interpretation or processing of interoceptive stimuli can result in distorted information being utilized during decision-making, potentially leading to risky behaviors such as substance use. However, the directionality (hyper-vs. hypoactivation of the insula) and causality of these relationships remain unclear in part due to methodological limitations, including small sample sizes, cross-sectional designs, and inconsistencies in age range, which may overlook variations in interoceptive processing as adolescents develop (i.e., changes in early vs. late adolescence). Despite these limitations, the emerging evidence highlights the relationship between interoception and substance and supports the hypothesis that altered interoceptive processing contributes to poor decision-making and substance use behaviors in adolescence.

5 Individual Differences in Interoceptive Processing

Interoception exhibits substantial variability across individuals, likely shaped by biological, hormonal, and sociocultural factors. In adolescence, these differences are particularly pronounced, as individuals undergo rapid physiological and psychological changes that impact interoceptive processing. Researchers have proposed that variability in interoceptive processing among adolescents can largely be attributed to differences in biological sex and the hormonal changes associated with puberty (Murphy et al. 2019).

The onset of puberty brings about significant hormonal shifts, including increases in testosterone, estrogen, and progesterone, which influence brain areas involved in interoceptive processing, such as the insula and anterior cingulate cortex (Blakemore et al. 2010). These hormonal changes may modulate interoceptive sensitivity, leading to fluctuations in bodily awareness and emotional responses. Research indicates that females often report higher interoceptive awareness yet poorer interoceptive accuracy than their male counterparts (Grabauskaitė et al. 2017; Prentice and Murphy 2022). This heightened sensitivity in females likely reflects an interaction between biological factors, such as physical changes and hormonal fluctuations, and sociocultural norms that encourage greater body awareness in females (Longarzo et al. 2021; Ma-Kellams et al. 2024).

Sociocultural factors can also influence interoceptive processing, though research in adolescence is limited. Studies in adult populations suggest that cultural and environmental factors can shape interoceptive accuracy. For example, adults from Western cultures have been found to demonstrate greater interoceptive accuracy than individuals from non-Western cultures, potentially reflecting cultural differences in bodily awareness and emotional expression (Ma-Kellams et al. 2024). Additionally, experiences of racial discrimination have also been proposed to negatively impact interoceptive processes with recent work revealing an association between racial discrimination and reduced functional connectivity in interoceptive brain regions, including the insula and medial prefrontal cortex, within a sample of 72 Black women (Elbasheir et al. 2024; Fani and Khalsa 2023). Early life adversity has also been linked to disruptions in interoception. Within a sample of 66 young adults, childhood trauma was negatively associated with interoceptive accuracy after experiencing an acute stressor (cold pressor; (Schaan et al. 2019)). Although these findings stem from adult samples, they highlight the potential influence of sociocultural and environmental factors on interoception and underscore the need for research focused on adolescents.

In summary, individual differences in interoceptive processing during adolescence likely arise from a complex interplay of biological, hormonal, and sociocultural factors with biological sex and pubertal hormones playing central roles, shaping sensitivity and awareness in distinct ways. However, sociocultural influences, such as cultural norms and experiences of adversity, may further shape interoceptive development, pointing to the need for tailored approaches to support adaptive interoceptive awareness in youth. Future research should explore these influences in adolescent samples to provide a more comprehensive understanding of how individual differences impact the development of interoception.

6 Potential Interventions and Applications: Leveraging Interoception for Emotional Resilience and Mental Health

Interventions targeting interoceptive awareness offer promising pathways to enhance emotional resilience and mental health in adolescents. Research suggests that interventions such as mindfulness-based practices, body-awareness training, and biofeedback lead to enhanced interoceptive accuracy and foster healthier relationships with internal bodily signals, providing individuals with tools to manage stress, anxiety, and other mental health challenges (Lazzarelli et al. 2024). By helping young people develop a more accurate and balanced awareness of their internal states, these approaches aim to support mental well-being and reduce the risk of affective disorders and risky behaviors during this critical developmental period.

Mindfulness-based interventions (MBIs) are among the most widely studied approaches to improving interoceptive awareness and emotion regulation although studies with adolescent samples are limited. Mindfulness involves cultivating a nonjudgmental awareness of present-moment experiences, including internal bodily sensations, emotions, and thoughts. There is evidence to suggest that MBIs, such as mindfulness-based stress reduction (MBSR) and mindfulness-based cognitive therapy (MBCT), enhance interoceptive accuracy by helping individuals attend to bodily sensations with curiosity and acceptance rather than avoidance or judgment (Hölzel et al. 2011). Specifically among adolescents, individuals who practice mindfulness appear to develop greater awareness and tolerance of physiological cues associated with emotional responses, such as a racing heart or muscle tension, i.e., enhanced interoceptive awareness, enabling them to adaptively respond to stressors (Tan and Martin 2015).

While direct examinations of changes in interoception following MBIs in adolescents are limited, research with adult samples suggests that MBIs can directly alter interoception. (See Chapter "Interceptive Awareness and Its Importance in Chronic Pain and Psychopathology" for a more in-depth discussion of interoceptive interventions with adults.) Research with adolescent samples has primarily focused on examining changes in mindfulness and psychological well-being following MBIs and has not directly measured interoceptive awareness or accuracy. A systematic review of 20 studies (*N*s ranged from 4 to 246) conducted with youth (<18 years of age) found that, compared to active control comparison conditions, MBIs significantly reduced psychological symptoms (*del* $= 0.37$; $p < 0.0001$) suggesting a potential link between enhanced interoceptive awareness (via mindfulness training) and improved mental health (Zoogman et al. 2015). Similarly, 108 adolescents (ages 13–18) recruited from community mental health clinics were randomly assigned to receive either treatment as usual (TAU) or TAU plus a 5-week mindfulness training (Tan and Martin 2015). Results revealed increased levels of mindfulness mediated the relationship between intervention and improved mental health (standardized $\beta = -0.42, p < 0.001$). Other work has delved into the underlying neural differences associated with mindfulness and emotional well-being. Among 82 adolescents, higher dispositional mindfulness was found to be associated with reduced thinning of the left anterior insula ($b = -0.263$, $p = 0.017$), a marker of enhanced functioning, from the age of 16–19 (Friedel et al. 2015). An association between changes in insular connectivity and self-reported emotional problems ($F = 5.43, p = 0.022$) was also observed within a sample of 100 healthy adolescents (ages 14–18) following a 12-week MBI (compared to wait-list control) as well as an overall increase in connectivity between interoceptive network brain regions (Tymofiyeva et al. 2024).

Overall, these studies demonstrate a small to medium effect of MBIs on mindfulness and emotional well-being. We suggest that change in interoceptive ability is the mechanism underlying this relationship, given research in adult samples demonstrating that MBIs lead to increased interoception (Lima-Araujo et al. 2022; Weng et al. 2021). Although further research with MBIs in adolescents directly measuring interoception are necessary, current findings suggest that MBIs can lead to beneficial outcomes for adolescents. Integrating such mindfulness practices into educational and clinical settings could aid adolescents in developing lifelong skills for managing emotional reactivity and building self-regulation capacities.

Body-awareness practices also hold promise for enhancing interoceptive sensitivity and emotional resilience. Body-awareness practices such as yoga emphasize attunement to physical sensations and the cultivation of a balanced awareness of one's body (Mehling et al. 2011). Research suggests that body-awareness practices can also lead to improvements in emotional regulation and mental well-being by helping individuals better understand the connection between their bodily states and emotional experiences (Felver et al. 2015; Menezes et al. 2015). Improvements in emotional regulation and psychopathology symptomatology following yoga interventions have been reported in adolescent samples. Among 27 adolescents (ages 15–17), those randomized to participate in a 16-week yoga intervention demonstrated increased emotion regulation compared to those who participated in a regular physical education class ($\eta^2 = 0.19$, $p = 0.01$; (Daly et al. 2015)). Significant increases in emotion regulation (Cohen's d = 0.12) have also been reported following completion of the Transformative Life Skills (TLS) program, a yoga-based social-emotional wellness program, compared to a control condition among 159 middle and high school-aged adolescent (Frank et al. 2017). Relatedly, a systematic review of 27 studies ranging from weak to moderate methodological quality (due to lack of randomization, blinding, and limited analyses), found that 70% of included studies reported an association between yoga-based interventions and reduced anxiety and depression (James-Palmer et al. 2020).

Yoga may also be effective at reducing risky behaviors among adolescents such as substance use. Researchers have examined the effects of yoga interventions on risky behaviors in adolescents, providing preliminary but promising evidence. In a pilot randomized controlled trial (RCT) with high-risk adolescents, decreased frequency of past month alcohol use was found for individuals who completed a 20-session mindful yoga intervention ($n = 49$) compared to a control group, though this finding was only trending ($\beta = -1.48$, $p < 0.10$) and the small sample size limited generalizability (Fishbein et al. 2016). A larger RCT ($N = 211$) evaluating a 32-session yoga intervention in middle schools found that post-intervention, students in the yoga group were less willing to try smoking cigarettes (partial $\eta^2 = 0.02$; $p = 0.03$). Although this study utilized a larger sample, this finding represents only a small effect and no intervention effect was found for other substances (i.e., alcohol, marijuana; (Butzer et al. 2017)).

Lastly, biofeedback is a therapeutic approach that leverages real-time physiological data to enhance interoceptive awareness and self-regulation. Biofeedback training typically involves monitoring physiological signals, such as heart rate, muscle tension, or skin conductance, and providing immediate feedback, allowing individuals to learn to regulate their physiological responses. Studies investigating the use of biofeedback in adolescent samples are limited. However, within adult populations, researchers have reported improvements in interoceptive accuracy ($p = 0.007$) and reduced state anxiety ($p = 0.004$) following completion of interoceptive training consisting of a modified heartbeat discrimination task (Sugawara et al. 2020). Increased interoceptive accuracy was also observed among 30 adults who received naturalistic haptic heartbeat feedback compared to adults who only received visual feedback (Dobrushina et al. 2024). Although conducted with adult samples, these findings suggest that biofeedback interventions may hold promise for improving interoceptive accuracy and emotional regulation in adolescents, though further research is needed to directly assess their efficacy within this specific age group.

The existing research on interventions for improving interoception in adolescents has several limitations that warrant consideration. Most studies rely on small, homogeneous samples, which limits the generalizability of findings to broader populations. Furthermore, there is considerable variability in how these interventions are defined and implemented across studies, making it challenging to compare results or draw definitive conclusions about their efficacy. For example, the duration, frequency, and specific components of yoga or mindfulness interventions vary widely, raising questions about the optimal design and dosage of such programs. Additionally, many studies focus on short-term outcomes, with little exploration of the sustained benefits of body-awareness training over time. The reliance on self-reported measures of interoceptive accuracy and emotional well-being also introduces potential biases, such as social desirability effects. Future research should address these limitations by recruiting larger, more diverse samples, standardizing intervention protocols, and incorporating objective measures to evaluate interoceptive processes and related outcomes.

Collectively, these interventions hold potential for enhancing interoceptive accuracy, developing emotional resilience, and reducing substance use in adolescents. By equipping young people with tools to recognize and respond adaptively to their internal states, these approaches help foster self-regulation and well-being during a crucial period of development. As we gain further insights into how these practices influence interoception, future research can refine and expand upon these strategies to better meet the diverse needs of adolescents, ultimately contributing to healthier developmental trajectories.

7 Key Insights, Limitations, and Future Research Directions

The nascent body of research on interoception in adolescence reveals a complex, multifaceted process essential to emotional regulation, behavioral choices, and mental health during a period of profound physiological and psychological change. This chapter has explored how interoceptive processing develops in adolescence, highlighting its potential influence on emotional and mental health outcomes and underscoring individual differences shaped by biological and sociocultural factors. Adolescence marks a critical developmental window where interoceptive sensitivity and accuracy mature in tandem with increased cognitive and emotional demands, positioning interoception as both a potential source of resilience and vulnerability. By understanding the role of interoception in adolescent well-being, researchers, educators, and clinicians can better identify at-risk individuals and develop targeted interventions to support adaptive interoceptive processes.

Several insights emerge from the research reviewed here. First, the developmental trajectory of interoception is closely tied to the maturation of brain regions involved in bodily awareness and emotional processing, such as the insula and anterior cingulate cortex. These changes provide adolescents with the tools to better interpret internal cues, yet the heightened sensitivity to interoceptive signals may also increase susceptibility to anxiety, mood fluctuations, and body image concerns. Developmental fluctuations in interoception also coincide with changes in decision-making processes, which may contribute to risky behaviors such as substance use during adolescence. Misinterpretation or heightened reactivity to bodily signals can impair adolescents' ability to evaluate risks and rewards, potentially driving impulsive or maladaptive choices. Moreover, individual differences in interoception highlight that interoceptive development is not a uniform process but is shaped by unique biological and psychosocial influences. These findings coincide with the proposed model in which variations in neural development underly fluctuations in interoception, which in turn influence emotion regulation and decision-making processes, leaving adolescents more susceptible to psychopathology and risk-taking. Lastly, interventions that improve interoceptive awareness, such as mindfulness practices, body-awareness training, and biofeedback, offer promising avenues for enhancing emotional resilience, particularly for adolescents who may struggle with interoception.

Our understanding of the development of interoception during adolescence remains incomplete. A major gap in the literature is the lack of longitudinal studies that map interoceptive accuracy and sensitivity from childhood through adolescence and into adulthood. This makes it challenging to definitively conclude how interoceptive processes evolve over time and how these changes relate to emotional, cognitive, and social outcomes. To our knowledge, only one cross-sectional study (May et al. 2014) has directly compared interoceptive abilities across different age groups, underscoring the need for more comprehensive developmental research. Additionally, there is a need for more nuanced examinations of how interoception interacts with specific disorders across adolescence. For example, differentiating between the roles of interoceptive hypersensitivity and hypo-awareness in anxiety versus depression could lead to more tailored interventions.

Continued research into the influence of individual factors on interoception is critical. Direct examinations of pubertal timing on interoceptive processing, particularly in relation to hormonal changes, could yield further insights into gender differences and inform personalized approaches to support adolescent mental health. Additionally, sociocultural factors, such as the influence of cultural norms, socioeconomic status, and experiences of adversity, remain underexplored, and many studies lack diversity in their samples, limiting the generalizability of findings.

Future studies should also explore the mechanisms by which therapeutic and educational interventions enhance interoceptive accuracy, particularly among adolescents at high risk for affective disorders. Investigating how different practices – such as mindfulness versus biofeedback – target specific facets of interoception could guide the design of programs tailored to varying interoceptive needs. Expanding school-based interventions that integrate interoceptive training as part of social-emotional learning programs offers a promising avenue for preventive mental health strategies. Experimental research is needed to establish the effectiveness of these interventions, measuring outcomes beyond symptom reduction to include resilience, academic performance, and social adaptation. In addition to intervention-focused research, studies examining the neurobiological underpinnings of interoceptive development are crucial. Limited research links interoceptive changes to the structural and functional maturation of brain regions like the insula and anterior cingulate cortex. Addressing these gaps through more inclusive and methodologically consistent research will advance our understanding of interoception and its critical role in adolescent development.

In sum, the study of interoception in adolescence represents a promising field with significant implications for understanding and supporting mental health during a critical period of development. Continued research is essential to refine our knowledge of interoceptive mechanisms, uncover factors that contribute to interoceptive variability, and identify interventions that can bolster emotional resilience. With a more comprehensive understanding of interoception, we can empower adolescents to develop a balanced and adaptive awareness of their internal states, laying the foundation for emotional well-being and resilience across the lifespan.

References

Baiano C, Santangelo G, Senese VP, Di Mauro G, Lauro G, Piacenti M, Conson M (2021) Linking perception of bodily states and cognitive control: the role of interoception in impulsive behaviour. Exp Brain Res 239:857–865

Barrett LF, Simmons WK (2015) Interoceptive predictions in the brain. Nat Rev Neurosci 16(7): 419–429. https://doi.org/10.1038/nrn3950

Berk L, Stewart JL, May AC, Wiers RW, Davenport PW, Paulus MP, Tapert SF (2015) Under pressure: adolescent substance users show exaggerated neural processing of aversive interoceptive stimuli. Addiction 110(12):2025–2036

Blakemore SJ, Mills KL (2014) Is adolescence a sensitive period for sociocultural processing? Annu Rev Psychol 65(1):187–207. https://doi.org/10.1146/annurev-psych-010213-115202

Blakemore SJ, Burnett S, Dahl RE (2010) The role of puberty in the developing adolescent brain. Hum Brain Mapp 31(6):926–933

Brand S, Lang K, Bellinghausen C, Witthöft M, Jungmann SM (2024) Interoceptive accuracy in children aged 8 to 13 and their parents: implications for mental health. Cogent Psychol 11(1): 2348043

Butzer B, LoRusso A, Shin SH, Khalsa SBS (2017) Evaluation of yoga for preventing adolescent substance use risk factors in a middle school setting: a preliminary group-randomized controlled trial. J Youth Adolesc 46:603–632

Casey BJ, Jones RM (2010) Neurobiology of the adolescent brain and behavior: implications for substance use disorders. J Am Acad Child Adolesc Psychiatry 49(12):1189–1201

Casey BJ, Getz S, Galvan A (2008) The adolescent brain. Dev Rev 28(1):62–77

Casey BJ, Jones RM, Somerville LH (2011) Braking and accelerating of the adolescent brain. J Res Adolesc 21(1):21–33

Craig AD (2002) How do you feel? Interoception: the sense of the physiological condition of the body. Nat Rev Neurosci 3(8):655–666. https://doi.org/10.1038/nrn894

Craig AD (2009) How do you feel – now? The anterior insula and human awareness. Nat Rev Neurosci 10(1):59–70. https://doi.org/10.1038/nrn2555

Critchley HD, Garfinkel SN (2017) Interoception and emotion. Curr Opin Psychol 17:7–14. https://doi.org/10.1016/j.copsyc.2017.04.020

Critchley HD, Wiens S, Rotshtein P, Öhman A, Dolan RJ (2004) Neural systems supporting interoceptive awareness. Nat Neurosci 7(2):189–195. https://doi.org/10.1038/nn1176

Daly LA, Haden SC, Hagins M, Papouchis N, Ramirez PM (2015) Yoga and emotion regulation in high school students: a randomized controlled trial. Evid Based Complement Alternat Med 2015(1):794928

De Witte NA, Sütterlin S, Braet C, Mueller SC (2016) Getting to the heart of emotion regulation in youth: the role of interoceptive sensitivity, heart rate variability, and parental psychopathology. PLoS One 11(10):e0164615

Dehne KL, Riedner G (2001) Adolescence – a dynamic concept. Reprod Health Matters 9(17): 11–15

Dobrushina O, Tamim Y, Wald IY, Maimon A, Amedi A (2024) Interoceptive training with real-time haptic versus visual heartbeat feedback. Psychophysiology 61(11):e14648

Domschke K, Stevens S, Pfleiderer B, Gerlach AL (2010) Interoceptive sensitivity in anxiety and anxiety disorders: an overview and integration of neurobiological findings. Clin Psychol Rev 30(1):1–11

Dunn BD, Dalgleish T, Ogilvie AD, Lawrence AD (2010) Heartbeat perception in depression. Behav Res Ther 48(11):1133–1138. https://doi.org/10.1016/j.brat.2006.09.008

Elbasheir A, Bond R, Harnett NG, Guelfo A, Karkare MC, Fulton TM et al (2024) Racial discrimination-related interoceptive network disruptions: a pathway to disconnection. In: Biological psychiatry: cognitive neuroscience and neuroimaging

Fani N, Khalsa SS (2023) The role of racial discrimination in dissociation and interoceptive dysfunction. Neuropsychopharmacology 48(1):225

Felver JC, Butzer B, Olson KJ, Smith IM, Khalsa SBS (2015) Yoga in public school improves adolescent mood and affect. Contemp Sch Psychol 19:184–192

Fishbein D, Miller S, Herman-Stahl M, Williams J, Lavery B, Markovitz L et al (2016) Behavioral and psychophysiological effects of a yoga intervention on high-risk adolescents: a randomized control trial. J Child Fam Stud 25:518–529

Frank JL, Kohler K, Peal A, Bose B (2017) Effectiveness of a school-based yoga program on adolescent mental health and school performance: findings from a randomized controlled trial. Mindfulness 8:544–553

Friedel S, Whittle SL, Vijayakumar N, Simmons JG, Byrne ML, Schwartz OS, Allen NB (2015) Dispositional mindfulness is predicted by structural development of the insula during late adolescence. Dev Cogn Neurosci 14:62–70

Garfinkel SN, Seth AK, Barrett AB, Suzuki K, Critchley HD (2015) Knowing your own heart: distinguishing interoceptive accuracy from interoceptive awareness. Biol Psychol 104:65–74. https://doi.org/10.1016/j.biopsycho.2014.11.004

Garfinkel SN, Tiley C, O'Keeffe S, Harrison NA, Seth AK, Critchley HD (2016) Discrepancies between dimensions of interoception in autism: implications for emotion and anxiety. Biol Psychol 114:117–126

Gennatas ED, Avants BB, Wolf DH, Satterthwaite TD, Ruparel K, Ciric R et al (2017) Age-related effects and sex differences in gray matter density, volume, mass, and cortical thickness from childhood to young adulthood. J Neurosci 37(20):5065–5073

Grabauskaitė A, Baranauskas M, Griškova-Bulanova I (2017) Interoception and gender: what aspects should we pay attention to? Conscious Cogn 48:129–137

Gross JJ (2015) Emotion regulation: current status and future prospects. Psychol Inq 26(1):1–26. https://doi.org/10.1080/1047840X.2014.940781

Hamm LL, Jacobs RH, Johnson MW, Fitzgerald DA, Fitzgerald KD, Langenecker SA et al (2014) Aberrant amygdala functional connectivity at rest in pediatric anxiety disorders. Biol Mood Anxiety Disorders 4:1–9

Herbert BM, Pollatos O (2014) Attenuated interoceptive sensitivity in overweight and obese individuals. Eat Behav 15(3):378–381

Herman AM, Rae CL, Critchley HD, Duka T (2019) Interoceptive accuracy predicts nonplanning trait impulsivity. Psychophysiology 56(6):e13339

Hölzel BK, Lazar SW, Gard T, Schuman-Olivier Z, Vago DR, Ott U (2011) How does mindfulness meditation work? Proposing mechanisms of action from a conceptual and neural perspective. Perspect Psychol Sci 6(6):537–559

James-Palmer A, Anderson EZ, Zucker L, Kofman Y, Daneault JF (2020) Yoga as an intervention for the reduction of symptoms of anxiety and depression in children and adolescents: a systematic review. Front Pediatr 8:78

Kelly AMC, Di Martino A, Uddin LQ, Shehzad Z, Gee DG, Reiss PT, Milham MP (2009) Development of anterior cingulate functional connectivity from late childhood to early adulthood. Cereb Cortex 19(3):640–657

Kerr KL, Moseman SE, Avery JA, Bodurka J, Zucker NL, Simmons WK (2016) Altered insula activity during visceral interoception in weight-restored patients with anorexia nervosa. Neuropsychopharmacology 41(2):521–528

Khalsa SS, Craske MG, Li W, Vangala S, Strober M, Feusner JD (2015) Altered interoceptive awareness in anorexia nervosa: effects of meal anticipation, consumption and bodily arousal. Int J Eat Disord 48(7):889–897

Khalsa SS, Adolphs R, Cameron OG, Critchley HD, Davenport PW, Feinstein JS et al (2018) Interoception and mental health: a roadmap. Biol Psychiatry Cogn Neurosci Neuroimaging 3(6): 501–513

Klabunde M, Juszczak H, Jordan T, Baker JM, Bruno J, Carrion V, Reiss AL (2019) Functional neuroanatomy of interoceptive processing in children and adolescents: a pilot study. Sci Rep 9(1):16184

Kleckner IR, Zhang J, Touroutoglou A, Chanes L, Xia C, Simmons WK et al (2017) Evidence for a large-scale brain system supporting allostasis and interoception in humans. Nat Hum Behav 1(5):69

Kruschwitz JD, Kausch A, Brovkin A, Keshmirian A, Paulus MP, Goschke T, Walter H (2019) Self-control is linked to interoceptive inference: craving regulation and the prediction of aversive interoceptive states induced with inspiratory breathing load. Cognition 193:104028

Lazzarelli A, Scafuto F, Crescentini C, Matiz A, Orrù G, Ciacchini R et al (2024) Interoceptive ability and emotion regulation in mind–body interventions: an integrative review. Behav Sci 14(11):1107

Li D, Zucker NL, Kragel PA, Covington VE, LaBar KS (2017) Adolescent development of insula-dependent interoceptive regulation. Dev Sci 20(5):e12438

Lima-Araujo GLD, de Sousa Júnior GM, Mendes T, Demarzo M, Farb N, Barros de Araujo D, Sousa MBCD (2022) The impact of a brief mindfulness training on interoception: a randomized controlled trial. PLoS One 17(9):e0273864

Longarzo M, Mele G, Alfano V, Salvatore M, Cavaliere C (2021) Gender brain structural differences and interoception. Front Neurosci 14:586860

Ma-Kellams C, Prentice F, Spooner R, Murphy J (2024) Demographic differences in interoception. In: Interoception: a comprehensive guide, pp 357–403

Mata F, Verdejo-Roman J, Soriano-Mas C, Verdejo-Garcia A (2015) Insula tuning towards external eating versus interoceptive input in adolescents with overweight and obesity. Appetite 93:24–30

May AC, Stewart JL, Tapert SF, Paulus MP (2014) The effect of age on neural processing of pleasant soft touch stimuli. Front Behav Neurosci 8:52

Mehling WE, Wrubel J, Daubenmier JJ, Price CJ, Kerr CE, Silow T et al (2011) Body awareness: a phenomenological inquiry into the common ground of mind-body therapies. Philos Ethics Humanit Med 6:1–12

Menezes CB, Dalpiaz NR, Kiesow LG, Sperb W, Hertzberg J, Oliveira AA (2015) Yoga and emotion regulation: a review of primary psychological outcomes and their physiological correlates. Psychol Neurosci 8(1):82

Migliorini R, Stewart JL, May AC, Tapert SF, Paulus MP (2013) What do you feel? Adolescent drug and alcohol users show altered brain response to pleasant interoceptive stimuli. Drug Alcohol Depend 133(2):661–668

Murphy J, Brewer R, Catmur C, Bird G (2017) Interoception and psychopathology: a developmental neuroscience perspective. Dev Cogn Neurosci 23:45–56. https://doi.org/10.1016/j.dcn.2016.12.006

Murphy J, Viding E, Bird G (2019) Does atypical interoception following physical change contribute to sex differences in mental illness? Psychol Rev 126(5):787

Nelson EE, Leibenluft E, McClure EB, Pine DS (2005) The social re-orientation of adolescence: a neuroscience perspective on the process and its relation to psychopathology. Psychol Med 35(2):163–174

Nord CL, Garfinkel SN (2022) Interoceptive pathways to understand and treat mental health conditions. Trends Cogn Sci 26(6):499–513

Paulus MP, Stein MB (2010) Interoception in anxiety and depression. Brain Struct Funct 214(5): 451–463

Paulus MP, Stewart JL (2014) Interoception and drug addiction. Neuropharmacology 76:342–350

Perkins NM, Ortiz SN, Smith AR, Brausch AM (2021) Suicidal ideation and eating disorder symptoms in adolescents: the role of interoceptive deficits. Behav Ther 52(5):1093–1104

Pfeifer JH, Blakemore SJ (2012) Adolescent social cognitive and affective neuroscience: past, present, and future. Soc Cogn Affect Neurosci 7(1):1–10

Pollatos O, Georgiou E (2016) Normal interoceptive accuracy in women with bulimia nervosa. Psychiatry Res 240:328–332

Pollatos O, Gramann K, Schandry R (2007) Neural systems connecting interoceptive awareness and feelings. Hum Brain Mapp 28(1):9–18

Pollatos O, Kurz AL, Albrecht J, Schreder T, Kleemann AM, Schöpf V et al (2008) Reduced perception of bodily signals in anorexia nervosa. Eat Behav 9(4):381–388

Prentice F, Murphy J (2022) Sex differences in interoceptive accuracy: a meta-analysis. Neurosci Biobehav Rev 132:497–518

Rakesh D, Lv J, Zalesky A, Allen NB, Lubman DI, Yücel M, Whittle S (2021) Altered resting functional connectivity patterns associated with problematic substance use and substance use disorders during adolescence. J Affect Disord 279:599–608

Roy AK, Fudge JL, Kelly C, Perry JS, Daniele T, Carlisi C et al (2013) Intrinsic functional connectivity of amygdala-based networks in adolescent generalized anxiety disorder. J Am Acad Child Adolesc Psychiatry 52(3):290–299

Schaan VK, Schulz A, Rubel JA, Bernstein M, Domes G, Schächinger H, Vögele C (2019) Childhood trauma affects stress-related interoceptive accuracy. Front Psych 10:750

Shenhav A, Botvinick MM, Cohen JD (2013) The expected value of control: an integrative theory of anterior cingulate cortex function. Neuron 79(2):217–240

Smith AR, Steinberg L, Chein J (2014) The role of the anterior insula in adolescent decision making. Dev Neurosci 36(3–4):196–209

Steinberg L (2005) Cognitive and affective development in adolescence. Trends Cogn Sci 9(2): 69–74. https://doi.org/10.1016/j.tics.2004.12.005

Sugawara A, Terasawa Y, Katsunuma R, Sekiguchi A (2020) Effects of interoceptive training on decision making, anxiety, and somatic symptoms. BioPsychoSoc Med 14:1–8

Tan L, Martin G (2015) Taming the adolescent mind: a randomised controlled trial examining clinical efficacy of an adolescent mindfulness-based group programme. Child Adolesc Mental Health 20(1):49–55

Tsakiris M, Critchley H (2016) Interoception beyond homeostasis: affect, cognition and mental health. Philos Trans R Soc B: Biol Sci 371(1708):20160002

Tymofiyeva O, Sipes BS, Luks T, Hamlat EJ, Samson TE, Hoffmann TJ et al (2024) Interoceptive brain network mechanisms of mindfulness-based training in healthy adolescents. Front Psychol 15:1410319

Voelker DK, Reel JJ, Greenleaf C (2015) Weight status and body image perceptions in adolescents: current perspectives. Adolesc Health Med Ther 6:149–158

Walter H, Kausch A, Dorfschmidt L, Waller L, Chinichian N, Veer I et al (2020) Self-control and interoception: linking the neural substrates of craving regulation and the prediction of aversive interoceptive states induced by inspiratory breathing restriction. NeuroImage 215:116841

Weng HY, Feldman JL, Leggio L, Napadow V, Park J, Price CJ (2021) Interventions and manipulations of interoception. Trends Neurosci 44(1):52–62

Zanetti T, Santonastaso P, Sgaravatti E, Degortes D, Favaro A (2013) Clinical and temperamental correlates of body image disturbance in eating disorders. Eur Eat Disord Rev 21(1):32–37

Zoogman S, Goldberg SB, Hoyt WT, Miller L (2015) Mindfulness interventions with youth: a meta-analysis. Mindfulness 6(2):290–302. https://doi.org/10.1007/s12671-013-0260-4

Subjective Time in Ordinary and Non-ordinary States of Consciousness: How Interoceptive Feelings Inform Us About the Passage of Time

Marc Wittmann and Sylvie Droit-Volet

Contents

Abstract In accordance with Bud (A.D.) Craig's theories, we maintain that ascending physiological signals in their temporal dynamics are a necessary prerequisite for human time judgments. Functional neuroimaging and psychophysiological evidence have increasingly demonstrated that the subjective judgment of time is based on the physical and emotional self. The psychological literature reveals how emotions and related body feelings shape subjective time. Empirical studies of altered states of consciousness, namely meditative states, are also of prime interest as the perception of the physical state is strongly modulated and thereby affects the subjective experience of time. Our conclusion is that the sense of time is strongly embodied.

M. Wittmann (✉)
Institute for Frontier Areas of Psychology and Mental Health, Freiburg, Germany
e-mail: wittmann@igpp.de

S. Droit-Volet
Université Clermont Auvergne, CNRS, LAPSCO, Clermont-Ferrand, France
e-mail: sylvie.droit-volet@uca.fr

Curr Topics Behav Neurosci (2025) 70: 199–221
https://doi.org/10.1007/7854_2024_520
Published Online: 02 November 2024

Keywords Altered states of consciousness · Embodiment · Emotions · Interoception · Time

1 Introduction

The brain with its neural oscillations and network dynamics is a temporal organ. It functions in time with its dynamic flow of information coded as duration and the order of events. We also consciously perceive the dynamics of our experienced self in time and space with the sensation of the acceleration and deceleration of the speed of time passage. Despite the importance of time for human beings at all descriptive levels, dominant theories of consciousness do not refer to time, to dynamic, continuous, and extended consciousness, features which constitute the phenomenal hallmark of experience (for an elaborate critique and overview, see Kent and Wittmann 2021; Montemayor and Wittmann 2022; Northoff and Lamme 2020). One exception is the Temporo-spatial Theory of Consciousness (TTC), which considers brain activity in explicit temporal and spatial terms, i.e. neural activity is characterized by temporal duration and spatial extension because time and space are basic dimensions of nature (Northoff and Huang 2017).

Another missing feature in the dominant theories of consciousness is its embodied basis. Physical sensations, visceral and proprioceptive input from the peripheral nervous system are fundamental for cognition, emotions, and self-consciousness (Damasio 1999). According to embodied models of subjectivity, the phenomenal first-person perspective depends on the physiological activity in the body and related bodily affects (Gallagher and Aguda 2015).

In phenomenological analyses of the twentieth century, (bodily) self-consciousness and time-consciousness were discussed as inseparable (especially by Husserl, Heidegger, and Merleau-Ponty). Metzinger (2004) equates the (bodily) self similarly with the experience of extended time. He summarizes this conception by stating that (1) the bodily self is the functional anchor of phenomenal experience and (2) phenomenal experience consists of an island of presence as related to what is happening right now in the continuous passage of time. Ergo, the bodily self is the functional anchor of time consciousness as extended experience within a dynamically changing present moment (Wittmann and Montemayor 2022). , In others words, the intrinsically dynamic experience of time could emerge through be manifestations of underlying self-representation processes (Bellingrath 2023).

A.D. (Bud) Craig's neuroscience research on how the interoceptive system is related to our sense of the bodily and emotional self combines the two missing elements – subjective time and body self – in theories of consciousness (Craig 2009, 2015). He maintains that the conscious awareness of our self is informed and constantly updated through ascending somatic signals integrated into the insular cortex. The insula is the primary interoceptive cortex which processes signals from all body organs. Accordingly, the consciously experienced passage of time evolves through awareness of successive moments of interoceptive and emotional self-realization constituting the living body. Craig links subjective time with the processing of the dynamically changing physiological signals through pathways that lead into the insula, the primary interoceptive cortex.

Until very recently, the question of how and where time is processed in the brain (mechanisms and neural substrates, respectively) remained an enigma for most researchers. Many competing theories concerning functional mechanisms exist, as well as identified brain regions which supposedly underlie the perception and judgment of temporal intervals (Buonomano 2017; Teghil et al. 2019; Wittmann and van Wassenhove 2009). One essential brain region underlying sensorimotor timing had been identified with the supplementary motor area (SMA; Coull et al. 2016). In 2023, and independently of each other, two research groups (Mondok and Wiener 2023; Naghibi et al. 2024) published two meta-analyses of the over 100 available neuroimaging experiments to identify brain structures related to the sense of time. The brain regions most consistently activated after inclusion of studies with stringent control tasks were the SMA and the insular cortex. Those two brain regions, which are functionally connected (Cauda et al. 2011), were shown to be activated across different time scales and independent of modality. Conceptually, the sense of time would thereafter be driven by sensorimotor timing processes (SMA) or physical action and by signal processing of body states (insula). This physical basis of subjective time had been anticipated by Craig (2009).

2 Interoceptive Time

2.1 *Body Time*

A seminal study testing the notion of the embodied mind was conducted by Strack et al. (1988). Cartoons were rated as funnier when participants held a pen in their mouths so that it facilitated muscle activity expressing a smile compared to when the pen inhibited smiling. Reporting more intense humor thus depended upon facial-muscle feedback. Many studies have shown that individuals overestimate the duration of emotional faces (e.g., Droit-Volet et al. 2004; Droit-Volet and Gil 2009). Building upon Strack et al.'s study, Effron et al. (2006) showed that participants overestimated the duration of emotional faces relative to neutral faces, but only when facial mimicry was possible. This outcome shows that the embodiment of emotions and that of subjective time are subject to feedback from physical activity, in this case signals from the facial muscles.

The realization that facial-muscle dynamics partly determine subjective duration in social interaction (explaining some variance) was confirmed by Fernandes and Garcia-Marques (2019). In their study, participants judged the temporal interval of neutral faces varying in familiarity, i.e., that had been previously presented in a series of 40 faces or not. The photos of the faces were presented for the duration of around 1 s while the time course of facial-muscle activity was monitored. Familiarity refers to the experience of ease that is associated with less mental effort, which leads to relatively longer subjective time estimates (Fernandes and Garcia-Marques 2020). The amplitude and duration of the corrugator supercilii muscle mediated subjective duration judgments, which led the researchers to suggest that subjective time is embodied through muscle feedback. The activity of the zygomaticus major muscle also predicted the familiarity effect on duration judgments in the expected direction, a relative expansion of duration. Accordingly, the authors suggested that duration perception of external events is compared with or simulated through processes of embodiment. Muscular activity therefore contributes to creating an internal representation of external duration.

Psychophysiological timing studies expand this notion of embodiment and confirm that signals from the body, most notably from the heart, are the basis of the sense of time in humans. In one study conducted by Meissner and Wittmann (2011), participants had to reproduce tone intervals with duration of 3, 9, and 18 s. Participants were first presented with a tone having one of these three durations (encoding phase). After a pause, a second tone was presented, and participants were asked to stop this tone by pressing a button when they thought that it had reached the duration of the first tone (reproduction phase). The simultaneous recording of psychophysiological measures revealed that the slowing down of the heartbeats observed in the encoding phase, with its positive linear trend, correlated significantly with duration reproduction performance. The steeper the slope, indicating the slowing of the heart rate, the longer the reproduced duration. In a following study, a similar deceleration of heartbeats and a decrease in skin-conduction levels over time were observed in a duration reproduction task using visual stimuli (Otten et al. 2015). This suggests that the interoceptive signatures were independent of the sensory modality of temporal stimuli to be reproduced. In a similar duration reproduction task, Pollatos et al. (2014) measured the cardiac cycle as it was phase-locked to the start and stop signals of to-be-timed intervals with durations ranging between 2 and 40 s. Using a synchronization index as a phase-locking value between the heart cycle and the start/stop signal, a significant synchronization between the heart period and duration reproduction accuracy was apparent for durations of up to 25 s. The authors conclude that, for this duration range, the sympathovagal tone measured by the heart rate mediates the accuracy of time perception.

Participants' abilities to be aware of somatic states helps them to be more accurate in their time judgments. This was demonstrated in studies carried out by a research group from Sapienza University in Rome, which also employed the duration reproduction methodology (Teghil et al. 2020a) and assessed individual personality traits of being aware of somatic states with the Self-Awareness Questionnaire (SAQ). A critical manipulation consisted of using two conditions, one with regular stimuli and the other with irregular stimuli. In the first condition, acoustically presented numbers were temporally separated into predictable, regular patterns. In the second condition, numbers were arranged by intervals of variable length (jittered) to create an irregular temporal pattern. Participants were therefore able to use the pattern of numbers to help them estimate the temporal duration of intervals in the first condition, but not in the second condition, where they had no external cues to rely on. Only timing accuracy in the second condition with irregular stimuli was predicted by trait-like bodily awareness as assessed by the SAQ. The general ability to be more aware of somatic states helped subjects to produce more accurate time judgments when no regular external signals were available. They had to rely on the dynamics of their bodily selves to reproduce the time intervals.

It has been proposed that the heart-rate evoked potential (HEP) in the electroencephalography (EEG) signal constitutes a physiological marker of the dynamics of the bodily self and thus of subjective time experience (Khoshnoud et al. 2022; Tallon-Baudry et al. 2018). In a study on the perception of long intervals (2 min) (Richter and Ibáñez 2021), an early negative component of the HEP amplitude was related to the accuracy in estimating the time interval, and the larger the HEP amplitude, the more accurate the duration estimate. In another study using shorter intervals (4, 8, and 12 s) and a control reaction-time task employing the same temporal intervals (Khoshnoud et al. 2024), the HEP amplitude within 130–270 ms (HEP1) and 470–520 ms (HEP2) after the R peak was smaller for the 4 s interval than for the 8 s and 12 s intervals. This is a timing-specific effect, as the HEP amplitude did not differ for intervals in the control reaction-time task. A ramp-like increase in HEP2 amplitudes was significant for the duration-encoding phase of the timing task, but not for the control reaction-time task. The HEP2 increase within the reproduction phase of the timing task also correlated significantly with the reproduced durations for the 8 s and the 12 s intervals. The larger the registered increase in the HEP2, the greater the under-reproduction of the estimated duration. In addition, the scores on the SAQ that we administered also correlated with the mean reproduced durations for 4 s, 8 s, and 12 s intervals, i.e. the higher the SAQ score (a stronger awareness of bodily signals), the longer the duration reproduced (which also meant greater timing accuracy). Therefore, interoceptive awareness (SAQ) as a trait and cortical responses to the heartbeat (HEP) predict duration reproductions and thus validate the hypothesis concerning the embodiment of time.

Two recent studies have demonstrated that the heart rate also plays a decisive role in the perception of very short durations of a few hundred milliseconds (Arslanova et al. 2023; Sadeghi et al. 2023). In these two studies, the cardiac fluctuations were studied between heartbeats or within a cardiac cycle. Sadeghi et al. (2023) showed that the subjective duration was relatively expanded when the interval between heartbeats before the stimulus was longer as compared to when it was shorter. Similarly, Arslanova et al. (2023) observed a relative overestimation of duration when the stimulus was presented during the diastole, i.e. the period of heart relaxation. According to these two studies on millisecond timing, performance is influenced by the participants' cardiac dynamics, beat by beat. To sum up, the conscious analysis of bodily signals in humans determines to some extent our temporal judgment of external stimuli at different duration scales.

2.2 *Insular Time*

The major neural processing route for ascending somatic signals in primates culminates in the dorsal insular cortex (Craig 2015; Strigo and Craig 2016; Evrard 2019; passim). This brain region was termed "primary sensory interoceptive cortex" because the ascending somatic pathways for the representation of the bodily states do not end in the classic primary somatosensory cortex, but in the posterior insula (Craig 2002, 2015). The insular cortex is highly connected to other brain areas. The anterior insula has anatomic-functional links to frontal and limbic regions, areas associated with the processing of cognitive tasks and affective responses. The middle-to-posterior insula has connections to many areas important for sensorimotor processing (Cauda et al. 2012; Wang et al. 2023). This substantiates the idea that interoceptive processes in the insular cortex are essential for the embodied mind, i.e. underlying perception, cognition, emotion, and motor action (Uddin et al. 2017).

Based on his knowledge as a functional neuroscientist, Bud Craig had theoretically predicted that the insula constitutes the decisive brain region for understanding time judgment (2009, 2015). He believed that the conscious awareness of complex feelings and the bodily self is based on a progression of processing in the posterior-to-mid-to-anterior direction in the insula, culminating in the anterior insula. An example related to the conscious perception of one's own heartbeat: activity in the right anterior insular predicted the participants' accuracy in the heartbeat detection task (Critchley et al. 2004). In the same study, the gray-matter volume in the right anterior insular region was associated with accuracy in detecting the heartbeat and in subjective ratings of awareness of the heartbeat. Progressive processing therefore integrates cognitive, motivational, and social information culminating in the anterior insula. This integration enables a perception of the homeostatic physical condition, the "material me," and is expressed as needs and desires. The anterior insula thus constitutes a unified meta-representation of homeostatic perception, which defines the experienced self at any given moment. A succession of such meta-representations of the self across time provides a continuity of conscious awareness as a series of emotional moments. Time perception accordingly evolves through the experienced progression of successive moments of bodily self-realization.

By combining the above-mentioned duration reproduction task with the functional magnetic resonance imaging (fMRI) method in two studies (Wittmann et al. 2010, 2011), ramp-like increases in fMRI activation were detected in the insular cortex which corresponded to stimulus duration. In contrast to a control task with the same durations marked by a continuous tone, but not related to timing, the recorded activity increased with the beginning of the acoustic signal and abruptly decreased with the termination of the signal in both the encoding (dorsal posterior insula) and the reproduction period (anterior insula). To our knowledge, these studies were the first to disclose and theoretically discuss empirical findings of insular cortex activation to explain the estimation of time duration. They were certainly the first to document a potential accumulation-type function in the insular cortex which was correlated with the processing of time intervals (Wittmann 2013).

In another more recent study employing the temporal reproduction task and the regularly and irregularly presented numbers conducted by the research group from Rome (Teghil et al. 2020b), the strength of functional connectivity of the right posterior insula with other regions was shown to vary as a function of duration reproduction accuracy in the irregular condition (i.e., the one where no external cues assisted in the timing performance and participants had to rely solely on internal bodily signals). Stronger connectivity of the posterior insula with other brain regions was also observed among participants who were better able to reproduce the intervals in the irregular condition and were more aware of their physical selves (SAQ scores). The above fMRI time-perception studies employed temporal intervals in the range of several seconds. However, according to Bud Craig, the insula is involved in the processing of much shorter durations too. Bueti and Macaluso (2011) also found the activation of the right mid-insula with visual stimuli of 600 and 1,000 ms during the encoding phase. This correlated with the reproduced times. These different fMRI studies confirm that the insular cortex plays a specific role in judgements of different durations, especially in individuals who are more sensitive to their own body signals.

Based on Bud Craig's theoretical considerations and the results of empirical research, we consider time judgement as embodied experience. We believe humans experience time through their physical sensations, which are integrated in the insular cortex. Bodily processes, such as the heart rate, are typically periodic physical patterns with specific frequencies on which individuals can base their explicit judgments of time. Studies of emotions (emotional time) and changes in the perception of the body and the self (altered time, meditative time) have provided additional results supporting the critical role of bodily states in time judgments.

3 Emotional Time

3.1 Emotional Arousal Time

Imaging studies have detected activation of the insular cortex during a wide variety of emotions, such as anger, fear, sadness, disgust, maternal love, and empathy (Gu et al. 2013; Craig 2009). Craig (2002, 2010) explained this activation of the insular cortex by the fact that emotions are associated with physiological changes, i.e. interoceptive information and body movements. Bud Craig follows in the footsteps of William James, who was convinced that emotion emerges from the perception of changes in bodily signals. When we are walking in the forest and see a wolf (a bear in James' 1884 example), we feel our heart rate accelerates, our breathing quickens, and our muscles tense up. Our whole body is automatically ready to flee for survival. According to William James, the perception of these changes in the autonomic and motor nervous systems helps us understand that we are afraid, i.e. to name what we have felt ("I am afraid"). We are afraid because our body has reacted in a typical way in this frightening situation, and not vice versa. It is not because we experience fear that the body reacts in this way. For Craig (2010) and James, emotion results from conscious interpretations of changes in bodily signals.

Studies on emotional time have provided data consistent with this conception of the relationship between physiological changes and emotions. Studies using a wide range of emotional stimuli (e.g., faces, pictures, movie clips, electric shocks) have found emotional distortions of time (temporal lengthening) when arousal levels increased as measured with a physiological (e.g., electrodermal activity) or subjective index (e.g., Self-Assessment Manikin Scale) (e.g., Fayolle et al. 2015; Mella et al. 2011; Ogden et al. 2019a; Piovesan et al. 2019). Pain that is an unpleasant subjective experience with a sensory and emotional component also produces distortions in the passage of time with the feeling of time slowing down when individuals feel physical pain (Ogden et al. 2023). A recent meta-analysis of study results on emotional time (Cui et al. 2022) confirmed that changes in physiological arousal with its physical manifestations (e.g., increased heart rate, muscle contractions) are the major factors responsible for emotional distortions of subjective time.

3.2 *Body Time Instead of Clock Time*

The exact role and place of the "time" component in the sequence of bodily changes and felt emotions remain obscure. The fact that felt emotions in a given situation follow bodily changes has been discussed in the literature (Smith and Lazarus 1990; Scherer 1982). However, few researchers have attempted to discuss the role of time in affective processes, with the exception of Craig and his cinemascopic model of awareness (the homeostatic model of self-awareness). This lack of discussion probably stems from researchers who initially imagined an independent and specific temporal mechanism (i.e., an internal clock) to account for the accuracy of time judgments in humans and animals (e.g., Gibbon et al. 1984; Treisman 1963). By analogy, this internal clock is conceived of as a kind of clock (stopwatch) which measures time independently of the characteristics of external events to be timed and the internal states of the person who has to make a temporal judgment. To account for time distortions, cognitive models incorporated a component that interferes with the internal clock mechanism. In the case of fear, which has been the most investigated emotion in time-perception studies, this component (i.e., arousal) is designed to speed up the internal clock (Block and Zakay 1996; Glicksohn 2001; for a review, see Wearden 2016). Most psychological time perception researchers adopt a modular approach to temporal information processing with the action of a module (arousal) on another (internal clock).

Craig, in contrast to this approach, defended in his cinemascopic model of awareness an integrative perspective by attributing a key role to the insula in the construction of an integrative and coherent representation of all bodily feelings at one moment. He then introduced the concept of "sentient self," which refers to a self-model across time with its temporal accelerations and decelerations (Craig 2009)

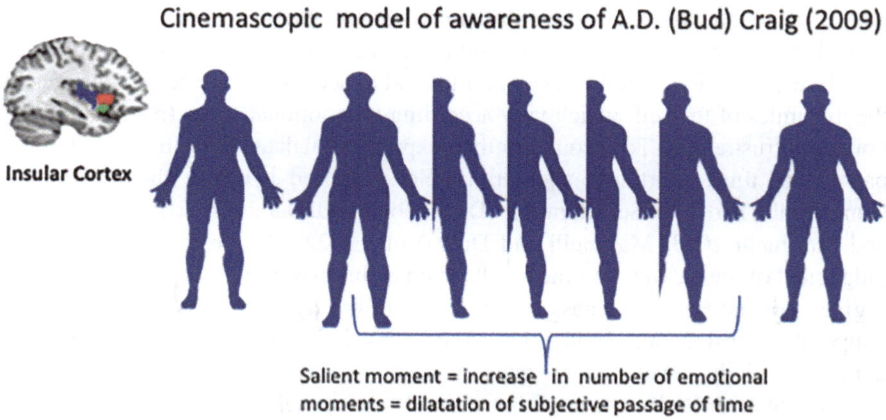

Fig. 1 A.D. (Bud) Craig's cinemascopic model of awareness (2009). The number of "global emotional moments," as derived from interoceptive information from the body, defines the subjective duration for a given objective time span

(see Fig. 1). He conceived that the temporal shifting in the number of "global emotional moments" for a given objective time span derived from interoceptive information underlies the perceived variations in the judgment of time passage, with the feeling of time dilating or contracting (see Fig. 1). According to Craig's theory, the judgement of time passage is not the result of a temporal mechanism separate from emotional or bodily states. On the contrary, time (conceived as temporal dynamics of body states) is embedded within emotional states, enabling the organism to construct a coherent representation of all subjective feelings at a given moment.

Recent studies on subjective time have provided evidence for this close association between the temporal dynamics of global emotional moments (sentient self) and the explicit judgment of the speed of the passage of time in the present. Martinelli and Droit-Volet (2022) recently conducted a study in which participants had to judge the speed of the passage of time for the sequence of images they saw with either positive or negative emotional valence. Participants judged that time passed faster in the positive than in the negative emotional condition, although the duration of the sequences of images was the same between the emotional conditions. In another experiment in that study, participants were asked to judge the speed at which time passed for an easy and a difficult task taking the same amount of time to perform. The passage of time was judged to pass faster with the difficult task than with the easy task. The judgment of the passage of time therefore depends on the context encountered and on its effect on the experienced self.

Additional evidence comes from young children aged 4 years and above, who, like adults, establish a link between emotions and the judgement of time passing, but not between duration estimation and the judgement of the passage of time (Martinelli and Droit-Volet 2023). For example, like adults, they reported that the happier they were, the faster time passed. But, unlike adults, children did not report that time passed more quickly for shorter durations because they did not understand the logical inverse relationship between duration and speed. The passage of time was therefore a metaphor used from early childhood to account for perceived changes in the dynamics of the self, which vary according to emotional states (e.g., happiness, boredom, frustration). To account for their experimental data on the judgement of the passage of time, Martinelli and Droit-Volet proposed a theory they called the Contextual Theory of Self-duration (Droit-Volet and Dambrun 2019; Droit-Volet and Martinelli 2023; Martinelli and Droit-Volet 2022). According to this theory, judgement of the passage of time results from an introspective analysis of the self in a given context and awareness of variations in the dynamics of the self with its temporal extension and contraction. They thus introduced the concept of self-duration initially developed by phenomenologists (Minkowski 1968). Craig did not use the term "self-duration," but rather "sentient self." The terms differ from author to author, but the ideas converge, attesting to the strength of Craig's theories. Thus, as suggested by Craig, awareness of the passage of time is generated by the conscious representation of the bodily self and global emotional moments, processed in the insular cortex.

3.3 Emotional Body Dynamics

Why do we use the metaphor of the speed of time passage to express our emotions? For example, why do we say that time passes quickly in the company of the loved one and that it passes slowly when we are bored? Is it because we have learned to translate our feelings into temporal terms or because we simply describe what we feel directly through the variations in the temporal dynamics of our body when active (e.g., slowing down during energy loss, effort during movement initiation, speed of movement execution) (Droit-Volet et al. 2013; Droit-Volet and Wittmann 2024)? In the latter case, time (duration, succession) would be an intrinsic characteristic of physiological behaviors. The judgment of duration would thus vary logically with physical activity (movement in space, energy, intensity) (Robbe 2023). This has been demonstrated in several studies. For example, it has been shown that duration judgments depended on the characteristics of the action perceived or performed by others, which we simulated internally (e.g., Hallez and Droit-Volet 2018; Nather et al. 2011; Sgouramani and Vatakis 2014), and temporal judgments were highly sensitive to the dynamics in emotional facial expressions (Fayolle and Droit-Volet 2014). Awareness of the passage of time and emotions then emerges from the dynamics of the body, from bodily time.

Craig did not specifically discuss this aspect of "bodily time" as one of the primary pieces of information underlying the awareness of emotional affect and the judgement of time passage. For him, the emotional state (number of emotional moments) takes precedence, deduced from interoceptive information, whatever its nature (e.g., the heart rate as discussed previously) (Fig. 1). Craig (2009) admitted the lack of data on the causal link between body, time, and emotions, which would enable this question to be solved, because he relied on existing, essentially correlational studies. However, when discussing variations in heart rate as a basis for judging time, for example, we are talking about the temporal properties (frequency) of bodily signals, and we can therefore state: "Our body is time." We can assume that emotions and time judgments arise from our awareness of bodily time. Several study results provide arguments supporting this idea that it is, above all, the variations in bodily time in a given context of which we become aware which determine our emotions, our explicit judgments of durations, and our discourse on the passage of time.

(1) Emotional distortions of time have mainly been observed with paradigms of explicit time judgments (for a review, see Droit-Volet et al. 2013) when time processing is conscious. However, two studies demonstrated the same type of emotional distortion in an implicit time-judgment paradigm when participants were not conscious that they were processing time (Droit-Volet 2016; Droit-Volet and Berthon 2017). This indicates that there is an automatic distortion of time (changes in physical dynamics) prior to any conscious process, i.e. before awareness of emotion or of the passage of time. (2) Consistently, emotional time distortions regarding the facial expression of anger have consistently been observed in very young children (aged 3) despite their limited language and executive-function abilities (Gil et al. 2007). (3) Some studies have also shown that the forced adoption of a specific emotional behavior, such as an emotional facial expression or an emotional posture, produces or modulates the emotion felt. Forcing yourself to smile, for example, increases the feeling of happiness. Time distortions observed during the presentation of emotional facial expressions (e.g., Droit-Volet et al. 2004; Visali et al. 2023) and emotional body posture (Droit-Volet and Gil 2016) have generally been explained by the fact that participants internally simulate the emotions perceived in the other person via automatic processes of sensorimotor imitation, through brain systems encompassing mirror neurons. (4) When sensorimotor imitation is prevented, e.g., by holding a pen in the mouth, which prevents smiling, simulation becomes difficult, and the temporal distortion classically observed in front of a facial expression of anger disappears (Effron et al. 2006). (5) In theories of embodied emotion (Niedenthal 2007), the capacity for empathy is also partially attributed to the ability to imitate the emotions of others. Mondillon et al. (2007) consistently demonstrated that the subjective distortion of time during the perception of facial expressions of anger was greater among the most empathic participants. (6) Cognitive control of physiological behaviors also impacts temporal judgments. Some studies have shown an improvement in the accuracy of duration judgements with mindfulness training, when participants were able to reduce their level of arousal by controlling their breathing rates or by slowly scanning different parts of their bodies (Droit-Volet and Héros 2017; Droit-Volet et al. 2018; Ogden et al. 2019b; Wittmann et al. 2014). (7) Having knowledge about the effect of emotions on time judgment (Droit-Volet et al. 2015a, b), awareness of the influence of emotions on time distortions (Lamotte et al. 2012), and being able to correctly perceive bodily information (Di Lernia et al. 2018) reduce the extent of temporal distortions.

Awareness of what is happening in the time body in emotional situations helps participants be more accurate in their time perceptions, as explained above.

In conclusion, a set of study data have supported the notion of embodied emotional time, bodily time as the temporal basis for the representation of the self at a given moment. What happens in the body during emotional situations therefore automatically influences temporal judgments. Awareness of bodily time would also help us become aware of our emotional affects and then be verbally translated in the form of a metaphor of time as flying or dragging. Further studies are still needed to better understand the complex body-time and emotion-consciousness interrelationship, and the role of the insular cortex in this complex interaction. Its role is still being debated. For example, although Mella et al. (2019) observed a reduced sensitivity to the duration of short intervals (300–1,500 ms) between patients with right insular lesions and those with left insular lesions and controls, they found no group differences in time perception for emotional auditory signals. These studies will enable us to accumulate data supporting the theory of embodied time (Droit-Volet et al. 2013; Droit-Volet and Wittmann 2024; Teghil 2024; Wittmann 2013; Wittmann and Droit-Volet 2024) and the idea that time awareness is underpinned by the integrative activity of the insular cortex relative to awareness of oneself across time and, therefore, of bodily time.

4 Altered Time

Insofar as the bodily self is involved in time judgement, as suggested previously, the degree to which attention is focused on the self necessarily influences time judgements. This is clearly sensed in the contrasting states of boredom and flow. When we are bored, time subjectively passes more slowly. This is true for durations in the range of minutes when people are forced to wait (Witowska et al. 2020), but also for the daily experience of time as reported during the Covid pandemic (e.g., Cellini et al. 2020; Droit-Volet et al. 2020a, b, 2023a; Ogden 2020; Kosak et al. 2022). When we are bored, we are typically highly self-aware. We complain about our lack of energy and our lack of mental stimulation. In this state, the time measured by the external clock has little influence on our judgment of the passage of time (Droit-Volet et al. 2023b). Because attention is particularly focused inward on our internal states, time seems to drag. By contrast, the flow state, when we are performing an exciting activity, is associated with pleasant emotional states. For example, in states of entertainment-induced flow, such as when individuals play a video game, time passes very quickly, and the more flow experienced, the faster the passage of time (Rutrecht et al. 2021). We are so absorbed in the activity that we no longer pay attention to ourselves, and time seems to pass quickly (Csikszentmihalyi and Csikszentmihalyi 1988). Since subjective time is closely linked to body feelings, the degree of attentional focus on the self in a given situation also influences the sense of time passage.

There are a few neuroimaging studies on neural correlates of boredom and states of flow pointing to the involvement of the insular cortex, but these have not assessed subjective time. In one study, fMRI-recorded neural activation in a group of neural regions including the insula was increased while individuals had to perform a boring, perceptual-decision task vs. a more pleasing control task (Dal Mas and Wittmann 2017). Higher activity during states of flow in the anterior insula has been interpreted as representing disengagement of task-irrelevant processes, such as those related to the default-mode network. The default-mode network, which in this context is typically attributed to representing activity of the narrative-self, was downregulated in flow states (Ulrich et al. 2016). Results regarding neural mechanisms underlying flow states are still quite inconsistent, probably due to the multifaceted concept of flow (Khoshnoud et al. 2020).

Altered states of consciousness, as induced through psychological techniques or psychopharmacological substances, logically cause strong distortions of self and, consequently, distortions in the judgments of the duration and passage of time (Wittmann 2015, 2018). We present below studies based on our embodied notion of self and time investigating altered states of consciousness (ASC) during meditative practice. Other induction methods, such as hypnosis (Cardeña and Terhune 2019), when immersed in a homogenous sensory environment (Glicksohn et al. 2017), in sensory deprivation of Floatation-REST (Reduced Environmental Stimulation Therapy; Hruby et al. 2024), and using psychedelics (Shanon 2001; Wittmann et al. 2007) have all demonstrated that the sense of time and self are often downregulated in ASC.

In addition, performing music is an activity for individuals and groups with a strong bodily component (Wöllner 2024). Listening to or performing music induces pleasant emotions which affect the judgment of durations (e.g., Droit-Volet et al. 2013) in everyone, but in some people can also induce strong temporal experiences ranging between the sensation of timelessness and time dilation, peak states being described by listeners as becoming one with the music and losing the sense of time (Schäfer et al. 2013).

Bud Craig was involved in neurological research on altered states of consciousness when, together with colleagues from Geneva and Linköping, he studied the auras preceding epileptic "ecstatic" seizures (Landtblom et al. 2011; Picard and Craig 2009). Neuroimaging during the studies showed that these ecstatic auras were related to hyperactivation in the anterior insula. Patients reported that they had experienced extremely heightened self-awareness, spiritual emotions, felt euphoric, and "orgasmic ecstasy" during these auras. Strong time distortions ranging from time dilation to the disappearance of time were reported. One person stated that she was "in a radiant sphere without any notion of time or space" (Picard and Craig 2009).

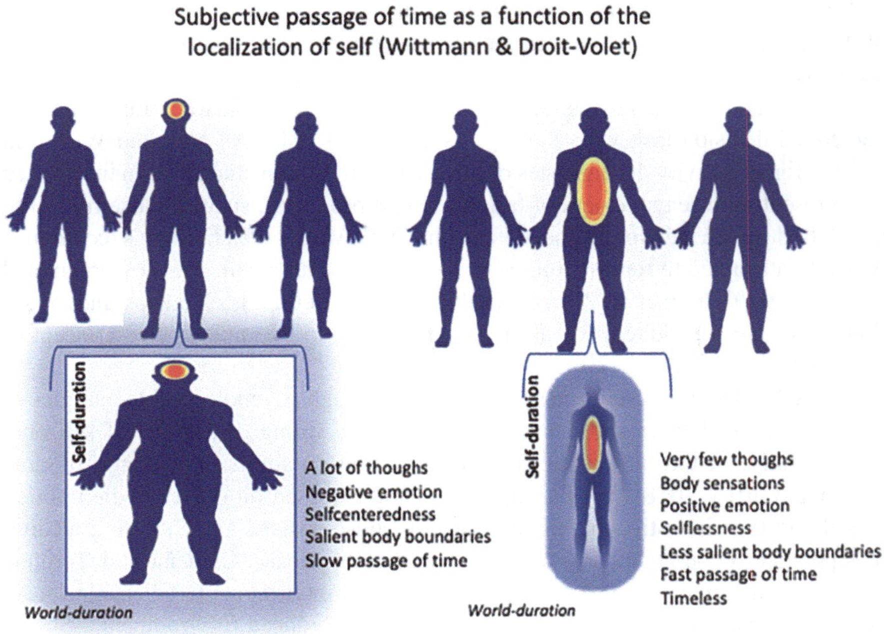

Fig. 2 Subjective time is positively related to body salience and the feeling of self-centeredness. More body awareness leads to a slowing of time passage and less body awareness to a faster time passage

4.1 Meditative Time

Experimental studies on meditative time are still rare. As we want to show, they are relevant for understanding the relationship of subjective time with the bodily self (Matko et al. 2021; Wittmann and Schmidt 2014). The results on time perception are rather inconsistent and even appear contradictory, as well as difficult to interpret (Morin and Grondin 2024). This is largely due to the participants tested (expert meditators, participants with different degrees of or without meditation training), different tasks and procedures used to assess time estimation, and the wide range of meditation techniques practiced (e.g., transcendental meditation, Zen-Buddhism meditation, secular-mindfulness meditation). For example, Berkovich-Ohana et al. (2011) found that time perception was altered after a session of mindfulness meditation, but not after a session of transcendental meditation. The quality of time judgment can also change between two techniques (breathing, body scan) although they both relate to mindfulness meditation. It is particularly more variable during breathing exercises than during the body-scan technique (Droit-Volet et al. 2018).

It has become well established that meditation improves abilities to regulate emotions and control attention and self-awareness. It also modifies the localization of the self in space as flexible (Dambrun et al. 2022), as demonstrated in studies on out-of-body experience (Droit-Volet et al. 2020b). During mindfulness meditation, participants reported a decrease in negative thoughts (associated with a decrease in anxiety and an increase in happiness) with a shift in the location of the self from the head (the seat of thoughts) to the parts of the body below the head (torso, abdomen, arms, hands) (Dambrun et al. 2022; Perestelo-Perez et al. 2017). Mindfulness meditation decreases ongoing thoughts and increases conscious awareness on the parts of the body (heart rate, breathing, small body movements) (Fig. 2).

As regard time, most studies on meditation and time have shown that practicing meditation exercises for several weeks or even a single 15- to 30-min session improves temporal performance in a prospective duration-judgment task performed after a meditation exercise when participants were instructed to judge the duration of presented stimuli (Droit-Volet et al. 2015a, b; Droit-Volet and Heros 2017). Indeed, regular meditation reduces anxiety when performing a task and increases the ability to control attention and to become aware of body signals, a set of factors that are important for the accurate processing of stimulus duration. The overestimation of stimulus duration found in novice meditators obtained in some studies cantherefore be explained by the fact that more attention was allocated to the processing of time in a temporal task during meditation (Kramer et al. 2013). Neuroimaging studies confirmed that mindfulness meditation induces structural and functional changes at an anatomical and functional level in different brain regions, including the prefrontal cortex, and especially in the insular cortex (Farb et al. 2013; Sharp et al. 2018; Wheeler et al. 2017). Experienced meditators show a higher concentration of gray matter (Hölzel et al. 2008) and greater cortical thickness (Lazar et al. 2005) in the right anterior insula compared to meditation-naïve control subjects.

Counterintuitively at first, meditation practice which increases the attentional focus on physical sensations decreases body salience, self-localization, and the feeling of self-centeredness (Dambrun and Richard 2011; Dambrun 2016). Self-centeredness is characterized by the sensation of a clear separation between the body and the rest of the world. In contrast, selflessness means a sense of profound connectedness between the body and the world, to the extent that the sense of self is attenuated or entirely dissolved (Dambrun and Richard 2011; Dambrun 2016; Dambrun et al. 2019). Meditators consistently reported less body salience, in other words, a greater dissolution of perceived body boundaries when they practiced mindfulness meditation as compared to a control exercise (Dambrun 2016). Linares Gutiérrez et al. (2022) also noted that the perception of body boundaries in experienced meditators was less salient during meditation when compared to a control activity, and found that this was associated with the feeling that time passed faster. Most studies have found that participants retrospectively judged that time had passed quickly during the meditation they had just completed, or that time had ceased to exist (e.g., Ataria et al. 2015; Droit-Volet et al. 2019; Liu et al. 2021; Thönes and Wittmann 2016), although some studies obtained a slowing down of time passage (e.g., Sucala and David 2013; for a review, see Morin and Grondin 2024). As explained by Berkovich-Ohana and Glicksohn (2014), meditation practice produces a shift from the narrative-self (conceptual thoughts about oneself) to the minimal-self, when the self is the object of immediate experience in the here and now.

Mindfulness is about attending to the present moment in a nonjudgmental way (Kabat-Zinn 2003). Neuroimaging studies have shown that the states of concentration induced by mindfulness meditation on "being in the present – here and now" (Farb et al. 2007), as well as focusing on momentary breathing sensations (Farb et al. 2013), were correlated with more activation of the brain region involved in the awareness of immediate body signals, i.e. the insular cortex.

Individuals with profound and prolonged meditative practice consistently reported that, during peak states, their sense of self and time were downregulated together or even non-existent with a dissolution of bodily boundaries (Costines et al. 2021; Droit-Volet and Dambrun 2019; Metzinger 2024). Some meditation experts have even reported the feeling of the disappearance of the self and being outside of time (timelessness) (Berkovich-Ohana et al. 2013; Droit-Volet and Dambrun 2019). Droit-Volet et al. (2018, 2019) have shown that all the participants' attention was captured by the meditation activity, i.e., an attention-demanding activity, so that the duration of the stimuli delivered during a meditation exercise was underestimated compared to a control exercise which also required attention. They paid little attention to the temporal dynamics of the outside world (external time) during meditation.

In summary, studies on meditative time have shown changes in the phenomenological experience of time during meditation practice (feeling that time has passed quickly, timelessness) and some short- or long-term effects of meditation practice on the judgement of durations with stimulus duration judged more accurately or as longer. This highlights the importance of disentangling the different types of temporal judgment and further examining the exact role of body signals and awareness for each type of judgments with different meditation techniques (Morin and Grondin 2024).

5 Conclusion

The question of how and where time is processed in the brain remained unsolved until very recently. There exists a strong consensus on basic neural processes regarding space perception and spatial navigation related to entorhinal and hippocampal structures (Moser et al. 2017), a work for which the Nobel Prize for Physiology was awarded in 2014. The notion of a link between ascending physiological signals processed in the insular cortex and subjective time was proposed on conceptual grounds over a decade ago by Bud Craig (2009, 2015). Ramp-like increases in fMRI activation (climbing activity) which terminate at the end of to-be-timed intervals in the range of several seconds were recorded in the insular cortex and the primary interoceptive cortex, providing first empirical evidence for Craig's theory (Wittmann et al. 2010, 2011). Psychophysiological studies directly reveal how body activity, i.e. the heart rate, modulates subjective time (Meissner and Wittmann 2011). Interoceptive awareness as measured with subjective and objective indices also predicts duration judgments in experimental settings (Teghil et al.

2020a; Teghil 2024). Subjective time is dependent upon body signals and related emotional states, suggesting that time is embodied (Droit-Volet and Gil 2009; Droit-Volet et al. 2013).

One theory of the consciousness-body problem proposed that subjectivity and recorded brain activity are treated as complementary, as dual aspects of the same underlying process (Velmans 2009). In such an approach, the phenomenal, behavioral, and neural dimensionalities are treated on a par. Related to the domain of time, these three dimensions reflect complementary perspectives of the subjective experience of time (e.g., felt as relatively extended), recorded behavioral activity (pressing a button relatively later than normal), and specific brain activity in the insular cortex (as correlated with behavioral duration judgment).

A strong version of such a body-consciousness complementarity was also revealed in an experimental study in which physical healing was affected by the experience of time, indicating a close connection between subjective time and body processes. For reasons not well established, Aungle and Langer (2023) were able to demonstrate that experimentally induced wounds healed faster when study participants thought that more time had passed than when in a different, within-subject condition the study participants thought that less time had passed and for whom wounds actually healed more slowly (clock time being equal for both conditions). This body-consciousness relationship goes beyond the neural domain and includes other cellular processing mechanisms (such as the immune system) for achieving successful adaptation and self-organization (Ciaunica et al. 2023). Cognition accordingly depends on multiscale information processing distributed across neuronal, immune, and other network systems. Temporal processes of the body, as well as conscious awareness of time, are then intrinsically linked to psychological well-being and physical health.

In conclusion, we have presented accumulating evidence of how interoception, related emotions, and associated brain networks surrounding the insular cortex determine the judgment of time. Based on current time-perception studies, we have also shown how the ongoing perception of an embodied self over time creates the feeling of time passage.

References

Arslanova I, Kotsaris V, Tsakiris M (2023) Perceived time expands and contracts within each heartbeat. Curr Biol 33(7):1389–1395

Ataria Y, Dor-Ziderman Y, Berkovich-Ohana A (2015) How does it feel to lack a sense of boundaries? A case study of a long-term mindfulness meditator. Conscious Cogn 37:133–147

Aungle P, Langer E (2023) Physical healing as a function of perceived time. Sci Rep 13(1):22432

Bellingrath JE (2023) The self-simulational theory of temporal extension. Neurosci Conscious 2023(1):niad015

Berkovich-Ohana A, Glicksohn J (2014) The consciousness state space (CSS) – a unifying model for consciousness and self. Front Psychol 5:341

Berkovich-Ohana A, Glicksohn J, Goldstein A (2011) Temporal cognition changes following mindfulness, but not transcendental meditation practice. Proc Fechner Day 27:245–250

Berkovich-Ohana A, Dor-Ziderman Y, Glicksohn J, Goldstein A (2013) Alterations in the sense of time, space, and body in the mindfulness-trained brain: a neurophenomenologically guided MEG study. Front Psychol 4:912

Block RA, Zakay D (1996) Models of psychological time revisited. Time Mind 33(9):171–195

Bueti D, Macaluso E (2011) Physiological correlates of subjective time: evidence for the temporal accumulator hypothesis. Neuroimage 57(3):1251–1263

Buonomano D (2017) Your brain is a time machine: the neuroscience and physics of time. WW Norton & Company, New York

Cardeña E, Terhune DB (2019) The roles of response expectancies, baseline experiences, and hypnotizability in spontaneous hypnotic experiences. Int J Clin Exp Hypn 67(1):1–27

Cauda F, D'Agata F, Sacco K, Duca S, Geminiani G, Vercelli A (2011) Functional connectivity of the insula in the resting brain. Neuroimage 55(1):8–23

Cauda F, Costa T, Torta DM, Sacco K, D'Agata F, Duca S, Geminiani G, Fox PT, Vercelli A (2012) Meta-analytic clustering of the insular cortex: characterizing the meta-analytic connectivity of the insula when involved in active tasks. Neuroimage 62(1):343–355

Cellini N, Canale N, Mioni G, Costa S (2020) Changes in sleep pattern, sense of time and digital media use during COVID-19 lockdown in Italy. J Sleep Res 29:e13074

Ciaunica A, Shmeleva EV, Levin M (2023) The brain is not mental! Coupling neuronal and immune cellular processing in human organisms. Front Integr Neurosci 17:1057622

Costines C, Borghardt TL, Wittmann M (2021) The phenomenology of "pure" consciousness as reported by an experienced meditator of the Tibetan Buddhist Karma Kagyu tradition. Analysis of interview content concerning different meditative states. Philosophies 6(2):50

Coull JT, Vidal F, Burle B (2016) When to act, or not to act: that's the SMA's question. Curr Opin Behav Sci 8:14–21

Craig AD (2002) How do you feel? Interoception: the sense of the physiological condition of the body. Nat Rev Neurosci 3:655–666

Craig AD (2009) Emotional moments across time: a possible neural basis for time perception in the anterior insula. Phil Trans R Soc B Biol Sci 364:1933–1942

Craig AD (2010) The sentient self. Brain Struct Funct 214:563–577

Craig AD (2015) How do you feel? Princeton University Press, Princeton

Critchley HD, Wiens S, Rotshtein P, Öhman A, Dolan RJ (2004) Neural systems supporting interoceptive awareness. Nat Neurosci 7(2):189–195

Csikszentmihalyi M, Csikszentmihalyi IS (1988) Optimal experience: psychological studies of flow in consciousness. Cambridge University Press, Cambridge

Cui X, Tian Y, Zhang L, Chen Y, Bai Y, Li D, Liu J, Gable P, Yin H (2022) The role of valence, arousal, stimulus type, and temporal paradigm in the effect of emotion on time perception: a meta-analysis. Psychon Bull Rev 30:1–21

Dal Mas DE, Wittmann BC (2017) Avoiding boredom: caudate and insula activity reflects boredom-elicited purchase bias. Cortex 92:57–69

Damasio A (1999) The feeling of what happens: body and emotion in the making of consciousness. Harcourt, San Diego

Dambrun M (2016) When the dissolution of perceived body boundaries elicits happiness: the effect of selflessness induced by a body scan meditation. Conscious Cogn 46:89–98

Dambrun M, Ricard M (2011) Self-centeredness and selflessness: a theory of self-based psychological functioning and its consequences for happiness. Rev Gen Psychol 15:138–157

Dambrun M, Berniard A, Didelot T, Chaulet M, Droit-Volet S, Corman M, Juneau C, Martinon L (2019) Unified consciousness and the effect of body scan meditation on happiness: alteration of inner-body experience and feeling of harmony as central processes. Mindfulness 10(8): 1530–1544

Dambrun M, Martinon L, Juneau C, Droit-Volet S, Corman M et al (2022) Changes in self-location during the practice of mindfulness meditation in novices. Mindfulness 14(1):174–191

Di Lernia D, Serino S, Pezzulo G, Pedroli E, Cipresso P, Riva G (2018) Feel the time. Time perception as a function of interoceptive processing. Front Hum Neurosci 12:74

Droit-Volet S (2016) Emotion and implicit timing. PLoS One 11(7):e0158474

Droit-Volet S, Berthon M (2017) Emotion and implicit timing: the arousal effect. Front Psychol 8: 176

Droit-Volet S, Heros J (2017) Time judgments as a function of mindfulness meditation, anxiety and mindfulness awareness. Mindfulness 8(2):266–275.

Droit-Volet S, Dambrun M (2019) Awareness of the passage of time and self-consciousness: what do meditators report? PsyCH J 8:51–65

Droit-Volet S, Gil S (2009) The time–emotion paradox. Phil Trans R Soc B Biol Sci 364:1943–1953

Droit-Volet S, Gil S (2016) The emotional body and time perception. Cogn Emotion 30(4):687–699

Droit-Volet S, Heros J (2017) Time judgments as a function of mindfulness meditation, anxiety and mindfulness awareness. Mindfulness 8(2):266–275

Droit-Volet S, Martinelli N (2023) The underpinnings of psychological feelings of time. In: Wöllner C, London J (eds) Performing time: synchrony and temporal flow in music and dance. Oxford University Press, Oxford, pp 67–76

Droit-Volet S, Wittmann M (2024) Le temps: des données en faveur d'un temps incarné. In: Coello Y (ed) Corps, cerveau et processus mentaux: les fondements sensortimoteurs de la cognition. Dunoud, Paris, pp 137–147

Droit-Volet S, Brunot S, Niedenthal PM (2004) Perception of the duration of emotional events. Cogn Emotion 18(6):849–858

Droit-Volet S, Fayolle S, Lamotte M, Gil S (2013) Time, emotion and the embodiment of timing. Timing Time Percept 1:99–126

Droit-Volet S, Fanget M, Dambrun M (2015a) Mindfulness meditation and relaxation training increases time sensitivity. Conscious Cogn 31:85–97

Droit-Volet S, Lamotte M, Izaute M (2015b) The conscious awareness of time distortions regulates the effect of emotion on the perception of time. Conscious Cogn 38:155–164

Droit-Volet S, Chaulet C, Dambrun M (2018) Time and meditation: when does the perception of time change with mindfulness exercise? Mindfulness 9(5):1557–1570

Droit-Volet S, Chaulet M, Dutheil F, Dambrun M (2019) Mindfulness meditation, time judgment and time experience: importance of the time scale considered (seconds or minutes). PlosOne 14(10):e0223567

Droit-Volet S, Gil S, Martinelli N, Andant N, Clinchamps M, Parreira L, Rouffiac K, Dambrun M, Huguet P, Dubuis B, Pereira B, COVISTRESS Network, Bouillon J-B, Dutheil F (2020a) Time and Covid-19 stress in the lockdown situation: time free, "dying" of boredom and sadness. PLoS One 15(8):e0236465

Droit-Volet S, Monceau S, Dambrun M, Martinelli N (2020b) Embodied time and the out-of-body experience of the self. PeerJ 8:e8565

Droit-Volet S, Martinelli N, Dezecache G, Belletier C, Gil S, Chevalère J, Huguet P (2023a) Experience and memory of time and emotions two years after the start of the COVID-19 pandemic. PLoS One 18(9):e0290697

Droit-Volet S, Monier F, Martinelli N (2023b) The feeling of the passage of time against the time of the external clock. Conscious Cogn 113:103535

Effron DA, Niedenthal PM, Gil S, Droit-Volet S (2006) Embodied temporal perception of emotion. Emotion 6(1):1–9

Evrard HC (2019) The organization of the primate insular cortex. Front Neuroanat 13:43

Farb NA, Segal ZV, Mayberg H, Bean J, McKeon D, Fatima Z, Adam K, Anderson AK (2007) Attending to the present: mindfulness meditation reveals distinct neural modes of self-reference. Soc Cogn Affect Neurosci 2:313–322

Farb NA, Segal ZV, Anderson AK (2013) Mindfulness meditation training alters cortical representations of interoceptive attention. Soc Cogn Affect Neurosci 8:15–26

Fayolle SL, Droit-Volet S (2014) Time perception and dynamics of facial expressions of emotions. PLoS One 9(5):e97944

Fayolle S, Gil S, Droit-Volet S (2015) Fear and time: fear speeds up the internal clock. Behav Process 120:135–140

Fernandes AC, Garcia-Marques T (2019) The perception of time is dynamically interlocked with the facial muscle activity. Sci Rep 9(1):18737

Fernandes AC, Garcia-Marques T (2020) A meta-analytical review of the familiarity temporal effect: testing assumptions of the attentional and the fluency-attributional accounts. Psychol Bull 146(3):187–217

Gallagher S, Aguda B (2015) The embodied phenomenology of phenomenology. J Conscious Studies 22(3–4):93–107

Gibbon J, Church RM, Meck WH (1984) Scalar timing in memory. In: Gibbon J, Allan L (eds) Timing and time perception. Annals of the New York Academy of Sciences, vol 423. New York Academy of Sciences, New York, pp 52–77

Gil S, Niedenthal PM, Droit-Volet S (2007) Anger and temporal perception in children. Emotion 7: 219–225

Glicksohn J (2001) Temporal cognition and the phenomenology of time: a multiplicative function for apparent duration. Conscious Cogn 10:1–23

Glicksohn J, Berkovich-Ohana A, Mauro F, Ben-Soussan TD (2017) Time perception and the experience of time when immersed in an altered sensory environment. Front Hum Neurosci 11: 487

Gu X, Hof PR, Friston KJ, Fan J (2013) Anterior insular cortex and emotional awareness. J Comp Neurol 521:3371–3388

Hallez Q, Droit-Volet S (2018) Young children embody the time of others in their time judgments: the role of the theory of mind. Infant Child Develop 26(6):e2101

Hölzel BK, Ott U, Gard T, Hempel H, Weygandt M, Morgen K, Vaitl D (2008) Investigation of mindfulness meditation practitioners with voxel-based morphometry. Social Cog Affect Neurosci 3(1):55–61

Hruby H, Schmidt S, Feinstein J, Wittmann M (2024) Induction of altered states of consciousness during floatation-REST is associated with the dissolution of body boundaries and the distortion of subjective time. Sci Rep 14(1):9316

James W (1884) What is emotion? Mind 9:188–205

Kabat-Zinn J (2003) Mindfulness-based interventions in context: past, present, and future. Clin Psychol Sci Pract 10(2):144–156

Kent L, Wittmann M (2021) Time consciousness: the missing link in theories of consciousness. Neurosci Conscious 2021(2):niab011

Khoshnoud S, Alvarez Igarzábal F, Wittmann M (2020) Peripheral-physiological and neural correlates of the flow experience while playing video games: a comprehensive review. PeerJ 8:e10520

Khoshnoud S, Alvarez Igarzábal F, Wittmann M (2022) Brain–heart interaction and the experience of flow while playing a video game. Front Hum Neurosci 16:819834

Khoshnoud S, Leitritz D, Cinar Bozdag M, Alvarez Igarzabal F, Noreika V, Wittmann M (2024) When the heart meets the mind: exploring the brain-heart interaction during time perception. J Neurosci 44(34)

Kosak F, Schelhorn I, Wittmann M (2022) The subjective experience of time during the pandemic in Germany: the big slowdown. PLoS One 17(5):e0267709

Kramer RSS, Weger UW, Sharma D (2013) The effect of mindfulness on time perception. Conscious Cogn 22:846–852

Lamotte M, Izaute M, Droit-Volet S (2012) The consciousness of time distortions and their effect on time judgment: a metacognitive approach. Conscious Cogn 21:835–842

Landtblom AM, Lindehammar H, Karlsson H, Craig AD (2011) Insular cortex activation in a patient with “sensed presence”/ecstatic seizures. Epilepsy Behav 20:714–718

Lazar SW, Kerr CE, Wasserman RH, Gray JR, Greve DN, Treadway MT, McGarvey M, Quinn BT, Dusek JA, Benson H, Rauch SL, Moore CI, Fischl B (2005) Meditation experience is associated with increased cortical thickness. Neuroreport 16:1893–1897

Linares Gutiérrez D, Schmidt S, Meissner K, Wittmann M (2022) Changes in subjective time and self during meditation. Biology 11(8):1116

Liu X, Liu Y, Shi H, Zheng M (2021) Effects of mindfulness meditation on musical aesthetic emotion processing. Front Psychol 12:648062

Martinelli N, Droit-Volet S (2022) What factors underlie our experience of the passage of time? Theoretical consequences. Psychol Res 86(2):522–530

Martinelli N, Droit-Volet S (2023) Development and relationship between the judgment of the speed of passage of time and the judgment of duration in young children. Front Psychol 14: 1160047

Matko K, Ott U, Sedlmeier P (2021) What do meditators do when they meditate? Proposing a novel basis for future meditation research. Mindfulness 12:1791–1811

Meissner K, Wittmann M (2011) Body signals, cardiac awareness, and the perception of time. Biol Psychol 86(3):289–297

Mella N, Conty L, Pouthas V (2011) The role of physiological arousal in time perception: psychophysiological evidence from an emotion regulation paradigm. Brain Cogn 75(2): 182–187

Mella N, Bourgeois A, Perren F, Viaccoz A, Kliegel M, Picard F (2019) Does the insula contribute to emotion-related distortion of time? A neuropsychological approach. Hum Brain Mapp 40(5): 1470–1479

Metzinger T (2004) Being no one. The self-model theory of subjectivity. MIT Press, Cambridge

Metzinger T (2024) The elephant and the blind. The experience of pure consciousness: philosophy, science, and 500+ experiential reports. MIT Press, Cambridge

Minkowski E (1968) Le temps vécu. Delachaux et Niestlé, Paris

Mondillon L, Niedenthal P, Gil S, Droit-Volet S (2007) Imitation of in-group versus out-group member's facial expression of anger. Soc Neurosci 2(3–4):223–237

Mondok C, Wiener M (2023) Selectivity of timing: a meta-analysis of temporal processing in neuroimaging studies using activation likelihood estimation and reverse inference. Front Hum Neurosci 16:1000995

Montemayor C, Wittmann M (2022) The illusions of time passage: why time passage is real. Philosophies 7(6):140

Morin A, Grondin S (2024) Mindfulness and time perception: a systematic integrative review. Neurosci Biobehav Rev:105657. In press

Moser E, Moser M-B, Mcnaughton B (2017) Spatial representation in the hippocampal formation: a history. Nat Neurosci 20:1448–1464

Naghibi N, Jahangiri N, Khosrowabadi R, Eickhoff CR, Eickhoff SB, Coull JT, Tahmasian M (2024) Embodying time in the brain: a multi-dimensional neuroimaging meta-analysis of 95 duration processing studies. Neuropsychol Rev 34:277–298

Nather FC, Bueno JLO, Bigand E, Droit-Volet S (2011) Time changes with the embodiment of another's body posture. PLoS One 6(5):e19818

Niedenthal P (2007) Embodying emotion. Science 18(316):1002–1005

Northoff G, Huang Z (2017) How do the brain's time and space mediate consciousness and its different dimensions? Temporo-spatial theory of consciousness (TTC). Neurosci Biobehav Rev 80:630–645

Northoff G, Lamme V (2020) Neural signs and mechanisms of consciousness: is there a potential convergence of theories of consciousness in sight? Neurosci Biobehav Rev 118:568–587

Ogden RS (2020) The passage of time during the UK Covid-19 lockdown. PLoS One 15(7): e0235871

Ogden RS, Henderson J, McGlone F, Richter M (2019a) Time distortion under threat: sympathetic arousal predicts time distortion only in the context of negative, highly arousing stimuli. PLoS One 14(5):e0216704

Ogden RS, Henderson J, Slade K, McGlone F, Richter M (2019b) The effect of increased parasympathetic activity on perceived duration. Conscious Cogn 76:102829

Ogden RS, Moore D, Piovesan A, Poole H (2023) Distortions to the passage of time during chronic pain: a mixed method study. Eur J Pain. https://doi.org/10.1002/ejp.2211

Otten S, Schötz E, Wittmann M, Kohls N, Schmidt S, Meissner K (2015) Psychophysiology of duration estimation in experienced mindfulness meditators and matched controls. Front Psychol 6:1215

Perestelo-Perez L, Barraca J, Penate W, Rivero-Santana A, Alvarez-Perez Y (2017) Mindfulness-based interventions for the treatment of depressive rumination: systematic review and meta-analysis. Int J Clin Health Psychol 17(3):282–295

Picard F, Craig AD (2009) Ecstatic epileptic seizures: a potential window on the neural basis for human self-awareness. Epilepsy Behav 16(3):539–546

Piovesan A, Mirams L, Poole H, Moore D, Ogden R (2019) The relationship between pain-induced autonomic arousal and perceived duration. Emotion 19(7):1148

Pollatos O, Laubrock J, Wittmann M (2014) Interoceptive focus shapes the experience of time. PLoS One 9(1):e86934

Richter F, Ibáñez A (2021) Time is body: multimodal evidence of crosstalk between interoception and time estimation. Biol Psychol 159

Robbe D (2023) Lost in time: relocating the perception of duration outside the brain. Neurosci Biobehav Rev 153:105312

Rutrecht H, Wittmann M, Khoshnoud S, Igarzábal FA (2021) Time speeds up during flow states: a study in virtual reality with the video game thumper. Timing Time Percept 9(4):353–376

Sadeghi S, Wittmann M, De Rosa E, Anderson AK (2023) Wrinkles in subsecond time perception are synchronized to the heart. Psychophysiology:e14270

Schäfer T, Fachner J, Smukalla M (2013) Changes in the representation of space and time while listening to music. Front Psychol 4:508

Scherer KR (1982) Emotion as process: function, origin and regulation. Soc Sci Inform 21:555–570

Sgouramani H, Vatakis A (2014) "Flash" dance: how speed modulates perceived duration in dancers and non-dancers. Acta Psychol (Amst) 147:17–24

Shanon B (2001) Altered temporality. J Conscious Stud 8:35–58

Sharp PB, Sutton BP, Paul EJ, Sherepa N, Hillman CH, Cohen NJ, Kramer AF, Shaurya Prakash R, Heller W, Telzer EH, Barbey AK (2018) Mindfulness training induces structural connectome changes in insula networks. Sci Rep 8:7929

Smith C, Lazarus R (1990) Emotion and adaptation. In: Pervin LA (ed) Handbook of personality theory and research. Guilford, New York, pp 609–637

Strack F, Martin LL, Stepper S (1988) Inhibiting and facilitating conditions of the human smile: a nonobtrusive test of the facial feedback hypothesis. J Pers Soc Psychol 54(5):768–777

Strigo IA, Craig AD (2016) Interoception, homeostatic emotions and sympathovagal balance. Phil Trans R Soc B Biol Sci 371(1708):20160010

Sucala M, David D (2013) Mindful about time in a fast forward world. The effects of mindfulness exercise on time perception. Transyl. J Psychol 14(2):243–253

Tallon-Baudry C, Campana F, Park HD, Babo-Rebelo M (2018) The neural monitoring of visceral inputs, rather than attention, accounts for first-person perspective in conscious vision. Cortex 102:139–149

Teghil A (2024) Interoceptive and bodily processing in prospective and retrospective timing. In: Wittmann M, Strigo I, Simmons A (eds) The bodily self, emotion, and subjective time: exploring Interoception through the contributions of A.D. (Bud) Craig, Current topics in behavioral neurosciences. Springer. In press

Teghil A, Boccia M, D'Antonio F, Di Vita A, de Lena C, Guariglia C (2019) Neural substrates of internally-based and externally-cued timing: an activation likelihood estimation (ALE) meta-analysis of fMRI studies. Neurosci Biobehav Rev 96:197–209

Teghil A, Boccia M, Nocera L, Pietranelli V, Guariglia C (2020a) Interoceptive awareness selectively predicts timing accuracy in irregular contexts. Behav Brain Res 377:112242

Teghil A, Di Vita A, D'Antonio F, Boccia M (2020b) Inter-individual differences in resting-state functional connectivity are linked to interval timing in irregular contexts. Cortex 128:254–269
Thönes S, Wittmann M (2016) Time perception in yogic mindfulness meditation – effects on retrospective duration judgments and time passage. Psychol Conscious Theory Res Practice 3(4):316–325
Treisman M (1963) Temporal discrimination and the indifference interval: implications for a model of the "internal clock". Psychol Monogr Gen Appl 77(13):1–31
Uddin LQ, Nomi JS, Hébert-Seropian B, Ghaziri J, Boucher O (2017) Structure and function of the human insula. J Clin Neurophysiol 34(4):300–306
Ulrich M, Keller J, Grön G (2016) Neural signatures of experimentally induced flow experiences identified in a typical fMRI block design with BOLD imaging. Social Cog Affect Neurosci 11(3):496–507
Velmans M (2009) Understanding consciousness. Routledge, Hove
Visalli A, Begliomini C, Mioni G (2023) The effect of emotion intensity on time perception: a study with transcranial random noise stimulation. Exp Brain Res 241:2179–2190
Wang R, Mo F, Shen Y, Song Y, Cai H, Zhu J (2023) Functional connectivity gradients of the insula to different cerebral systems. Hum Brain Mapp 44(2):790–800
Wearden JH (2016) The psychology of time perception. Palgrave Macmillan
Wheeler MS, Arnkoff DB, Glass CR (2017) The neuroscience of mindfulness: how mindfulness alters the brain and facilitates emotion regulation. Mindfulness 8:1471–1487
Witowska J, Schmidt S, Wittmann M (2020) What happens while waiting? How self-regulation affects boredom and subjective time during a real waiting situation. Acta Psychol (Amst) 205: 103061
Wittmann M (2013) The inner sense of time: how the brain creates a representation of duration. Nat Rev Neurosci 14(3):217–223
Wittmann M (2015) Modulations of the experience of self and time. Conscious Cogn 38:172–181
Wittmann M (2018) Altered states of consciousness: experiences out of time and self. MIT Press, Cambridge
Wittmann M, Droit-Volet S (2024) Les théories et modèles du temps: les récentes théories du temps incarné. In: Coello Y (ed) Corps, cerveau et processus mentaux: les fondements sensortimoteurs de la cognition. Dunoud, Paris, pp 147–158
Wittmann M, Montemayor C (2022) Reinterpreting the Einstein-Bergson debate through contemporary neuroscience. In: Campo A, Gozzano S (eds) Einstein vs. Bergson: an enduring quarrel of time. De Gruyter, Berlin, pp 349–374
Wittmann M, Schmidt S (2014) Mindfulness meditation and the experience of time. In: Schmidt S, Walach H (eds) Meditation – neuroscientific approaches and philosophical implications. Springer, New York, pp 199–209
Wittmann M, van Wassenhove V (2009) The experience of time: neural mechanisms and the interplay of emotion, cognition and embodiment. Phil Trans R Soc B Biol Sci 364:1809–1813
Wittmann M, Carter O, Hasler F, Cahn BR, Grimberg U, Spring P, Hell D, Flohr H, Vollenweider FX (2007) Effects of psilocybin on time perception and temporal control of behaviour in humans. J Psychopharmacol 21(1):50–64
Wittmann M, Simmons AN, Aron JL, Paulus MP (2010) Accumulation of neural activity in the posterior insula encodes the passage of time. Neuropsychologia 48(10):3110–3120
Wittmann M, Simmons AN, Flagan T, Lane SD, Wackermann J, Paulus MP (2011) Neural substrates of time perception and impulsivity. Brain Res 1406:43–58
Wittmann M, Peter J, Gutina O, Otten S, Kohls N, Meissner K (2014) Individual differences in self-attributed mindfulness levels are related to the experience of time and cognitive self-control. Pers Individ Differ 64:41–45
Wöllner C (2024) Music in the air and in the body: an interoceptive system's perspective on musical emotions, awareness and time. In: Wittmann M, Strigo I, Simmons A (eds) The bodily self, emotion, and subjective time: exploring Interoception through the contributions of A.D. (Bud) Craig, Current topics in behavioral neurosciences. Springer. In press

Interoceptive and Bodily Processing in Prospective and Retrospective Timing

Alice Teghil

Contents

Abstract This chapter reviews some directions along which Craig's proposal of subjective time as emergent from interoceptive and bodily dynamics allows to frame recent findings on prospective and retrospective time processing. Behavioral and neuroimaging evidence from prospective timing studies demonstrates that an interoceptive-insular system may support the development of a primary representation of time in the context of large-scale networks involved in duration processing. Studies showing a tight link between episodic memory and interoceptive, emotional, and sensorimotor states further provide insights on processes supporting retrospective timing. These lines of evidence show that acknowledging its dependence on bodily states is most likely a crucial step toward a mechanistic understanding of time perception.

Keywords Embodiment · Hippocampus · Insula · Memory · Time perception

A. Teghil (✉)
Department of Psychology, Sapienza University of Rome, Rome, Italy
e-mail: alice.teghil@uniroma1.it

Curr Topics Behav Neurosci (2025) 70: 223–245
https://doi.org/10.1007/7854_2024_516
Published Online: 23 October 2024

1 Introduction

A temporal dimension is intrinsic to all behaviors and cognitive processes. Integration of sensory inputs in the range of a few milliseconds results in our perception of a dynamically changing environment, while the generation of precise temporal patterns is essential to respond to this perceived flow of change. Moving along the continuum, representing the timing and duration of events from seconds to years is a prerequisite to high-level cognitive functions such as planning and memory (Block and Gruber 2014). Finally, the perception of ourselves as entities continuously existing over time provides the basis for autonoetic consciousness, which is intimately intertwined with our sense of self and personal identity (Nelson and Fivush 2004; Markowitsch and Staniloiu 2011).

A growing body of results has accumulated supporting the notion that our perception of time is fundamentally rooted in the experience of our bodies. Evidence for such a causal link between the processing of information arising from the body and the emergence of our sense of time has been collected all along the continuum of processing levels and temporal scales. At the level of sensory processing, phases of cardiac dynamics influence the perceived duration of events in the subsecond range (Sadeghi et al. 2023; Arslanova et al. 2023). Abstract representations of time as progressing along a back-to-front axis (with the past at our back and the future in front of us) also develop at least partially as a consequence of the sensorimotor experience of our body movements in the environment (Miles et al. 2010; Rinaldi et al. 2016; Teghil et al. 2021).

The idea that subjective time emerges from the integration of bodily information into a unified meta-representation of "the sentient self" was first systematized by A.D. (Bud) Craig. In his cinemascopic model of awareness, Craig (2009a, 2010, 2015) proposed that the anterior insular cortex – the convergence point of an ascending pathway signaling the state of tissues of the whole body, as well as of multiple channels conveying information about environmental events (Craig 2009b) – supports the development of a representation of all relevant conditions for the organism at a given time, in the form of subjective feelings with homeostatic significance ("global emotional moments"). These moments of awareness are continuously replicated, so that their sequential progression allows for the subjective continuity of the self across time and supports awareness for time passage. Craig's model thus posits a causal dependence of the sense of time on mechanisms supporting the representation of the physiological conditions of the body.

The following sections will review findings from different lines of research in the field of time perception. These results highlight some of the directions along which Craig's work opened new roads to understand the relation between time perception and bodily processing when time is "in the foreground" (MacDonald 2014) as a focus of attention (*prospective timing*, Sect. 2), and when time is "in the background," being encoded incidentally in memory (*retrospective timing*, Sect. 3).

2 Time in the Foreground: Interoceptive Signals and Prospective Timing

2.1 Neural Networks for Prospective Timing

Multiple lines of evidence show that time perception depends on an interrelated set of mental processes and neural systems. Lesions to different brain structures, including the cerebellum (Mangels et al. 1998; Gooch et al. 2010), the dorsolateral prefrontal (Koch et al. 2002) and parietal (Coslett et al. 2009) cortex, the precentral gyrus (Gooch et al. 2011), and the insula (Monfort et al. 2014; Mella et al. 2019) produce alterations in temporal processing. Although these lesions may increase variability (Gooch et al. 2011; Coslett et al. 2009), reduce accuracy (Harrington et al. 1998; Koch et al. 2002), and distort perceived durations (Monfort et al. 2014; Mella et al. 2019; see also Coelho et al. 2022, for a review) in specific timing tasks, none of them completely abolishes the sense of time (Merchant et al. 2013; Wiener et al. 2011). These findings are consistent with neuroimaging literature. Meta-analyses synthetizing the results of several functional magnetic resonance imaging (fMRI) studies have highlighted the involvement of a distributed network in time perception, including the dorsal striatum, the cerebellum, the supplementary motor area (SMA), the motor cortex, the inferior parietal cortex, the inferior frontal gyrus, the superior temporal gyrus, and the insula (Wiener et al. 2010; Nani et al. 2019; Teghil et al. 2019; Naghibi et al. 2024).

Different factors interact in modulating the likelihood that specific brain regions or networks are recruited in timing. Subcortical structures such as the cerebellum and the striatum are more likely involved in timing durations approximately below 1 s, whereas processing suprasecond durations entails activation of cortical regions, including the middle frontal gyrus and the superior parietal cortex (Wiener et al. 2010; Lewis and Miall 2003; Nani et al. 2019). Modality-specific mechanisms are also involved in processing time in the range of milliseconds, with evidence of neural population responses tuned to specific durations in primary visual (Harvey et al. 2020; Hendrikx et al. 2022) and auditory (van Ackooij et al. 2022) regions. The range of durations tested further interacts with the sensorimotor demands of the task, as not completely overlapping networks are activated when intervals to be timed are defined by a motor response, and when the task requires a perceptual duration judgment (Bueti et al. 2008; Wiener et al. 2010; Nani et al. 2019).

Both intrinsic and dedicated models have been proposed to explain how we process and perceive time (see Ivry and Schlerf 2008; Wittmann 2013; Merchant et al. 2013, for reviews). However, "intermediate" accounts, which do not necessarily posit the existence of a unique amodal timing system, still acknowledge the involvement of specific brain networks and neural mechanisms in timing, have gained credibility over time. This type of accounts appears in fact to capture better the dynamic and multifaceted nature of neural networks involved in the perception of time. Time perception thus would be supported by specific, yet multiple and overlapping, neural systems. These procedures likely operate at different temporal scales and serve different purposes (Paton and Buonomano 2018). Nonetheless, separate timing mechanisms may also be involved simultaneously or in an exchangeable way, based on task features, previous context, and dimensions of individual variability (Wiener et al. 2011; Murai and Yotsumoto 2016; Grahn and McAuley 2009). Merchant et al. (2013) proposed that timing is supported by a "core" network, interacting with multiple brain regions depending on the specific task demands. The core timing network was identified with the cortico–thalamic–basal ganglia circuit, including the SMA (Merchant et al. 2013; Merchant and Yarrow 2016).

Consistent with this framework, a recent quantitative meta-analysis of fMRI experiments (Naghibi et al. 2024) assessed the impact of several task dimensions on the likelihood of specific brain regions to be activated during timing. Specific patterns of activation were found to be associated with different combinations of task dimensions and stimulus features. However, the only two regions with significant activation probability across *all* included task dimensions were the more rostral part of the SMA (pre-SMA) and the insular cortex. The same two regions were highlighted in an analysis restricted to fMRI experiments using more stringent control tasks. An independent study confirmed these findings using a reverse inference approach (Mondok and Wiener 2023). On the one hand, these results align with evidence that several factors determine which brain networks are recruited for timing, consistent with intermediate accounts on brain correlates of time processing. On the other hand, they point to the SMA and the insular cortex as to core timing nodes in humans.

The SMA is a functionally and anatomically heterogeneous region broadly involved in motor processing and cognition, with functions spanning from supporting sequential aspects of movements to cognitive control (Nachev et al. 2008). This region represents both space and time, along a gradiental organization which may reflect the tight coupling of these two metrics in service of action (Cona et al. 2021). Accordingly, it has been proposed that the consistent involvement of the SMA in time perception is due to the rooting of abstract time representation into sensorimotor processing (Naghibi et al. 2024).

The insular cortex is similarly a complex and multifaceted brain structure, implicated in several features of sensorimotor, affective, and cognitive processing, including action control, processing of exteroceptive and interoceptive sensory information, and decision-making (Zhang et al. 2024; Craig 2009b). The anterior insula is a crucial node of the salience network, which supports the representation of homeostatically relevant stimuli and outcomes, and the maintenance of an appropriate cognitive set to respond to perceived salience (Seeley et al. 2007; Seeley 2019). Different findings show that the insula may have an integral role in supporting the perception of time through the building of a continuous representation of the internal environment, and that this mechanism provides a basis for the estimation of durations independently from external inputs.

2.2 *The Insula and Prospective Timing*

The hypothesis that time perception crucially relies on the insular cortex has been first put forward by A.D. (Bud) Craig (2009a, 2015), based on the convergence of his research on neural pathways signaling the conditions of the body with evidence from behavioral, neurophysiological, and neuroimaging studies. The dorsal posterior insula is the cortical termination point of an ascending pathway signaling the physiological state of all bodily tissues and organs, and is the first cortical site at which a somatotopically organized sensory representation is developed of distinct sensations (Craig 2002). In the middle insula, this primary interoceptive image is integrated with information arising from proprioceptive and mechanoreceptive activity, signaling the effects of one's own action, as well as with exteroceptive information concerning salient features of the environment (Craig 2015). Finally, the anterior insula supports the development of a second-order representation of all perceived feelings and salient conditions in a given moment, allowing the subjective awareness of such feelings. Consistent with this integrative progression, the dorsal posterior insula is activated by several categories of stimuli with homeostatic significance, such as noxious stimuli, heat, and affective touch (Craig 2002; Björnsdotter et al. 2009), whereas neural activity in the anterior insula is associated with accuracy in perceiving one's own heartbeats (Critchley et al. 2004). As the process responsible for the development of a representation of the material self through the integration of salience across relevant conditions is replicated at each moment in time, according to the rate of endogenous brain oscillations, this mechanism supports the continuity of the sentient self across time and allows for the experience of the duration of events (Craig 2009a, 2015).

The proposal that the experience of time is rooted in interoceptive processing and the insula has been confirmed by several lines of research (Wittmann and Droit-Volet 2024). Behavioral and physiological studies show that different dimensions of interoception, including afferent signaling and its preconscious impact, neural processing of interoceptive signals, interoceptive accuracy and sensibility (Nord and Garfinkel 2022), are associated with measures of temporal processing. At the level of afferent signaling, the perceived duration of stimuli in the millisecond range is affected by moment-by-moment cardiac dynamics (Sadeghi et al. 2023; Arslanova et al. 2023) and measures of vagal activity are associated with accuracy in duration reproduction (Pollatos et al. 2014a) and bisection (Cellini et al. 2015). Other studies (Meissner and Wittmann 2011; Otten et al. 2015) demonstrated a correlation between the slope of cardiac periods during the encoding of time intervals of multiple seconds and their reproduction accuracy. At the neural level, the amplitude of heartbeat-evoked potentials, a marker of cortical processing of afferent cardiac signals, is correlated with the perceived length of time intervals both in the range of seconds (Khoshnoud et al. 2024) and minutes (Richter and Ibáñez 2021). Accuracy in reproducing second-long time intervals further correlates with the ability to track one's own heartbeats (Meissner and Wittmann 2011), as well as with individual differences in trait-like self-reported dispositions toward bodily signals (Teghil et al.

2020a; Khoshnoud et al. 2024). Thus, involuntary bodily dynamics and the conscious perception of internal signals both affect time processing, most likely with variations in the weight of different interoceptive dimensions according to the timescale of processed durations.

Results of some carefully conducted task-based fMRI studies point to a role for the encoding of bodily changes in the insular cortex in the perception of time. Livesey et al. (2007) manipulated the difficulty of the control task to identify brain activations associated with timing independently of other cognitive demands. The bilateral anterior insula was one of the few brain regions surviving the task difficulty manipulation. Across two experiments, Wittmann et al. (2010, 2011) showed ramping neural activity in the insula. The authors asked participants to reproduce time intervals of multiple seconds marked by acoustical signals. A secondary memory task (holding a series of numbers in memory for the duration of the trial) was also included to prevent chronometric counting. In the encoding phase of the reproduction task, a ramping pattern of activity, peaking at the end of the presented duration, was observed in the posterior insula, and in a set of regions including the precentral gyrus and the posterior superior temporal gyrus. Accumulating neural activation, peaking before the button press, was also observed in the reproduction phase in the anterior insula, as well as in the medial and inferior frontal cortex. These findings are consistent with the role of the posterior insula as the primary interoceptive cortex, and of the anterior insula in the awareness of bodily feelings, suggesting that the accumulation of physiological changes during duration encoding provides a basis for the subjective experience of time, while the explicit use of this representation is supported by the anterior insula (Wittmann et al. 2010, 2011).

Other findings contribute further insights into the functional significance of this mechanism in the context of distributed networks supporting time processing. In a behavioral study (Teghil et al. 2020a), the authors assessed the relation between interoceptive sensibility (corresponding to beliefs on how well one perceives their own bodily signals, Critchley and Garfinkel 2017) and duration reproduction accuracy. Interoceptive sensibility was evaluated using the Self-Awareness Questionnaire (SAQ) (Longarzo et al. 2015), a self-report instrument that assesses the frequency of sensations arising from the whole body, including visceral feelings, pain, and skin temperature. Participants reproduced time intervals between ~8 and 19 s, marked by two auditory signals. Following Wittmann et al. (2010, 2011), a secondary task was added to prevent chronometric counting: a series of numbers was presented auditorily during both the encoding and reproduction phases, and participants had to indicate whether a number displayed at the end of the intervals was present in the series. This secondary task allowed to set two task conditions: in a first condition, numbers were arranged into a regular pattern that was maintained between the encoding and reproduction phase of a single trial. The condition with a regular pattern aimed to provide participants with an external sensory cue to aid time estimation. Conversely, in a second condition, numbers for the secondary task were presented at random intervals, thus being arranged in an irregular pattern that did not provide any reliable cue on elapsed time. Supporting the effectiveness of this manipulation, participants reproduced time intervals more accurately in the

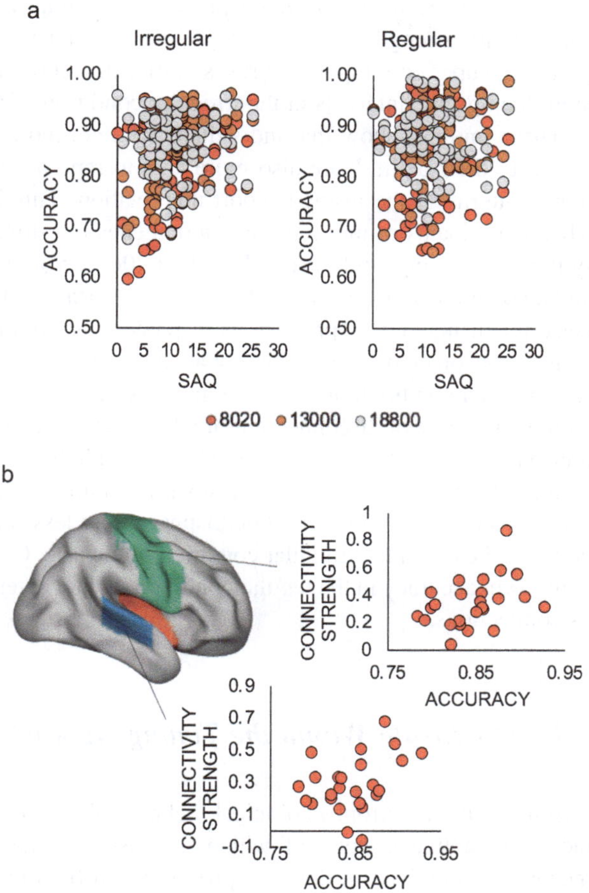

Fig. 1 (**a**) Correlation between reproduction accuracy for the three target durations (8,020, 13,000, 18,800 ms) and SAQ scores in the regular and irregular conditions of the duration reproduction task. (**b**) Correlation between accuracy in the irregular condition of the duration reproduction task and connectivity patterns of the right posterior insula (shown in red). The uppermost scatterplot refers to connectivity with the precentral gyrus (shown in green); the lowermost scatterplot refers to connectivity with the posterior superior temporal gyrus (shown in blue)

condition with a regular pattern. More importantly, SAQ scores predicted accuracy in duration reproduction specifically in the second (irregular) condition, while no significant relation was apparent between interoceptive sensibility and timing performance when a regular pattern was provided (Fig. 1a).

The relation between interoceptive sensibility and time perception when not relying on external cues is supported by specific neural correlates, as shown by the results of a resting-state fMRI study (Teghil et al. 2020b) which investigated the association between individual differences in intrinsic functional connectivity and behavioral performance. Participants performed the duration reproduction task in regular/irregular conditions described above, and underwent resting-state fMRI in a separate session. Within the brain network involved in prospective timing, only individual differences in the functional coupling between the putamen, the insula, and the precentral gyrus were associated with timing performance in the irregular condition. As expected, individual differences in interoceptive sensibility (assessed using the SAQ) predicted the connectivity of the insula with a large set of brain regions involved in sensorimotor processing. Within this latter brain network, the connectivity strength of the right posterior insula with the precentral gyrus and posterior superior temporal gyrus significantly correlated only with accuracy in reproducing time intervals in the irregular condition of the timing task (Fig. 1b).

These findings show that individuals who are more confident in their ability to "listen to their bodies" are also more able to reproduce elapsed time, and connectivity patterns of the insula support this relation. This link is specifically apparent when participants cannot rely on external cues to estimate elapsing time. A lesion-symptom mapping study (Teghil et al. 2020c), involving a sample of patients with unilateral brain damage, provides further, causal evidence for a key role of the posterior insula and its projections in processing durations when reliable environmental cues on elapsing time are not available. Reduced accuracy in both the regular and irregular condition of the duration reproduction task mentioned above correlated with the severity of disconnection of white matter tracts connecting the dorsolateral prefrontal cortex, the superior frontal gyrus including the SMA, and lateral parietal regions. However, patients with lesions including the posterior insula and fronto-insular connections in the right hemisphere were less accurate in duration reproduction specifically in the irregular condition of the task. Conversely, lesions of patients with lower accuracy in the regular condition did not extend to the posterior insula nor its connections.

2.3 The Insula Within the Timing Network

Taken together, neurophysiological, behavioral, and neuroimaging studies point to the interoceptive system, of which the insula is the crucial cortical node, as a mechanism involved in duration processing in both the range of milliseconds and seconds. Thus, the insula represents a main hub within networks supporting prospective timing (Naghibi et al. 2024; Mondok and Wiener 2023), likely through the development of a representation of time based on the integration of ascending bodily signals. Evidence for a selective involvement of interoceptive processing and the insula in timing when the environment provides no reliable duration cues further suggests that this mechanism is preferentially active when a representation of durations has to be derived independently from external inputs.

Neural pathways relaying homeostatic information on the conditions of bodily tissues provide ongoing inputs to the insular cortex. Individuals immersed in an environment with minimal auditory, visual, and tactile stimulation can produce accurate estimates of elapsed time, although they also experience time distortions (Hruby et al. 2024), as a continuous experience of bodily information is warranted even in sensory deprivation conditions. Findings reviewed in this chapter demonstrate that bodily processing is a primary route through which humans develop a representation of time. Nonetheless, they suggest that the interoceptive-insular pathway is more or less relied upon depending on task conditions, consistent with models acknowledging the existence of multiple timing systems (Merchant et al. 2013; Wiener et al. 2011).

Several factors associated with the features of external sensory signals, as well as with inter-individual variability (see, e.g., Grahn and McAuley 2009) may in principle affect the balance between interoceptive processing and other sources of information when timing durations, and investigations in this field are still lacking. Recent studies suggest that different sensory information channels are weighted according to their reliability during time perception to improve the precision of estimates (de Kock et al. 2021a, 2023). A possibility, consistent with studies reported above, is that the reliance on the interoceptive-insular timing system is decreased in presence of exteroceptive signals that act as predictable cues on event duration. From such a perspective, this could be especially true for individuals who are less able to rely on their bodily feelings. As people greatly differ in their ability and propensity to use interoceptive signals (Ainley et al. 2016; Murphy 2022), assessing individual differences in interoception in relation to time perception in different environmental conditions would provide a first test to this hypothesis.

Although a relative consensus has been reached on brain regions part of the prospective timing network (Wiener 2024), task-dependent interactions between nodes of this network are far from being well understood. In the context of spatial cognition, Ekstrom et al. (2017) proposed that environmental navigation depends on a continuum of interaction between multiple regions of a network, centered on specific nodes. These nodes work as functional hubs, with a bias toward processing particular categories of information. This framework could be usefully borrowed to conceptualize time perception as a network phenomenon, in which the processing of bodily and environmental information continuously interacts, resulting in our experience of time. Results reviewed above show that the insular cortex is a key node within the timing network. An important challenge for future research will be to characterize factors allowing to predict its interactions with other brain regions to support different temporal demands.

3 Time in the Background: The Bodily Self in Memory

3.1 Retrospective Timing

Most evidence on psychological and neural mechanisms involved in perceiving time comes from studies asking participants to attend to the duration of stimuli or events. Nonetheless, estimates of durations in everyday life are typically given in hindsight, without people paying deliberate attention to the unfolding of time (Riemer et al. 2021; Balcı et al. 2023). This type of duration processing is referred to as *retrospective timing*, in contrast with *prospective timing*, involving deliberate attention to elapsing time (Block and Zakay 1997). Paradigms to test retrospective timing entail participants performing a task unrelated to time or duration. Then, after task completion, volunteers are asked unexpectedly to provide a judgment concerning the duration of experienced events, stimuli, or empty time intervals.

As the passage of time is not attended to, estimating the duration of events in retrospect relies on the retrieval of information incidentally encoded in episodic memory (Zakay and Block 2004). According to one of the most popular accounts, the *contextual change hypothesis* (Block 1978; Block and Reed 1978), retrospective duration estimates depend on the remembered amount of change in cognitive context during the intervals to be estimated (Block and Reed 1978; Block 1992). Episodic information associated with a sequence of events is assumed to be encoded in memory incidentally; when a retrospective duration judgment is required, contextual information is retrieved together with other features of memories for events occurred during the interval (Zakay and Block 2004). The perceived duration of events in memory thus depends on the amount of change in cognitive context retrieved when providing the retrospective duration judgment (Block 1982; Zakay and Block 2004). Contextual elements affecting retrospective judgments of durations include changes in cognitive processes required by tasks performed during the intervals (Block and Reed 1978; Block 1986), environmental conditions (Block 1982), and stimulus features (Faber and Gennari 2015a, b), as well as emotional reactions.

3.2 Emotional Effects on Retrospective Timing

Different findings suggest that emotional experiences affect retrospective time estimates. Music inducing positive affect is judged to last longer in retrospect compared with music inducing less positive affect (Kellaris and Kent 1992) and with an identical time interval spent performing a backward counting task (Bisson et al. 2008). The degree of focus on bodily sensations also contributes to the effect of emotion on retrospective duration judgments: in a study by Pollatos et al. (2014b), participants were shown frightful, funny, and neutral movie clips after having been instructed to either focus on their physiological reactions or on movie details. When estimated in hindsight, the duration of fun and frightening clips was, respectively, under- and over-estimated compared to that of neutral clips, and interoceptive attention increased this effect.

Although Craig's model did not explicitly discuss emotional effects on long-term memory and on the experience of elapsed time in hindsight, the proposal that interoceptive and sensorimotor feelings create structural units of awareness in the form of global emotional moments offers a potential framework to understand the effect of emotion on retrospective duration judgments. In the study mentioned above, the duration of the movies was subjectively expanded in memory for frightful stimuli, and compressed for funny ones (Pollatos et al. 2014b). The authors suggested that increased arousal while volunteers were watching the frightful movie led to an expansion in remembered time, through a faster accumulation of global emotional moments. In the funny movie condition, instead, the contraction of remembered time may have resulted from global emotional moments progressing more slowly, due to the decrease in sympathetic salience (Pollatos et al. 2014b).

Results showing that the emotional context during an event affects the remembered duration of the same experience suggest that long-term memory representations supporting retrospective time processing embed original feeling states. Recent studies support the possibility that retrospective timing relies on a reconstructive process, which involves the reinstatement of patterns of neural activity corresponding to the unfolding of the original experience, including a representation of affective and sensorimotor states.

3.3 *Retrospective Timing and the Hippocampus*

Substantial evidence points to a crucial role of the hippocampus in memory for time. Patients with Alzheimer's Disease, in which medial temporal lobe atrophy is a pathological hallmark, show alterations in retrospective timing (Mioni et al. 2021), even in early and prodromal stages (Teghil et al. 2023a). Neuroimaging studies further demonstrate the involvement of the medial temporal lobes in the incidental encoding of elapsed time. Thavabalasingam et al. (2018) showed participants two sequences of pictures in each trial of a task-based fMRI study. Pictures within each sequence were separated by a variable time interval in the range of seconds, and participants had to discriminate if the order of pictures was the same or different between the first and second sequence. Unbeknownst to volunteers, time intervals between pictures could also be the same or different from the first to the second sequence. Despite information about the duration of the intervals between pictures being task-irrelevant, patterns of activation in the hippocampus were more similar between the first and second sequence of a pair when the intervals between pictures were not different. Other studies showed that experienced events judged to be closer in time are also represented by more similar activation patterns in the hippocampus (Ezzyat and Davachi 2014; Deuker et al. 2016).

This body of findings complements results from neural recording in non-human animals, showing the existence of populations of "time cells" in the hippocampus. These cells fire at specific moments in time during intervals embedded in non-temporal tasks, so that their population activity spans across the whole intervals (e.g., MacDonald et al. 2011, 2013; Kraus et al. 2013). The existence of time cells in medial temporal lobes has been recently shown also in humans (Reddy et al. 2021; Schonhaut et al. 2023), and the stability of their firing patterns at encoding predicts the clustering of elements of the original experience upon retrieval (Umbach et al. 2020). These findings show that the hippocampus encodes a representation of duration as a component of event context, and that activity patterns of time cells likely provide a mechanism to encode the passage of time, representing the evolving temporal context during the encoding of event sequences (Eichenbaum 2014; Tsao et al. 2022).

3.4 Retrospective Timing and Episodic Reinstatement

Tsao et al. (2022) recently proposed a model of how retrospective estimates of durations may be accomplished at the neural level. According to this model, segmentation of the continuous stream of experience into "chunks," corresponding to discrete events, happens at different timescales, from seconds to minutes. Since each event is represented as a unique neural population state, the long-term representation of segmented events may be used for retrospective timing. Retrospective timing thus depends on the possibility to reinstate population states characterizing the initial experience.

The reinstatement of the neural representation of past events – that is, the reactivation of distributed brain activation patterns present at encoding – is generally acknowledged as a central mechanism of episodic memory. Through the reactivation of neural patterns corresponding to the encoding of the original events, episodic memory allows to "re-experience" such events, supporting the subjective feeling of re-living the original episodes (Staresina et al. 2012; Horner et al. 2015).

The proposal that retrospective processing of time requires the reinstatement of the original experience including its contextual details is consistent with the results of a study by Brunec et al. (2017). The authors asked participants to retrospectively discriminate the duration of periods of waiting at intersections during a spatial navigation task. Participants' performance was enhanced when they could re-experience the events through the reinstatement of their spatio-temporal context, which depends on the hippocampus (Howard and Eichenbaum 2013). In the same study, Brunec et al. (2017) showed that retrospective duration judgments tended to be more accurate when participants performed a movement during waiting, compared to a passive, no-movement condition.

Multiple sources of information are integrated during the reinstatement of original memories to support episodic re-experiencing. Medial temporal lobe regions, including the parahippocampal and hippocampal cortex, guide the reinstatement of discrete events, including details on people, places, objects, and animals (Staresina et al. 2012; Liang and Preston 2017; Tompary et al. 2016; Horner et al. 2015). This is achieved through pattern completion mechanisms that allow the recollection of multidimensional episodes upon presentation of partial cues (Rolls 2013; Brunec et al. 2020). Pattern completion thus affords a means to achieve a holistic recollection of events based on neocortical reactivation of the component features of the original experience (Horner et al. 2015).

A growing number of findings demonstrate that information conveyed by bodily signals is part of the multimodal representation of episodic memories for one's own life events (episodic autobiographical memories). Episodic autobiographical memories are recollected faster when face and bodily expressions upon retrieval are congruent with the emotional context at encoding (Riskind 1983) and when participants adopt the same bodily posture as during the original event (Dijkstra et al. 2007). These findings show that episodic autobiographical memories embed information on sensorimotor and affective features of the original experience, which thus act as effective cues to trigger memory reinstatement. A representation of previously experienced physiological states also appears to be incorporated into episodic autobiographical memories, and this effect is modulated by individual variations in how well one perceives somatic signals. McCall et al. (2015) recorded heart rate and skin conductance while participants experienced a series of threatening situations in an immersive virtual reality environment. Volunteers then watched a playback of their experience and rated their arousal during different moments of the original experience. Individuals more accurately perceiving their own's heartbeat showed higher coherence between remembered arousal levels and physiological measures recorded during the original experience. These findings are in line with fMRI studies showing that emotional memories are reinstated both in the anterior insula and in the hippocampus (Hennings et al. 2022; Kobelt et al. 2024; Crombie et al. 2024).

A different line of research supports the importance of bodily self-representation in the retrieval and re-experiencing of episodic autobiographical memories. Using fMRI, Bergouignan et al. (2014) showed reduced hippocampal responses at retrieval when an out-of-body experience was experimentally induced during the original event, leading to lower episodic memory performance. Viewing one's own body during an experience affects connectivity patterns of the hippocampus, with a positive association between subjective measures of re-experiencing, and connectivity between the hippocampus and the SMA and pre-/post-central gyrus (Gauthier et al. 2020). Recent data show even more compellingly that neural reinstatement of an experience is related with bodily self-awareness at encoding. Iriye et al. (2024) used a full-body illusion to manipulate the degree of body ownership during the observation of naturalistic events in a virtual environment. Participants retrieved memories for the same events 1 week later; at both sessions, their brain responses were assessed using fMRI. The hippocampus not only distinguished between events encoded with lower or higher body ownership, but showed increased neural signatures of reinstatement when higher body ownership was present at encoding. These results are consistent with those of an independent study (Meyer et al. 2023) showing that reinstatement in the hippocampus, precentral gyrus, and SMA is higher for scenes encoded in conditions of visuomotor as well as perspectival congruency. Episodic memory for life events thus likely requires interactions between the hippocampus and brain networks supporting the sense of one's own body.

As described in the next paragraph, there is evidence that retrospective duration judgments may rely on similar interactions between brain regions supporting episodic retrieval and those involved in bodily self-representation and sensorimotor processing.

3.5 Retrospective Timing and Bodily Processing

fMRI studies discussed in Sect. 3.3 support the notion that the hippocampus incidentally encodes elapsed time (Thavabalasingam et al. 2018; Ezzyat and Davachi 2014; Deuker et al. 2016). Nonetheless, most fMRI studies assessing the relation between hippocampal activation and judgments of temporal distance between experienced events used sequences composed of unrelated pictures. These laboratory stimuli are very different from complex, semantically meaningful events experienced in everyday life.

A recent study used resting-state fMRI and a retrospective timing task to provide insights on connectivity patterns supporting memory for durations. Stimuli were developed to resemble multimodal, everyday experiences, and consisted of a set of video clips, shot from a first-person perspective (Teghil et al. 2023b). Video clips displayed common situations, such as ordering a drink at the vending machine, or calling the elevator, and had a duration between 5 and 29 s. Participants performed a retrospective timing task outside the scanner: in a first phase, they were asked to observe the video clips, and to answer questions on their content, without any mention of the need to estimate time. Then, they were unexpectedly presented with pairs of screenshots taken from the video clips, and had to decide which of the two corresponding video clips had lasted longer. Accuracy with which participants retrospectively discriminated the duration of the video clips predicted the strength of resting-state connectivity of the bilateral hippocampus with a cluster of voxels in the right hemisphere, mainly spanning the pre- and extending to the postcentral gyrus (Fig. 2a). The hippocampus was more strongly coupled with the precentral gyrus at rest in participants with better duration discrimination performance.

The precentral gyrus is part of the network supporting body representation and body ownership (Di Vita et al. 2016; Grivaz et al. 2017) and is involved in the so-called *body schema*, a dynamic representation of body parts that interacts with the motor system during action generation (Boccia et al. 2020). Information conveyed by proprioceptive and vestibular inputs on the position of one's own body parts in space and time, that is represented into body schema, is a central aspect of body self-representation (Corradi-Dell'Acqua et al. 2009; Blanke et al. 2015).

The right precentral gyrus is also activated by prospective timing tasks, independent of actual motor demands (Nani et al. 2019) and is a shared activation site between brain networks supporting time perception and interoception (Richter and Ibáñez 2021). This region is more strongly functionally coupled with the posterior insula in individuals more sensitive toward their bodily signals, and the strength of the same connectivity patterns correlates with timing accuracy when there are no reliable external duration cues (Teghil et al. 2020b, see also Sect. 1).

A role of primary motor regions in the perception of time is not surprising considering the tight link between time processing and action dynamics. Timing precision is higher when participants are free to move while encoding an auditory stimulus (Wiener et al. 2019), and sounds are perceived as shorter when movements are slowed down (De Kock et al. 2021b). This is in line with results that the activation of facial muscles during a perceptual timing task predicts duration judgments (Fernandes and Garcia-Marques 2019). Droit-Volet et al. (2013) further reported that the duration of pictures of body postures suggesting different movements is affected by the content of the same pictures: pictures showing a running posture are perceived as lasting longer than pictures depicting a standing posture. Prospective time processing has thus been proposed to involve the reactivation and

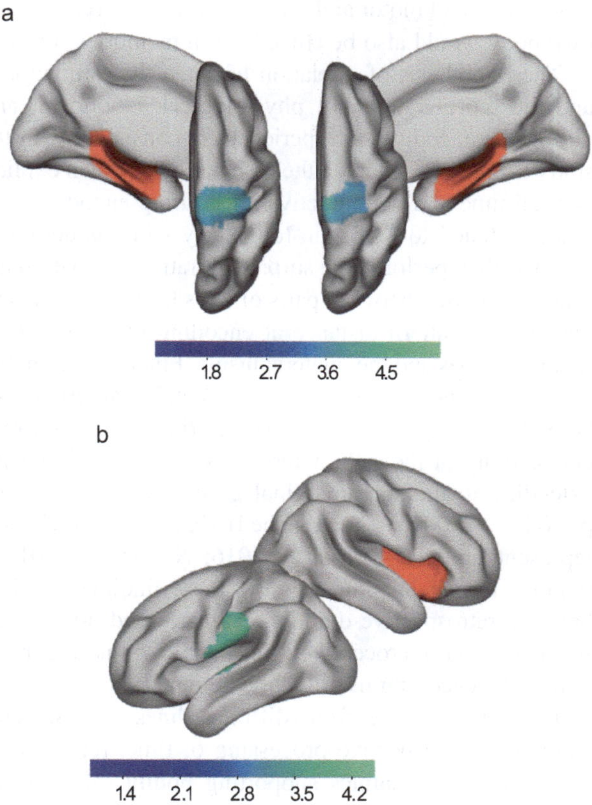

Fig. 2 (**a**) Clusters of voxels (in blue-to-green) showing a positive association between connectivity with the hippocampus (shown in red) and retrospective duration discrimination. Results for the left and right hippocampus are, respectively, displayed on the left and right side of the panel. (**b**) Clusters of voxels (in blue-to-green) showing a positive association between connectivity with the right anterior insula (shown in red) and retrospective duration discrimination

simulation of sensorimotor states, including action dynamics (Droit-Volet et al. 2013; Fernandes and Garcia-Marques 2019), through the involvement of motor structures of the brain (Coull and Droit-Volet 2018; Naghibi et al. 2024).

A possible interpretation of the relation between hippocampal-precentral functional coupling and retrospective duration discrimination in the resting-state fMRI study described above (Teghil et al. 2023b) is that participants with better retrospective duration discrimination performance are those in which a stronger hippocampal-motor coupling allows a more effective reinstatement of the sensorimotor simulation involved in the original experience, supporting increased re-living of the original event. This effect could be further related to the nature of the stimuli used in the retrospective timing task, i.e. video clips of actions shot from a first-person perspective that may have favored sensorimotor simulation. Such an interpretation is consistent with results showing that the strength of intrinsic connectivity of the hippocampus with pre/postcentral regions correlates with the experience of re-living, and is affected by the degree of bodily self-consciousness during the original experience (Gauthier et al. 2020). Results mentioned above further suggest that mechanisms supporting prospective and retrospective timing may at least in part overlap, as sensorimotor simulations would be involved during prospective duration processing, and motor and sensory dynamics contributing to the experience of one's own body would also be embedded in memories for the original events.

The hypothesis of a relation between the estimation of durations from memory and the representation of physiological, emotional, and sensorimotor conditions during the encoding of experiences is consistent with the results of the only fMRI study to date that assessed the association between explicit retrospective estimates of elapsed time and neural activations during encoding (Lositsky et al. 2016). Participants listened to a 25-min-long story while undergoing fMRI. Then, outside the scanner, they performed a surprise duration estimation task asking to say how much time has elapsed between pairs of clips from the story. Duration estimates correlated with neural pattern distance at encoding in a set of regions, including the medial temporal lobes and the bilateral insula. Finally, a re-analysis of data from the resting-state fMRI study described above (Teghil et al. 2023b) shows a positive association between retrospective duration discrimination performance and the strength of connectivity of the right anterior insula with a cluster in the left postcentral gyrus, extending to the supramarginal gyrus (peak at MNI -66, −18, + 34, cluster-size p-FWE = 0.0016) (unpublished) (Fig. 2b), which are both implicated in body representation (Di Vita et al. 2016; Naito et al. 2016). Although this possibility requires testing in future studies, these findings would be compatible with a link between retrospective duration estimates and differences in the integration of sensorimotor and interoceptive information within a representation of one's own body during the encoding of original experiences.

In sum, evidence from different lines of research supports the notion that, similarly to prospective processing of time, retrospective timing involves at least in part neural dynamics supporting bodily self-representation, and their trace in memory. A recent magnetoencephalography (MEG) study showed that spontaneous alpha activity recorded during wakeful rest predicts retrospective duration estimates at the scale of minutes (Azizi et al. 2023). In line with the proposal that individual units of representation of the physiological condition of the body across time correspond to oscillation cycles in the alpha rhythm (Craig 2015), the authors suggested that alpha bursts may represent an intrinsic mechanism parsing experience into discrete states of awareness, possibly indexed to provide an estimate of elapsed time when time is not attended to. From this perspective, the reinstatement of such states of awareness could be a candidate process to explain the link between retrospective duration estimates and the effect of physiological, emotional, and sensorimotor conditions present at encoding. Importantly, this possibility is compatible with current models of retrospective timing based on event segmentation and the reinstatement of population states (Tsao et al. 2022).

4 Conclusion

Within his model of human awareness, Craig theorized a mechanism which, through the moment-by-moment integration of salient internal and environmental conditions in the insula, supports both the extended experience of the self across time and the perception of durations in the millisecond-to-second range. Since this proposal has been put forward more than 10 years ago, convergent findings are increasingly providing support to the notion that the understanding of psychological and neural dynamics responsible for our experience of time – in the perception of present events, as well as in memory for past happenings – cannot be separated by the study of how we process information conveyed by channels signaling the state of our bodies.

The studies reviewed here suggest a framework in which the experience of time and the processing of durations extending from present to future rely on the interaction between ascending signals from our bodies and the current demands of the external environment. These same bodily dynamics may be an important component of the reconstruction in memory of the content of our experience, thus contributing to explain subjective features of the temporal dimension of our past. This can necessarily only be a starting point, as many issues need to be addressed. Among several others, these include the impact of different dimensions of interoception at scales of temporal processing, the need to characterize network interactions between the interoceptive-insular timing system and other brain regions involved in time perception, as well as the understanding of dynamics putatively allowing the reinstatement of previous states of awareness in service of retrospective judgments of time. Nonetheless, acknowledging the causal relation between bodily states and time perception is likely a crucial step toward a mechanistic understanding of timing at different timescales and processing levels.

Acknowledgments I never had the chance to know A.D. (Bud) Craig. Nonetheless, I owe to his ideas and groundbreaking conceptualizations most of my research agenda up to now. I am grateful to Marc Wittmann for the opportunity to contribute to this book, as well as for all the insightful exchanges and deeply stimulating discussions I was so lucky to have with him over the last year.

I am thankful to Maddalena Boccia for her supervision and help during many of the studies described here, and to Matilde Conti for her invaluable support and for providing feedback on a first version of this chapter. I am also grateful to both for their long-lasting friendship.

References

Ainley V, Apps MAJ, Fotopoulou A, Tsakiris M (2016) 'Bodily precision': a predictive coding account of individual differences in interoceptive accuracy. Philos Trans R Soc B 371(1708): 20160003

Arslanova I, Kotsaris V, Tsakiris M (2023) Perceived time expands and contracts within each heartbeat. Curr Biol 33(7):1389–1395.e4

Azizi L, Polti I, Van Wassenhove V (2023) Spontaneous α Brain Dynamics Track the Episodic "When." J Neurosci 43(43):7186–7197

Balcı F, Ünübol H, Grondin S, Sayar GH, Van Wassenhove V, Wittmann M (2023) Dynamics of retrospective timing: a big data approach. Psychon Bull Rev 30(5):1840–1847

Bergouignan L, Nyberg L, Ehrsson HH (2014) Out-of-body–induced hippocampal amnesia. Proc Natl Acad Sci U S A 111(12):4421–4426

Bisson N, Tobin S, Grondin S (2008) Remembering the duration of joyful and sad musical excerpts: assessment with three estimation methods. Neuroquantology 7(1)

Björnsdotter M, Löken L, Olausson H, Vallbo Å, Wessberg J (2009) Somatotopic organization of gentle touch processing in the posterior insular cortex. J Neurosci 29(29):9314–9320
Blanke O, Slater M, Serino A (2015) Behavioral, neural, and computational principles of bodily Self-consciousness. Neuron 88(1):145–166
Block RA (1978) Remembered duration: effects of event and sequence complexity. Mem Cognit 6(3):320–326
Block RA (1982) Temporal judgments and contextual change. J Exp Psychol Learn Mem Cogn 8(6):530–544
Block RA (1986) Remembered duration: imagery processes and contextual encoding. Acta Psychol (Amst) 62(2):103–122
Block RA (1992) Prospective and retrospective duration judgment: the role of information processing and memory. In: Macar F, Pouthas V, Friedman WJ (eds) Time, action and cognition. Springer, Dordrecht, pp 141–152
Block RA, Gruber RP (2014) Time perception, attention, and memory: a selective review. Acta Psychol (Amst) 149:129–133
Block RA, Reed MA (1978) Remembered duration: evidence for a contextual-change hypothesis. J Exp Psychol Hum Learn Mem 4(6):656–665
Block RA, Zakay D (1997) Prospective and retrospective duration judgments: a meta-analytic review. Psychon Bull Rev 4(2):184–197
Boccia M, Raimo S, Di Vita A, Battisti A, Matano A, Guariglia C, Grossi D, Palermo L (2020) Topological and hodological aspects of body representation in right brain damaged patients. Neuropsychologia 148:107637
Brunec IK, Ozubko JD, Barense MD, Moscovitch M (2017) Recollection-dependent memory for event duration in large-scale spatial navigation. Learn Mem 24(3):104–114
Brunec IK, Robin J, Olsen RK, Moscovitch M, Barense MD (2020) Integration and differentiation of hippocampal memory traces. Neurosci Biobehav Rev 118:196–208
Bueti D, Walsh V, Frith C, Rees G (2008) Different brain circuits underlie motor and perceptual representations of temporal intervals. J Cogn Neurosci 20(2):204–214
Cellini N, Mioni G, Levorato I, Grondin S, Stablum F, Sarlo M (2015) Heart rate variability helps tracking time more accurately. Brain Cogn 101:57–63
Coelho P, Rodrigues JA, Nascimento Alves P, Fonseca AC (2022) Time perception changes in stroke patients: a systematic literature review. Front Neurol 13:938367
Cona G, Wiener M, Scarpazza C (2021) From ATOM to GradiATOM: cortical gradients support time and space processing as revealed by a meta-analysis of neuroimaging studies. Neuroimage 224:117407
Corradi-Dell'Acqua C, Tomasino B, Fink GR (2009) What is the position of an arm relative to the body? Neural correlates of body schema and body structural description. J Neurosci 29(13): 4162–4171
Coslett HB, Shenton J, Dyer T, Wiener M (2009) Cognitive timing: neuropsychology and anatomic basis. Brain Res 1254:38–48
Coull JT, Droit-Volet S (2018) Explicit understanding of duration develops implicitly through action. Trends Cogn Sci 22(10):923–937
Craig AD (2002) How do you feel? Interoception: the sense of the physiological condition of the body. Nat Rev Neurosci 3(8):655–666
Craig AD (Bud) (2009a) Emotional moments across time: a possible neural basis for time perception in the anterior insula. Philos Trans R Soc B 364(1525):1933–1942
Craig AD (Bud) (2009b) How do you feel – now? The anterior insula and human awareness. Nat Rev Neurosci 10(1):59–70
Craig AD (Bud) (2010) The sentient self. Brain Struct Funct 214(5–6):563–577
Craig AD (2015) How do you feel? An interoceptive moment with your neurobiological self. Princeton University Press, Princeton
Critchley HD, Garfinkel SN (2017) Interoception and emotion. Curr Opin Psychol 17:7–14

Critchley HD, Wiens S, Rotshtein P, Öhman A, Dolan RJ (2004) Neural systems supporting interoceptive awareness. Nat Neurosci 7(2):189–195

Crombie KM, Azar A, Botsford C, Heilicher M, Jaeb M, Gruichich TS, Schomaker CM, Williams R, Stowe ZN, Dunsmoor JE, Cisler JM (2024) Decoding context memories for threat in large-scale neural networks. Cereb Cortex 34(2):bhae018

De Kock R, Gladhill KA, Ali MN, Joiner WM, Wiener M (2021a) How movements shape the perception of time. Trends Cogn Sci 25(11):950–963

De Kock R, Zhou W, Joiner WM, Wiener M (2021b) Slowing the body slows down time perception. eLife 10:e63607

De Kock R, Zhou W, Datta P, Mychal Joiner W, Wiener M (2023) The role of consciously timed movements in shaping and improving auditory timing. Proc R Soc B 290(1992):20222060

Deuker L, Bellmund JL, Navarro Schröder T, Doeller CF (2016) An event map of memory space in the hippocampus. eLife 5:e16534

Di Vita A, Boccia M, Palermo L, Guariglia C (2016) To move or not to move, that is the question! Body schema and non-action oriented body representations: an fMRI meta-analytic study. Neurosci Biobehav Rev 68:37–46

Dijkstra K, Kaschak MP, Zwaan RA (2007) Body posture facilitates retrieval of autobiographical memories. Cognition 102(1):139–149

Droit-Volet S, Fayolle S, Lamotte M, Gil S (2013) Time, emotion and the embodiment of timing. Timing Time Percept 1(1):99–126

Eichenbaum H (2014) Time cells in the hippocampus: a new dimension for mapping memories. Nat Rev Neurosci 15(11):732–744

Ekstrom AD, Huffman DJ, Starrett M (2017) Interacting networks of brain regions underlie human spatial navigation: a review and novel synthesis of the literature. J Neurophysiol 118(6): 3328–3344

Ezzyat Y, Davachi L (2014) Similarity breeds proximity: pattern similarity within and across contexts is related to later mnemonic judgments of temporal proximity. Neuron 81(5): 1179–1189

Faber M, Gennari SP (2015a) Representing time in language and memory: the role of similarity structure. Acta Psychol (Amst) 156:156–161

Faber M, Gennari SP (2015b) In search of lost time: reconstructing the unfolding of events from memory. Cognition 143:193–202

Fernandes AC, Garcia-Marques T (2019) The perception of time is dynamically interlocked with the facial muscle activity. Sci Rep 9(1):18737

Gauthier B, Bréchet L, Lance F, Mange R, Herbelin B, Faivre N, Bolton TAW, Ville DVD, Blanke O (2020) First-person body view modulates the neural substrates of episodic memory and autonoetic consciousness: a functional connectivity study. Neuroimage 223:117370

Gooch CM, Wiener M, Wencil EB, Coslett HB (2010) Interval timing disruptions in subjects with cerebellar lesions. Neuropsychologia 48(4):1022–1031

Gooch CM, Wiener M, Hamilton AC, Coslett HB (2011) Temporal discrimination of sub- and Suprasecond time intervals: a voxel-based lesion mapping analysis. Front Integr Neurosci 5

Grahn JA, McAuley JD (2009) Neural bases of individual differences in beat perception. Neuroimage 47(4):1894–1903

Grivaz P, Blanke O, Serino A (2017) Common and distinct brain regions processing multisensory bodily signals for peripersonal space and body ownership. Neuroimage 147:602–618

Harrington DL, Haaland KY, Knight RT (1998) Cortical networks underlying mechanisms of time perception. J Neurosci 18(3):1085–1095

Harvey BM, Dumoulin SO, Fracasso A, Paul JM (2020) A network of topographic maps in human association cortex hierarchically transforms visual timing-selective responses. Curr Biol 30(8): 1424–1434.e6

Hendrikx E, Paul JM, Van Ackooij M, Van Der Stoep N, Harvey BM (2022) Visual timing-tuned responses in human association cortices and response dynamics in early visual cortex. Nat Commun 13(1):3952

Hennings AC, McClay M, Drew MR, Lewis-Peacock JA, Dunsmoor JE (2022) Neural reinstatement reveals divided organization of fear and extinction memories in the human brain. Curr Biol 32(2):304–314.e5

Horner AJ, Bisby JA, Bush D, Lin W-J, Burgess N (2015) Evidence for holistic episodic recollection via hippocampal pattern completion. Nat Commun 6(1):7462

Howard MW, Eichenbaum H (2013) The hippocampus, time, and memory across scales. J Exp Psychol Gen 142(4):1211–1230

Hruby H, Schmidt S, Feinstein JS, Wittmann M (2024) Induction of altered states of consciousness during floatation-REST is associated with the dissolution of body boundaries and the distortion of subjective time. Sci Rep 14(1):9316

Iriye H, Chancel M, Ehrsson HH (2024) Sense of own body shapes neural processes of memory encoding and reinstatement. Cereb Cortex 34(1):bhad443

Ivry RB, Schlerf JE (2008) Dedicated and intrinsic models of time perception. Trends Cogn Sci 12(7):273–280

Kellaris JJ, Kent RJ (1992) The influence of music on consumers' temporal perceptions: does time fly when you're having fun? J Consum Psychol 1(4):365–376

Khoshnoud S, Leitritz D, Çinar Bozdağ M, Alvarez Igarzábal F, Noreika V, Wittmann M (2024) When the heart meets the mind: exploring the brain-heart interaction during time perception. J Neurosci:e2039232024. https://doi.org/10.1523/JNEUROSCI.2039-23.2024

Kobelt M, Waldhauser GT, Rupietta A, Heinen R, Rau EMB, Kessler H, Axmacher N (2024) The memory trace of an intrusive trauma-analog episode. Curr Biol 34(8):1657–1669.e5

Koch G, Oliveri M, Carlesimo GA, Caltagirone C (2002) Selective deficit of time perception in a patient with right prefrontal cortex lesion. Neurology 59(10):1658–1658

Kraus BJ, Robinson RJ, White JA, Eichenbaum H, Hasselmo ME (2013) Hippocampal "Time Cells": time versus path integration. Neuron 78(6):1090–1101

Lewis PA, Miall RC (2003) Distinct systems for automatic and cognitively controlled time measurement: evidence from neuroimaging. Curr Opin Neurobiol 13(2):250–255

Liang JC, Preston AR (2017) Medial temporal lobe reinstatement of content-specific details predicts source memory. Cortex 91:67–78

Livesey AC, Wall MB, Smith AT (2007) Time perception: manipulation of task difficulty dissociates clock functions from other cognitive demands. Neuropsychologia 45(2):321–331

Longarzo M, D'Olimpio F, Chiavazzo A, Santangelo G, Trojano L, Grossi D (2015) The relationships between interoception and alexithymic trait. The self-awareness questionnaire in healthy subjects. Front Psychol 6

Lositsky O, Chen J, Toker D, Honey CJ, Shvartsman M, Poppenk JL, Hasson U, Norman KA (2016) Neural pattern change during encoding of a narrative predicts retrospective duration estimates. eLife 5:e16070

MacDonald CJ (2014) Prospective and retrospective duration memory in the hippocampus: is time in the foreground or background? Philos Trans R Soc B 369(1637):20120463

MacDonald CJ, Lepage KQ, Eden UT, Eichenbaum H (2011) Hippocampal "Time Cells" bridge the gap in memory for discontiguous events. Neuron 71(4):737–749

MacDonald CJ, Carrow S, Place R, Eichenbaum H (2013) Distinct hippocampal time cell sequences represent odor memories in immobilized rats. J Neurosci 33(36):14607–14616

Mangels JA, Ivry RB, Shimizu N (1998) Dissociable contributions of the prefrontal and neocerebellar cortex to time perception. Cogn Brain Res 7(1):15–39

Markowitsch HJ, Staniloiu A (2011) Memory, autonoetic consciousness, and the self. Conscious Cogn 20(1):16–39

McCall C, Hildebrandt LK, Bornemann B, Singer T (2015) Physiophenomenology in retrospect: memory reliably reflects physiological arousal during a prior threatening experience. Conscious Cogn 38:60–70

Meissner K, Wittmann M (2011) Body signals, cardiac awareness, and the perception of time. Biol Psychol 86(3):289–297

Mella N, Bourgeois A, Perren F, Viaccoz A, Kliegel M, Picard F (2019) Does the insula contribute to emotion-related distortion of time? A neuropsychological approach. Hum Brain Mapp 40(5): 1470–1479

Merchant H, Yarrow K (2016) How the motor system both encodes and influences our sense of time. Curr Opin Behav Sci 8:22–27

Merchant H, Harrington DL, Meck WH (2013) Neural basis of the perception and estimation of time. Annu Rev Neurosci 36(1):313–336

Meyer NH, Gauthier B, Stampacchia S, Boscheron J, Babo Rebelo M, Potheegadoo J, Herbelin B, Lance F, Alvarez V, Franc E, Esposito F, Morais Lacerda M, Blanke O (2023) Embodiment in episodic memory through premotor-hippocampal coupling. bioRxiv:2023.09.23.559108. https://doi.org/10.1101/2023.09.23.559108

Miles LK, Karpinska K, Lumsden J, Macrae CN (2010) The meandering mind: vection and mental time travel. PLoS One 5(5):e10825

Mioni G, Román-Caballero R, Clerici J, Capizzi M (2021) Prospective and retrospective timing in mild cognitive impairment and Alzheimer's disease patients: a systematic review and meta-analysis. Behav Brain Res 410:113354

Mondok C, Wiener M (2023) Selectivity of timing: a meta-analysis of temporal processing in neuroimaging studies using activation likelihood estimation and reverse inference. Front Hum Neurosci 16:1000995

Monfort V, Pfeuty M, Klein M, Collé S, Brissart H, Jonas J, Maillard L (2014) Distortion of time interval reproduction in an epileptic patient with a focal lesion in the right anterior insular/inferior frontal cortices. Neuropsychologia 64:184–194

Murai Y, Yotsumoto Y (2016) Context-dependent neural modulations in the perception of duration. Front Integr Neurosci 10

Murphy J (2022) Propensity to use interoceptive signals: an important individual difference. Biol Psychol 171:108326

Nachev P, Kennard C, Husain M (2008) Functional role of the supplementary and pre-supplementary motor areas. Nat Rev Neurosci 9(11):856–869

Naghibi N, Jahangiri N, Khosrowabadi R, Eickhoff CR, Eickhoff SB, Coull JT, Tahmasian M (2024) Embodying time in the brain: a multi-dimensional neuroimaging meta-analysis of 95 duration processing studies. Neuropsychol Rev 34(1):277–298

Naito E, Morita T, Amemiya K (2016) Body representations in the human brain revealed by kinesthetic illusions and their essential contributions to motor control and corporeal awareness. Neurosci Res 104:16–30

Nani A, Manuello J, Liloia D, Duca S, Costa T, Cauda F (2019) The neural correlates of time: a meta-analysis of neuroimaging studies. J Cogn Neurosci 31(12):1796–1826

Nelson K, Fivush R (2004) The emergence of autobiographical memory: a social cultural developmental theory. Psychol Rev 111(2):486–511

Nord CL, Garfinkel SN (2022) Interoceptive pathways to understand and treat mental health conditions. Trends Cogn Sci 26(6):499–513

Otten S, Schötz E, Wittmann M, Kohls N, Schmidt S, Meissner K (2015) Psychophysiology of duration estimation in experienced mindfulness meditators and matched controls. Front Psychol 6

Paton JJ, Buonomano DV (2018) The neural basis of timing: distributed mechanisms for diverse functions. Neuron 98(4):687–705

Pollatos O, Yeldesbay A, Pikovsky A, Rosenblum M (2014a) How much time has passed? Ask your heart. Front Neurorobot 8

Pollatos O, Laubrock J, Wittmann M (2014b) Interoceptive focus shapes the experience of time. PLoS One 9(1):e86934

Reddy L, Zoefel B, Possel JK, Peters J, Dijksterhuis DE, Poncet M, Van Straaten ECW, Baayen JC, Idema S, Self MW (2021) Human hippocampal neurons track moments in a sequence of events. J Neurosci 41(31):6714–6725

Richter F, Ibáñez A (2021) Time is body: multimodal evidence of crosstalk between interoception and time estimation. Biol Psychol 159:108017

Riemer M, Wolbers T, Van Rijn H (2021) Age-related changes in time perception: the impact of naturalistic environments and retrospective judgements on timing performance. Q J Exp Psychol 74(11):2002–2012

Rinaldi L, Locati F, Parolin L, Bernardi NF, Girelli L (2016) Walking on a mental time line: temporal processing affects step movements along the sagittal space. Cortex 78:170–173

Riskind JH (1983) Nonverbal expressions and the accessibility of life experience memories: a congruence hypothesis. Soc Cogn 2(1):62–86

Rolls ET (2013) The mechanisms for pattern completion and pattern separation in the hippocampus. Front Syst Neurosci 7

Sadeghi S, Wittmann M, De Rosa E, Anderson AK (2023) Wrinkles in subsecond time perception are synchronized to the heart. Psychophysiology 60(8):e14270

Schonhaut DR, Aghajan ZM, Kahana MJ, Fried I (2023) A neural code for time and space in the human brain. Cell Rep 42(11):113238

Seeley WW (2019) The salience network: a neural system for perceiving and responding to homeostatic demands. J Neurosci 39(50):9878–9882

Seeley WW, Menon V, Schatzberg AF, Keller J, Glover GH, Kenna H, Reiss AL, Greicius MD (2007) Dissociable intrinsic connectivity networks for salience processing and executive control. J Neurosci 27(9):2349–2356

Staresina BP, Henson RNA, Kriegeskorte N, Alink A (2012) Episodic reinstatement in the medial temporal lobe. J Neurosci 32(50):18150–18156

Teghil A, Boccia M, D'Antonio F, Di Vita A, De Lena C, Guariglia C (2019) Neural substrates of internally-based and externally-cued timing: an activation likelihood estimation (ALE) meta-analysis of fMRI studies. Neurosci Biobehav Rev 96:197–209

Teghil A, Boccia M, Nocera L, Pietranelli V, Guariglia C (2020a) Interoceptive awareness selectively predicts timing accuracy in irregular contexts. Behav Brain Res 377:112242

Teghil A, Di Vita A, D'Antonio F, Boccia M (2020b) Inter-individual differences in resting-state functional connectivity are linked to interval timing in irregular contexts. Cortex 128:254–269

Teghil A, Di Vita A, Pietranelli V, Matano A, Boccia M (2020c) Duration reproduction in regular and irregular contexts after unilateral brain damage: evidence from voxel-based lesion-symptom mapping and atlas-based hodological analysis. Neuropsychologia 147:107577

Teghil A, Marc IB, Boccia M (2021) Mental representation of autobiographical memories along the sagittal mental timeline: evidence from spatiotemporal interference. Psychon Bull Rev 28(4): 1327–1335

Teghil A, Boccia M, Di Vita A, Zazzaro G, Sepe Monti M, Trebbastoni A, Talarico G, Campanelli A, Bruno G, Guariglia C, De Lena C, D'Antonio F (2023a) Multidimensional assessment of time perception along the continuum of Alzheimer's disease and evidence of alterations in subjective cognitive decline. Sci Rep 13(1):22117

Teghil A, Bonavita A, Procida F, Giove F, Boccia M (2023b) Intrinsic hippocampal connectivity is associated with individual differences in retrospective duration processing. Brain Struct Funct 228(2):687–695

Thavabalasingam S, O'Neil EB, Lee ACH (2018) Multivoxel pattern similarity suggests the integration of temporal duration in hippocampal event sequence representations. Neuroimage 178:136–146

Tompary A, Duncan K, Davachi L (2016) High-resolution investigation of memory-specific reinstatement in the hippocampus and perirhinal cortex. Hippocampus 26(8):995–1007

Tsao A, Yousefzadeh SA, Meck WH, Moser M-B, Moser EI (2022) The neural bases for timing of durations. Nat Rev Neurosci 23(11):646–665

Umbach G, Kantak P, Jacobs J, Kahana M, Pfeiffer BE, Sperling M, Lega B (2020) Time cells in the human hippocampus and entorhinal cortex support episodic memory. Proc Natl Acad Sci U S A 117(45):28463–28474

Van Ackooij M, Paul JM, Van Der Zwaag W, Van Der Stoep N, Harvey BM (2022) Auditory timing-tuned neural responses in the human auditory cortices. Neuroimage 258:119366

Wiener M (2024) Coordinate-based meta-analyses of the time perception network. In: Merchant H, De Lafuente V (eds) Neurobiology of interval timing. Springer, Cham, pp 215–226

Wiener M, Turkeltaub P, Coslett HB (2010) The image of time: a voxel-wise meta-analysis. Neuroimage 49(2):1728–1740

Wiener M, Matell MS, Coslett HB (2011) Multiple mechanisms for temporal processing. Front Integr Neurosci 5

Wiener M, Zhou W, Bader F, Joiner WM (2019) Movement improves the quality of temporal perception and decision-making. eNeuro 6(4):ENEURO.0042-19.2019

Wittmann M (2013) The inner sense of time: how the brain creates a representation of duration. Nat Rev Neurosci 14(3):217–223

Wittmann M, Droit-Volet S (2024) Subjective time in ordinary and non-ordinary states of consciousness: how interoceptive feelings inform us about the passage of time. In: Wittmann M, Strigo I, Simmons A (eds) The bodily Self, emotion, and subjective time: exploring Interoception through the contributions of A.D. (Bud) Craig, Current topics in behavioral neurosciences. Springer. In press

Wittmann M, Simmons AN, Aron JL, Paulus MP (2010) Accumulation of neural activity in the posterior insula encodes the passage of time. Neuropsychologia 48(10):3110–3120

Wittmann M, Simmons AN, Flagan T, Lane SD, Wackermann J, Paulus MP (2011) Neural substrates of time perception and impulsivity. Brain Res 1406:43–58

Zakay D, Block R (2004) Prospective and retrospective duration judgments: an executive-control perspective. Acta Neurobiol Exp 64(3):319–328

Zhang R, Deng H, Xiao X (2024) The insular cortex: an interface between sensation, emotion and cognition. Neurosci Bull. https://doi.org/10.1007/s12264-024-01211-4

Insights from Ecstatic Epilepsy: From Uncertainty to Metacognitive Feelings

Nina M. Sooter, Federico Seragnoli, and Fabienne Picard

Contents

Abstract Ecstatic epilepsy is a rare form of focal epilepsy linked to the anterior insula in which patients experience a blissful state with a unique set of symptoms, including a feeling of physical well-being, mental clarity, a sense of oneness with the universe, and time dilation. In this chapter, we reflect on how these symptoms

N. M. Sooter
Geneva School of Economics and Management, University of Geneva, Geneva, Switzerland

F. Seragnoli
Institute of Psychology, Lausanne University, Lausanne, Switzerland

Psychiatry Department, Geneva University Hospital, Geneva, Switzerland

F. Picard (✉)
Department of Clinical Neurosciences, Geneva University Hospitals and Medical School, Geneva, Switzerland
e-mail: fabienne.picard@hug.ch

Curr Topics Behav Neurosci (2025) 70: 247–267
https://doi.org/10.1007/7854_2024_528
Published Online: 23 October 2024

coincide with our current knowledge of the insula's functions and explore how this stunning natural model can further inform our understanding of the insula's role in the sentient self, uncertainty and surprise monitoring, and metacognitive feelings.

Keywords Ecstatic epilepsy · Insula · Metacognitive feeling · Mystical experience · Self-awareness · Uncertainty

1 Introduction

> *The immense joy that fills me is above physical sensations. It is a feeling of total presence, an absolute integration of myself, a feeling of unbelievable harmony of my whole body and myself with life, with the world, with the 'All'.*
>
> Quote from one of the patients (Picard and Craig 2009)

In a rare subset of epileptic seizures, aptly named ecstatic, patients describe a unique, blissful experience. The sense of self is "magnified," characterized by an intense and enhanced sense of self-awareness, and accompanied by a feeling of certainty, of having found the truth. Research with patients who experience ecstatic epilepsy has identified the anterior insula (AIns) as the main structure at its genesis, providing an exceptional opportunity to use the ecstatic epilepsy model and its rich symptomatology as a means of furthering our understanding of the AIns.

As proposed by Bud Craig: "In order to address the functions of the insula, new techniques for combining imaging and electrophysiological recordings will be needed, perhaps involving patients with ecstatic epileptic seizures" (Luo et al. 2010). It is in this spirit that we engage in an in-depth reflection of the model of ecstatic epilepsy as a stepping stone in understanding the broader role of the anterior insula.

We begin by defining and describing ecstatic epilepsy, followed by a section on the insula and its known role in a myriad of affective and cognitive functions. We then link the symptoms experienced in ecstatic epilepsy with our general knowledge of the insula by exploring two hypotheses that could explain this experience and the insights they provide into the overall function of this fascinating brain area.

2 Ecstatic Epilepsy: Definition, Description, and Origin

2.1 Definition and Description

Epilepsy is a common neurological disorder, affecting nearly 1% of the population. It is divided into "generalized" epilepsies – mostly of genetic origin – and the more frequent "focal" epilepsies, in which the electrical epileptic discharge begins in part of one cerebral hemisphere and whose cause is most often structural (i.e., related to a cortical lesion). Epileptic seizures are due to abnormal hyperactivity in a group of neurons, which activate synchronously and spontaneously without being guided by any voluntary action or physiological adaptation.

Ecstatic epilepsy is a rare form of focal epilepsy in which patients experience a blissful state at the very beginning of their seizures. This state comprises a feeling of physical well-being, enhanced self-awareness, and a feeling that everything makes sense (a feeling of certainty, of having insight into depths of truth), as well as a sense of union with everything. Patients also report a sense of time dilation, as if they were living in an eternal now. These extremely intense, positive experiences typically occur at the onset of a seizure and may be immediately followed by a secondarily generalized, tonic-clonic (convulsive) seizure with all its co-occurring risks and discomfort.

The first known written description of this form of epilepsy comes from Dostoevsky, both in letters to friends, as well as through the experiences of his fictional characters, such as Prince Mychkine in *The Idiot* (Dostoevsky 1869). To date, about 60 patients with ecstatic epilepsy have been reported in the literature (Bartolomei et al. 2019; Gschwind and Picard 2016). However, the frequency of this form of epilepsy is likely underestimated as people can be unable or reluctant to speak about the ecstatic symptoms; they often cannot find the words to express what they experienced and are afraid of being judged/misunderstood.

According to the phenomenological report of the patients we interviewed (Gschwind and Picard 2016; Picard 2013; Picard and Craig 2009), the experience comprises:

1. A sense of physical well-being.
2. A sense of bliss.
3. A feeling of (nonconceptual) knowledge as if everything makes sense (a kind of mental clarity, sense of certainty, epiphany, or "revelation").
4. A feeling of unity, of being part of a unified, interconnected whole (with the environment and with others).
5. A sense of time dilation, the feeling of an eternal now.

Interestingly, one patient suffering from ecstatic epilepsy described the ecstatic phenomenon as "a continuous series of profound 'aha!' moments" (Picard 2013).

The ecstatic experience of epileptic origin is very similar to the mystical-type experience described by William James, which can be accessed through ritual practices, such as meditation or prayer, or through the use of psychedelic substances. He defined the mystical experience by four characteristics: ineffability, "noetic" quality, transiency, and passivity (James 1985). The "noetic" quality describes a state of knowledge or insight into depths of truth, unplumbed by discursive intellect, an illumination, enlightenment or revelation. The three other characteristics also apply to the epileptic ecstatic experience, which appears ineffable to the patient, is transient, and is not guided by the will.

Such ecstatic/mystical experiences can have colossal transformative effects (Barros and Schultz 2023; Picard and Craig 2009) which can last for years and even permanently change a person's life and existential narrative. Hence, the interest in better understanding their underlying neural basis, which could ultimately lead to therapeutic use in neuropsychiatric disorders such as depression or anxiety.

2.2 The Involvement of the Insula in Ecstatic Epilepsy

Ecstatic epilepsy was first related to the temporal lobe as patients with ecstatic epilepsy presented with temporal lobe lesions or had scalp EEG abnormalities in the surface region of the temporal lobe. However, the symptoms of an epileptic seizure may not be due to a disruption of the first region of the discharge (epileptogenic area), but rather to the propagation of the epileptic discharge to another region, the symptomatogenic region, if the first region does not produce any symptoms. Thus, although early associations were made between ecstatic epilepsy and the temporal lobe, in collecting patients' detailed descriptions of their experiences over the last 15 years and reading neuroscientific literature with their symptoms in mind, we were guided toward the anterior insula as a potential source of this phenomenon (Picard and Craig 2009), given its postulated role in interoception and self-awareness in Bud Craig's papers (Craig 2002, 2003). Based on this context, the major role of the anterior insula in this experience was demonstrated in recent years (Bartolomei et al. 2019; Nencha et al. 2022; Picard et al. 2013).

In 2002, Craig proposed that the re-representation of the body's (interoceptive) state from the posterior to the anterior insula, contextualized by external signals, serves as a sensory substrate for the development of subjective feelings and emotions (Craig 2002). Thus, the AIns would be at the basis of the "subjective awareness" of feelings. Craig also proposed the concept of "global emotional moments," each a unified, final meta-representation of the sentient self at one moment in time (produced in the cortical area near the junction of the AIns and the frontal operculum, Craig 2009a). The AIns thus constituted the brain region best able to explain a phenomenon comprising physical well-being, positive emotion, and a complete change in the feeling of the sentient self.

The first evidence of the involvement of the AIns in ecstatic epileptic symptoms was obtained thanks to nuclear imaging (ictal Single-Photon Emission Computed Tomography) studies in three patients: a first published case in Picard and Craig (2009, case 2) reporting increased blood flow in the right anterior insula during an ecstatic seizure; a second case with an ictal "sensed presence"/ecstatic feeling associated with increased blood flow in the left anterior insula (Landtblom et al. 2011); and a third case in which the ictal SPECT showed an increased blood flow maximal at the junction of the right dorsal mid-insula and the central operculum (Picard 2013, patient 1). The ictal SPECT, which reflects the blood flow during the first 30 s to 1 min after injection, often relates to a region of propagation of the epileptic discharge.

Stronger evidence for the AIns's involvement in the ecstatic phenomenon came from a study with intracerebral electrodes (stereo-electroencephalogram or S-EEG) implanted in a patient with drug-resistant ecstatic epilepsy. Using intracerebral electrodes within the right temporal lobe and insular cortex, the recording of a spontaneous seizure showed a right hippocampal onset of the discharge with a very rapid propagation (within 1 s) from the hippocampus to the ipsilateral dorsal AIns (Picard et al. 2013). What is more, the ecstatic auras were reproduced via electrical stimulation in the dorsal AIns outside of a seizure (being part of the standard presurgical evaluation of epilepsy), while electrical stimulation of all the other electrodes did not induce any ecstatic phenomenon.

Since then, the reproduction of ecstatic auras through electrical stimulation of the dorsal AIns was reported in several other patients with ecstatic epilepsy (Bartolomei et al. 2019; Tyrand et al. 2024), as well as the identification of the fast propagation of the discharge to the AIns during spontaneous seizures recorded in several patients not reported in the literature (personal communications). Fascinatingly, we observed the induction of ecstatic phenomena by electrical stimulation of the dorsal AIns in a patient suffering from temporal lobe epilepsy, who had never had ecstatic symptoms before (Nencha et al. 2022), that is, without any potential underlying modified networks prone to this phenomenon. This novel induction of ecstatic symptoms following electrical stimulations was also observed in other patients, in Sweden and the United Kingdom (personal communications).

Yet, the induction of an ecstatic phenomenon through electrical stimulation of the dorsal AIns is not the rule, as there are many epileptic patients in whom stimulations of this region during presurgical evaluation do not provoke ecstatic experiences. Thus, there appears to be some predisposing factors to experiencing this state of consciousness (possibly anatomical or functional factors, such as alterations in structural and/or functional connectivity of the dorsal anterior insula, or a difference in the autonomic nervous system, with a higher parasympathetic baseline tone), which are as yet unknown.

Regarding hemispheric differences, no lateralization of ecstatic symptoms has yet been shown, with no clear differences between patients with right-sided or left-sided epilepsy (Gschwind and Picard 2016). However, there are connections between both insulae, allowing rapid propagation of the electrical activity from one insula to the other (Lacuey et al. 2016). The effect of an imbalance between the left and right insulae has not yet been elucidated, despite old hypotheses of a trend of a sympathetic dominance (associated with avoidance behavior) for the right insula and parasympathetic dominance (associated with approach behavior) for the left insula (Harrison 2015; Xavier et al. 2018).

3 The Insula: Functional Heterogeneity

The insula is a multifaceted hub extensively connected to a range of cortical and subcortical brain areas (Gogolla 2017), placing it in an ideal position for its involvement in a range of sensory, affective, and cognitive processes. As such, it responds both to the external world and to emerging feelings and cognitions used in executive functions (Sheldon et al. 2024). We highlight some of these functions below.

3.1 Salience Detection

The AIns belongs to the "salience network" which also includes the adjacent frontal operculum and the dorsal anterior cingulate cortex (Seeley et al. 2007). Salience refers to the quality of being noticeable or standing out, and it is defined by both low-level sensory features (e.g., color or intensity) and high-level cognitive and affective processes (e.g., emotional relevance) (Parr and Friston 2019). Salience detection involves the processing of internal and external stimuli to prioritize information that is meaningful or behaviorally relevant – whether aversive or appetitive – and can constitute an epistemic gain ("epistemic" meaning "of or relating to knowledge"). As part of its crucial role in detecting salient events, the AIns initiates appropriate cognitive responses, acting as a hub to redirect attention and mediate interactions between networks.

3.2 Interoception, Multisensory Integration, and Feelings

The insula is considered to be the primary interoceptive cortex, perceiving the physiological condition of the body and activating for stimuli including heartbeat, respiration, thirst, shortness of breath, itch, temperature, pain, and others (for a review, see Craig 2002, 2009a). The dynamic processing of interoceptive signals can be understood in the context of predictive coding (see Box 1).

Box 1 The Predictive Coding Theory of Brain Function

The brain sits in its cranial box with no direct means of capturing the outside world. Instead, it must piece together an "image" of the environment by synthetizing the indirect information it receives via the eyes, ears, skin, and other sensory modalities. However, waiting for this image to be built anew before acting would be slow and suboptimal in most environments. Thus, the brain relies on predictions to both speed up the process and fill in any perceptual gaps. In this way, what we experience as the world is a mixture of (1) the brain's predictions and (2) the new information, that is, the actual input from the outside world (Friston et al. 2017). In a sense, we are all operating and interacting with the world based on our own private fantasies.

According to the predictive coding theory, the brain is a prediction machine, constantly anticipating what the world, "our reality," will be like in the next moment. In the different hierarchical regions of the brain that manage sensory modalities (e.g., vision, hearing, touch, and internal body sensations) or cognitive processes, there are predictions, or "top-down" signals, based on what the brain knows from both the previous moment and the accumulated knowledge of all past experiences. When a new signal arrives from the environment – at any level of the brain system/module hierarchy – it is

(continued)

Box 1 (continued)

compared to the prediction (made by the brain in the layer above), weighted by "precision" or level of uncertainty. In case of a mismatch, a prediction error is produced. The prediction error enables the next prediction to be updated so that it is as close as possible to the next real signal that comes in (Clark 2013), allowing the brain to constantly correct its "database" through learning. The brain's aim is to reduce the uncertainty in one's encounter with the world via prediction error minimization.

Predictions are made simultaneously in all the different brain systems (e.g., interoception, vision, audition, touch, speech listening[1], and cognition at varied abstraction levels), at multiple timescales, and are combined to form a structured representation of the world at each moment. In other words, the brain uses a hierarchical hypothesis-testing mechanism, predicting sensory input on the basis of prior experience and updating predictions on the basis of prediction errors. The interpretation of the external world depends on selecting the predictions made in different brain systems at every given moment that best fit together to form a coherent experience.

Importantly, the precision weights given to top-down predictions versus bottom-up environmental and interoceptive signals are dependent on beliefs about the volatility of the environment, which influence levels of uncertainty. There are constantly fluctuating, instantaneous prediction errors depending on the tuning of precision weighting of the predictions, but also on the temporal dynamics of prediction error reduction (or increase) over longer time scales.

Our reality thus "emerges as the consequence of an ongoing, continually modified constructive process during which stored knowledge within an experiencer (as the reactivation and recombination of prior experience, referred to as "top-down" influence) makes incoming sensory inputs meaningful as *situated conceptualizations*" (Barrett 2014, page 3).

This predictive functioning of the brain explains how the same external world can be seen and experienced differently by different people, depending on the predictions that each has made. In short, what each person thinks and anticipates may be more important than what they see or hear, and clearly contributes to their perception of the world.

The posterior insula receives ascending interoceptive signals from the body, which are transmitted through the mid and into the anterior insula in a series of steps, each with the integration of additional polymodal sensory and cognitive information. As the signals move anteriorly, they are increasingly compared to and modified by top-down interoceptive predictions (Barrett and Simmons 2015). Mismatches generate prediction errors that correspond to conflict/ambiguity in the

[1] For an example using speech listening, see Caucheteux et al. (2023).

physiological state of the body, which inform subsequent interoceptive predictions and trigger adaptation in autonomic systems.

Functional MRI studies have identified activation in the insula for interoceptive predictions for breathing-related signals (Harrison et al. 2021) and heartbeat and related cardiorespiratory signals (Hassanpour et al. 2016, 2018) in humans. Interoceptive predictions in the insula were also nicely demonstrated in mice models (Livneh et al. 2020; Prilutski and Livneh 2023). Additionally, the anterior insula is vital for interoceptive attention and accuracy, showing increased activation during tasks requiring attention to internal signals, while lesions in this area impair interoceptive discrimination and sensitivity (Wang et al. 2019).

In addition, the insula plays a multimodal role in analyzing the constant interaction between interoceptive and exteroceptive signals, allowing for a coherent representation of the body's internal state in the context of external environmental changes (Salomon et al. 2016). By generating interoceptive predictions that account for exteroceptive predictions, the insula allows for better anticipation of interoceptive changes related to the external world. This provides a multisensory representation of the world from the perspective of an *embodied* person, i.e. a person with a body. The dorsal AIns appears especially crucial for integrating different types of information to create this coherent experience (Kurth et al. 2010). Interestingly, interoceptive processing may also affect exteroceptive processing as, for instance, images of emotional faces were processed differently if presented during the systolic or the diastolic phase of heartbeats (Critchley and Garfinkel 2017).

The AIns is also greatly implicated in emotional processing, with most neuroimaging studies reporting AIns activation for affective feelings across the spectrum (for a review, see Craig 2009a). Thus, the AIns is not only involved in subjective "bodily feelings" (Craig et al. 2000; Zhang et al. 2019), but subjective feelings more generally, which is consistent with theories on the close link between emotions and somatic markers (Damasio 1994; James 1890). In support of this, one study showed a specific cluster in the AIns activated both during heart rate monitoring (bodily awareness) and emotional processing (Zaki et al. 2012).

Interestingly, according to Seth and Friston, feelings, emotions, and possibly self-consciousness (embodied selfhood) are related to the succession of interoceptive *predictions* rather than to the succession of real interoceptive signals (Seth and Friston 2016), that is, our felt experience is primarily about what goes on in our anticipations.

3.3 Metacognitive Feelings

The dorsal AIns is consistently involved in cognitive tasks, reflecting a role in high-level cognitive processing (Chang et al. 2013). Recent converging evidence showed that some prefrontal regions interact with the interoceptive (insular) cortex to support metacognition and contextualize its inferences.

Metacognition is the cognitive function that enables a person to reflect on their own state and react from a subjective perspective (Fleming and Dolan 2012). It is composed of two facets: a declarative and a procedural one (Proust 2019). The declarative aspects of metacognition are related to the knowledge one has of one's own mental states, the strategies one knows on how to regulate them, and the epistemic beliefs one holds about how knowledge is acquired. The procedural aspects are instead nonconceptual and related to one's capacity to monitor and control their own cognitive activity. They correspond to the competence of being aware of what is going on in one's mind and being able to regulate and manipulate it through conscious attentional and cognitive effort (Goupil and Proust 2023, Menon and Uddin 2010). For instance, metacognition is what produces the awareness of mind wandering, possibly by detecting the targeted mismatch between a goal (e.g., if the goal was to maintain focused attention on the breath) and current mental state (e.g., thinking about an upcoming meeting). This awareness is at the origin of switching between the default mode network and executive networks through the salience network, in particular, the bilateral anterior insula (and dorsal anterior cingulate cortex) (Hasenkamp et al. 2012; Tang et al. 2012).

"We do not just think about ideas, we feel them too" (Laukkonen et al. 2023). Through the process of monitoring and manipulating one's own states of mind, a person is guided by metacognitive feelings related to ongoing cognitive action (Goupil and Proust 2023). Metacognitive feelings provide an effective model of the situation. They are affective experiences with positive or negative valence which are part of a cognitive evaluative process and guide behavior, allowing people to deal with uncertainty about the world (Fernández Velasco and Loev 2024). In accordance with insular functioning, bodily changes under the control of the autonomic nervous system serve as cues to predict the dynamics of prediction error reduction relating to a particular mental process. The insula can then be understood as having a role in metacognition from the subjective experience perspective, likely in part through conscious detection of errors (Cosentino et al. 2015; Vaccaro and Fleming 2018).

One example of the phenomenology of the metacognitive feelings is the "tip-of-the-tongue" feeling, where a person knows they know something but cannot retrieve it (e.g., "*I know I know this person's name, let's retrieve it*"). This feeling is defined as the feeling accompanying temporary inaccessibility of a target word that a person is trying to retrieve (Schwartz and Metcalfe 2011). In the more general context of a problem-solving failure, attentional and cognitive resources are recruited to resolve an inconsistency, a gap, between an internal model of the world and a salient, surprising set of cognitive affordances that contradict it. If, at a given point, this contradiction is suddenly resolved through a flash of insight – seemingly out of the blue and not via analytical strategies – the individual experiences an "aha!" or "Eureka!" moment (Stuyck et al. 2021). This experience of transition to a state of resolution of uncertainty (or resolution of chronic surprise) is classically characterized by a sudden sense of having understood or a sudden certainty of being correct, and a positive-value feeling of joy or pleasure (Danek and Wiley 2017). The insight phenomenon comprises thus a cognitive component (sudden change in the problem representation with identification of a solution) and an affective component. The insight feeling can be defined as a metacognitive feeling of epistemic gain (Laukkonen et al. 2023).

3.4 Uncertainty, Risk, and Surprise

The AIns's role in integrating interoceptive signals with external information and top-down predictions puts it in an ideal position to track uncertainty. Following the predictive coding hypothesis (see Box 1), uncertainty is computed on different hierarchical levels in each sensory and interoceptive region, and then passed up the chain for higher-order processing (for an example in the visual cortex, see Rao and Ballard 1999). The insula is at the link in the chain where these signals are combined to give one the sense of global awareness via a global emotional moment (Craig 2009a), which comprises a level of uncertainty.

The AIns's integrated role in global uncertainty processing, regardless of domain, was first suggested by Singer et al. (2009). Activation of the AIns has since been shown in uncertainty processing related to strategic thinking (Nagel et al. 2018) and more recently in the conscious perceptual uncertainty and surprises related to the different perceptions of the Necker cube (Loued-Khenissi et al. 2020). The insula was further implicated in a measure of reported subjective confidence (Geurts et al. 2022) which could be seen as an alternative measure for the perceived uncertainty of the environment. Additionally, the AIns was identified as a key region in tracking risk prediction and risk prediction errors in probabilistic tasks (Christopoulos et al. 2009; Knutson and Huettel 2015; Kuhnen and Knutson 2005; Preuschoff et al. 2008; Rudorf et al. 2012) and was shown to belong to a neural network associated with general error awareness (Bossaerts 2010; Craig 2009a; Dali et al. 2022).

Our current postulate is that the AIns encodes the overall uncertainty moment-by-moment by using predictive processing; that is, comparing the brain's virtual model of the world to the real-world model created using affordances that it receives on different levels: interoceptive, multisensory perceptual, semantic, and cognitive (see Box 1). Moreover, depending on the level of uncertainty in the immediate environment, the brain predicts not only what real-world signals will be, but also provides expectations about the global magnitude of mismatches. A violation of these expectations (a magnitude of mismatch above the expected level of mismatch) would result in surprise.

Given the AIns's role in error awareness and uncertainty, recent theories posit that it acts as the global surprise monitor, taking into account the global uncertainty prediction error (Bossaerts 2018; Picard et al. 2021; Picard 2023). If surprised, that is if the overall amplitude or frequency of prediction errors exceeds (or is below[2]) the

[2]One should allow for the possibility that surprise includes the situation where the world is less uncertain than in one's "virtual model of the world." This may prompt the agent to change the world to have more "excitement."

expected global amplitude of the prediction errors (subjective uncertainty prediction), the AIns sends feedback to the brain's attentional and learning modules on the need for increased control,[3] which may then update their learning protocol, such as by adjusting the learning rate speed, in order to better control the environment (Liakoni et al. 2022; Picard 2023). In addition, the global surprise signal and interoception should be completely entangled, with surprise causing an autonomic visceromotor response, which in turn affects interoceptive predictions and subsequent surprise.

Surprise related to prediction errors can be either expected or unexpected. Expected surprise does not exceed the overall expected level of mismatches in each brain system, e.g., in a volatile environment one expects one's predictions to be less accurate. Contrarily, unexpected surprise is experienced when surprise differs from what one predicted based on beliefs about the uncertainty in the environment, that is, when the size of prediction errors or the "overall" prediction error differs from expectations. There is no unexpected surprise as long as the continual increases/decreases of prediction errors follow our expectations.

In environments with low uncertainty, predictions/reference models are expected to be accurate (weighted with high precision), so when there is a large mismatch between them and environmental inputs, unexpected surprises may occur, and the brain learns from this and adapts its predictions/reference model. In environments with high uncertainty, predictions/reference models are expected to be less accurate, and the learning rate is kept low, so that cognitive systems do not overreact to noise in the environment. (In these cases where control is lower, a better strategy is to increase attention, thereby improving mediation of environmental effects.)

Early evidence for the AIns's role in surprise monitoring was provided by a task studying stochastic events in environments with different levels of uncertainty (d'Acremont and Bossaerts 2016). Using fMRI, activation was observed in the AIns only following outliers in the more volatile condition where participants had difficulty controlling their mistakes. A separate study showed that surprise was computed separately from prediction error valence, that is, whether the outcome was better or worse than expected (Fouragnan et al. 2018).

We discuss below how a neurocognitive hypothesis for the ecstatic epilepsy experience (Sect. 4.2) provides additional evidence for the AIns's role in uncertainty and surprise monitoring.

[3]As such, the AIns operates like the adapter in the Model Reference Based Adaptive Control (MRAC) model in engineering (Nguyen 2018), sending modulatory signals when it receives input that is surprising compared to what is expected based on the reference model.

4 Neurocognitive Models for the Ecstatic Experience

Based on our knowledge of the insula's multiple functions and the patients' descriptions of ecstatic epilepsy, we propose two neurocognitive models for the ecstatic experience.

4.1 Suppression of Interoceptive Prediction Errors

As described above (Sect. 3.2), the insula integrates moment-by-moment internal bodily signals (interoception), environmental information (external multisensory perceptions), and emotional states, creating a comprehensive, global awareness that guides behavior and cognitive processes (Craig 2009a; Simmons et al. 2013).

Our first neurocognitive model for the ecstatic phenomenon is that the epileptic discharge within the anterior insula interrupts the processing of interoceptive predictions and/or of interoceptive prediction errors. A state without interoceptive prediction errors would mimic perfect prediction of the body's physiological state (Picard 2013; Picard and Friston 2014). As interoception is consistently processed and integrated with multisensory information in the insula, and as the self and one's representation of the external world continuously co-emerge (Colman 2015; Thompson and Varela 2001), we hypothesize that a bodily state which seems perfectly predicted, devoid of any conflict or ambiguity at all internal levels for a few consecutive seconds – a state never achieved under normal insular functioning – could be projected onto one's perception of the outside world and in that respect, could explain the ecstatic quality of the experience and the feeling of mental clarity and "revelation" (Picard 2023).

4.2 Suppression of Surprise Processing (and Comparison with the Aha! Experience)

Postulating that the insula acts as a multimodal surprise and uncertainty monitor, we hypothesize that a sudden dysfunction of the anterior insula could interrupt the integration of predictions with incoming signals not just in the interoceptive domain, but more globally, suspending *all* uncertainty monitoring and updating. The consequent absence of surprise and error awareness would result in a sense that everything is under control (regardless of whether it actually is). As a result, no adaptation signals would be sent to the learning modules, freeing the individual to simply experience bottom-up inputs as they occur – a state never experienced in everyday life. In the absence of surprise signals, a feeling of being in complete control of the situation, of mental clarity and certainty, could originate where the world appears self-evident, interconnected, and blissful without doubts or conflict.

If we further consider the insula's role in surprise, the interruption of predictions/bottom-up conflict and error awareness could result in an intuitive sense of "just knowing," a sense of something being revealed, mimicking a sustained "aha!" moment, where one feels as if they have suddenly solved the lifelong prediction puzzle, ending the cognitive effort and frustration involved and resolving the need to provide a semantic interpretation of the world. The transitory absence of semantic and conceptual interpretations could explain the feeling of unity with the "All" observed in patients with ecstatic epilepsy. This non-dual state (so called in the context of spirituality or theology) refers to the idea that there is no fundamental distinction or separation between the self, and the rest of existence, e.g., no separation between self and other or between subject and object.[4] Previous theories posit that categorization and reification are required to delineate this separation (Josipovic 2019; Laukkonen and Slagter 2021), ergo, a suspension of the need for semantic interpretations could result in the feeling of oneness with the universe described.

In short, due to the suspension of uncertainty processing, the phenomenology of ecstatic epilepsy could be an amplification of the intensity and duration of an aha metacognitive feeling of epistemic gain. Interestingly, one fMRI study of aha! moments demonstrated an activation affecting the anterior insula in particular (Aziz-Zadeh et al. 2009).

Of note, in ecstatic epilepsy, this intense "aha!"/"Eureka!" experience is not triggered by a coherently intense epistemic gain resulting from the resolution of a specific mismatch between predictive processing and the actual external or cognitive affordances. Instead, given the disruption of insular activity and the interruption of interoceptive and/or uncertainty prediction errors, this feeling seems to be activated without any specific independent epistemic justification. The phenomenology of insight was reported by McGovern and colleagues to occur without epistemic gain under psychedelics, which may cause false insights (illusions mimicking veridical insights) and false beliefs (McGovern et al. 2024). These authors wondered "how do we improve the likelihood that the insights and subsequent belief changes engendered by psychedelics result in beliefs that move one closer to reality." Could the subjectively meaningful narratives that are expressed by people following an ecstatic epileptic seizure be a post-hoc justification conjured by the mind to make sense of such an intense metacognitive feeling of epistemic gain (e.g., a feeling of having insight into depth of the truth, of understanding of the meaning of life)? Following an ecstatic experience, people might tend to attribute their intense experienced feelings to something equally as intense semantically and may orient their interpretations to great existential and spiritual themes. This interpretation may assimilate the new experience into pre-existing assumptions or may completely change the person's world and life view. In our opinion, during the epileptic ecstatic experience, the

[4]In meditation, non-dualistic traditions assert that ultimate reality is a unified whole, and any apparent separations or distinctions are illusion. This view seems to have been shared by Albert Einstein who wrote "A human being experiences himself, his thoughts and feelings as something separated from the rest – a kind of optical delusion of his consciousness" (1950's letter to Robert Marcus).

shutdown of uncertainty (and conflict related to one's own anticipations) could unveil an unfiltered reality, with the first one-of-a-kind experience of all the sensory inputs unchanged by predictions, i.e., a new view of reality, which justifies an epistemic gain.

4.3 Bridging the Hypotheses

While both hypotheses discussed above could account for the ecstatic epilepsy experience on their own, there is no reason to believe these processes are not interconnected and operating simultaneously. In fact, as the epileptic discharge disrupts the AIns's major functions, one would expect a disruption in both interoception and uncertainty/surprise monitoring, which could jointly contribute to the intensity of the experience. We briefly discuss here how disruption in each of these processes could affect the other.

We begin with a disruption in uncertainty monitoring and an acute cessation of surprise signals. Firstly, with the uncertainty monitor switched off, the discrepancy between top-down predictions and bottom-up information in the interoceptive domain would not reach awareness. Secondly, interoceptive changes triggered by top-down modulation would be reduced, thereby reducing bottom-up signals of new interoceptive states. Jointly, these effects could mimic the sense of perfect physiological bodily prediction described in the interoception hypothesis.

Likewise, a feeling of perfect prediction in interoception would greatly please the uncertainty monitor. On the one hand, we project our experience of the self onto the world, so that a self that is perfectly predicted could temporarily give rise to the illusion that the world is perfectly predicted. On the other hand, conflict or ambiguity experienced in the external world often induces a bodily reaction, so that the sense of a perfect physiological state of the body could feel as if there was an absence of uncertainty in the experienced world, which also reconciles the two assumptions.

As we can see, these processes likely build on each other and could jointly cause the phenomenological experiences our patients described.

4.4 A Note on Time Dilation

The time dilation experienced by patients during their ecstatic epileptic seizures could be related to a change in the sampling rate of global emotional moments, which, as discussed above, are also synthesized in the AIns (Craig 2009a). The representation of each moment was previously evaluated to advance as a moving window of present time at a probable frame rate of around 8 Hz (i.e., lasting 125 ms), thus producing a continuous cinemascopic image of the sentient self (Craig 2009b; Picard and Craig 2009). When there is an increase in salient information, as in the case of increased attentional processes, the representation of moments through neural instantiations of feelings could be filled up more rapidly, causing the sampling rate to increase (Tse et al. 2004).

During the ecstatic experience, we propose that every internal and external signal appears salient and reaches awareness, as they are not filtered by predictions. This may enhance sensory information at low levels, with real incoming bottom-up signals being fully transmitted without prediction errors from the lower to the highest hierarchical levels. An increase in global emotional moments during the time when salient moments occur rapidly would give the brain the capacity to receive and encode these internal/environmental signals in greater detail, resulting in the feeling of time dilation reported (Picard and Kurth 2014). Moreover, with prediction updating and adaptation put on hold, subjective time is allowed to stretch without the need for integration or mental modification.

There is also evidence to suggest that time dilation could occur without increased saliency if regular insular functioning is disrupted. A study with patients showed that those with right insular lesions were impaired in their discrimination of duration, demonstrating reduced temporal sensitivity (Mella et al. 2019). Our interpretation was that the insula compares features of a stimulus such as its temporal processing, to a general representation of the same feature stored in long-term memory. Thus, a disruption in the insula's role in integration of new input could also result in a sense of time dilation, regardless of saliency.

5 The Insula's Involvement in Other Altered States of Consciousness

Given the involvement of the insula in subjectively felt consciousness processes and its role in ecstatic epilepsy, we can consider how its activity could be related to other similar kinds of altered states of consciousness.

The discovery of the insula as the main region involved in the specific pattern of symptoms found in ecstatic epilepsy, resulted from a phenomenological and clinical-centered approach. By collecting similar descriptions of people who experienced this particular set of symptoms, we identified the core symptomatology of the ecstatic phenomenon and predicted the location of the epileptic focus thanks to the subjective description of the symptoms (Picard and Craig 2009). Crucially, fundamentally proving our forecasting, we were able to electrically induce ecstatic epileptic symptoms in patients implanted with intracerebral electrodes (Nencha et al. 2022).

This approach could possibly be used to specifically track the involvement of the insula in similar altered states of consciousness. The main altered states of consciousness displaying a potential common core of experience, with a feeling of unity, of revelation, alteration in time perception, and intense positive emotion, are the use of psychedelics, non-dual meditative practices, and near-death experiences (Martial et al. 2020; Timmermann et al. 2023).

Considering a neurophenomenological approach (Timmermann et al. 2023; Varela 1996), the subjective descriptions in reported experiences of altered states of consciousness could lead to the creation of testable hypotheses to examine the insula's potential implication in these different altered states of consciousness. Some studies already pointed to functional alterations of the insula, such as a decreased insula/salience network functional connectivity under MDMA, also commonly known as ecstasy (Walpola et al. 2017), yet this theme not being the focus of this chapter, we leave this discussion to future work.

6 Conclusion

In conclusion, ecstatic epilepsy is a stunning natural model for understanding the function of the insula, corroborating its role in physical well-being and the sentient self. The most powerful and blissful experience of self-awareness, whether called ecstatic, mystical, or transcendental, can be related to a dysfunction of the anterior insula with scientifically relevant arguments. Specifically, this model confirms the AIns's cardinal role in interoception and the sense of self and provides support for a hypothesis on the AIns's function as a surprise monitor. We look forward to seeing how future research into insular functioning and other altered states of consciousness can continue to be informed by the natural ecstatic epilepsy model to increase our understanding of this fascinating, heterogenous brain region.

Acknowledgments We thank Prof. Peter Bossaerts (University of Cambridge) for very fruitful discussions, and Dr. Nathalie Mella (University of Geneva) for helpful comments.

Funding Sources This research did not receive any specific grant from funding agencies in the public, commercial, or not-for-profit sectors.

Competing Interests Statement The authors declare that they have no known competing financial interests or personal relationships that could have appeared to influence the work reported in this manuscript. By declaring no competing interests, the authors affirm that their research findings and interpretations are presented without any undue influence from personal or financial relationships.

Author Contributions **FP**: Conceptualization; Writing – original draft, review, and editing; Supervision. **NS** Writing – original draft, review, and editing. **FS:** Conceptualization; Writing – original draft, review, and editing.

References

Aziz-Zadeh L, Kaplan JT, Iacoboni M (2009) "Aha!": the neural correlates of verbal insight solutions. Hum Brain Mapp 30(3):908–916. https://doi.org/10.1002/hbm.20554

Barrett LF (2014) The conceptual act theory: a Précis. Emot Rev 6(4):292–297. https://doi.org/10.1177/1754073914534479

Barrett LF, Simmons WK (2015) Interoceptive predictions in the brain. Nat Rev Neurosci 16(7): 419–429. https://doi.org/10.1038/nrn3950

Barros M, Schultz L (2023) The transformative potential of religious, spiritual, and mystical experiences. 868201 Bytes. https://doi.org/10.6084/M9.FIGSHARE.22225213.V2

Bartolomei F, Lagarde S, Scavarda D, Carron R, Bénar CG, Picard F (2019) The role of the dorsal anterior insula in ecstatic sensation revealed by direct electrical brain stimulation. Brain Stimul 12(5):1121–1126. https://doi.org/10.1016/j.brs.2019.06.005

Bossaerts P (2010) Risk and risk prediction error signals in anterior insula. Brain Struct Funct 214(5):645–653. https://doi.org/10.1007/s00429-010-0253-1

Bossaerts P (2018) Formalizing the function of anterior insula in rapid adaptation. Front Integr Neurosci 12. https://doi.org/10.3389/fnint.2018.00061

Caucheteux C, Gramfort A, King J-R (2023) Evidence of a predictive coding hierarchy in the human brain listening to speech. Nat Hum Behav 7(3):430–441. https://doi.org/10.1038/s41562-022-01516-2

Chang LJ, Yarkoni T, Khaw MW, Sanfey AG (2013) Decoding the role of the insula in human cognition: functional parcellation and large-scale reverse inference. Cereb Cortex 23(3): 739–749. https://doi.org/10.1093/cercor/bhs065

Christopoulos GI, Tobler PN, Bossaerts P, Dolan RJ, Schultz W (2009) Neural correlates of value, risk, and risk aversion contributing to decision making under risk. J Neurosci 29(40): 12574–12583. https://doi.org/10.1523/JNEUROSCI.2614-09.2009

Clark A (2013) Whatever next? Predictive brains, situated agents, and the future of cognitive science. Behav Brain Sci 36(3):181–204. https://doi.org/10.1017/S0140525X12000477

Colman W (2015) Bounded in a nutshell and a king of infinite space: the embodied self and its intentional world. J Anal Psychol 60(3):316–335. https://doi.org/10.1111/1468-5922.12152

Cosentino S, Brickman AM, Griffith E, Habeck C, Cines S, Farrell M, Shaked D, Huey ED, Briner T, Stern Y (2015) The right insula contributes to memory awareness in cognitively diverse older adults. Neuropsychologia 75:163–169. https://doi.org/10.1016/j.neuropsychologia.2015.05.032

Craig AD (2002) How do you feel? Interoception: the sense of the physiological condition of the body. Nat Rev Neurosci 3(8):655–666. https://doi.org/10.1038/nrn894

Craig AD (2003) Interoception: the sense of the physiological condition of the body. Curr Opin Neurobiol 13(4):500–505. https://doi.org/10.1016/S0959-4388(03)00090-4

Craig AD (2009a) How do you feel – now? The anterior insula and human awareness. Nat Rev Neurosci 10(1):59–70. https://doi.org/10.1038/nrn2555

Craig AD (2009b) Emotional moments across time: a possible neural basis for time perception in the anterior insula. Philos Trans R Soc Lond B Biol Sci 364(1525):1933–1942. https://doi.org/10.1098/rstb.2009.0008

Craig AD, Chen K, Bandy D, Reiman EM (2000) Thermosensory activation of insular cortex. Nat Neurosci 3(2):184–190. https://doi.org/10.1038/72131

Critchley HD, Garfinkel SN (2017) Interoception and emotion. Curr Opin Psychol 17:7–14. https://doi.org/10.1016/j.copsyc.2017.04.020

d'Acremont M, Bossaerts P (2016) Neural mechanisms behind identification of leptokurtic noise and adaptive behavioral response. Cereb Cortex 26(4):1818–1830. https://doi.org/10.1093/cercor/bhw013

Dali G, Brosnan M, Tiego J, Johnson BP, Fornito A, Bellgrove MA, Hester R (2022) Examining the neural correlates of error awareness in a large fMRI study. Cereb Cortex 33(2):458–468. https://doi.org/10.1093/cercor/bhac077

Damasio A (1994) Descartes' error: emotion, reason and the human. Brain.

Danek AH, Wiley J (2017) What about false insights? Deconstructing the Aha! Experience along its multiple dimensions for correct and incorrect solutions separately. Front Psychol 7. https://www.frontiersin.org. https://doi.org/10.3389/fpsyg.2016.02077

Dostoevsky F (1869) The idiot

Fernández Velasco P, Loev S (2024) Metacognitive feelings: a predictive-processing perspective. Perspect Psychol Sci:17456916231221976. https://doi.org/10.1177/17456916231221976

Fleming SM, Dolan RJ (2012) The neural basis of metacognitive ability. Philosophical Transactions of the Royal Society B: Biological Sciences 367(1594):1338–1349. https://doi.org/10.1098/rstb.2011.0417

Fouragnan E, Retzler C, Philiastides MG (2018) Separate neural representations of prediction error valence and surprise: evidence from an fMRI meta-analysis. Hum Brain Mapp 39(7): 2887–2906. https://doi.org/10.1002/hbm.24047

Friston K, FitzGerald T, Rigoli F, Schwartenbeck P, Pezzulo G (2017) Active inference: a process theory. Neural Comput 29(1) Article 1. https://doi.org/10.1162/NECO_a_00912

Geurts LS, Cooke JRH, van Bergen RS, Jehee JFM (2022) Subjective confidence reflects representation of Bayesian probability in cortex. Nat Hum Behav 6(2):294–305. https://doi.org/10.1038/s41562-021-01247-w

Gogolla N (2017) The insular cortex. Curr Biol 27(12):R580–R586. https://doi.org/10.1016/j.cub.2017.05.010

Goupil L, Proust J (2023) Curiosity as a metacognitive feeling. Cognition 231:105325. https://doi.org/10.1016/j.cognition.2022.105325

Gschwind M, Picard F (2016) Ecstatic epileptic seizures: a glimpse into the multiple roles of the insula. Front Behav Neurosci 10. https://doi.org/10.3389/fnbeh.2016.00021

Harrison DW (2015) Parasympathetic and sympathetic tone. In: Harrison IDW (ed) Brain asymmetry and neural systems. Springer, pp 441–453. https://doi.org/10.1007/978-3-319-13069-9_26

Harrison OK, Köchli L, Marino S, Luechinger R, Hennel F, Brand K, Hess AJ, Frässle S, Iglesias S, Vinckier F, Petzschner FH, Harrison SJ, Stephan KE (2021) Interoception of breathing and its relationship with anxiety. Neuron 109(24):4080–4093.e8. https://doi.org/10.1016/j.neuron.2021.09.045

Hasenkamp W, Wilson-Mendenhall CD, Duncan E, Barsalou LW (2012) Mind wandering and attention during focused meditation: a fine-grained temporal analysis of fluctuating cognitive states. Neuroimage 59(1):750–760. https://doi.org/10.1016/j.neuroimage.2011.07.008

Hassanpour MS, Yan L, Wang DJJ, Lapidus RC, Arevian AC, Simmons WK, Feusner JD, Khalsa SS (2016) How the heart speaks to the brain: neural activity during cardiorespiratory interoceptive stimulation. Philos Trans R Soc B Biol Sci 371(1708):20160017. https://doi.org/10.1098/rstb.2016.0017

Hassanpour MS, Simmons WK, Feinstein JS, Luo Q, Lapidus RC, Bodurka J, Paulus MP, Khalsa SS (2018) The insular cortex dynamically maps changes in cardiorespiratory interoception. Neuropsychopharmacology 43(2):426–434. https://doi.org/10.1038/npp.2017.154

James W (1890) The principles of psychology. http://psychclassics.yorku.ca/James/Principles/

James W (1985) The varieties of religious experience. Harvard University Press. https://www.hup.harvard.edu/books/9780674932258

Josipovic Z (2019) Chapter 12 – nondual awareness: consciousness-as-such as non-representational reflexivity. In: Srinivasan N (ed) Progress in brain research, vol 244. Elsevier, pp 273–298. https://doi.org/10.1016/bs.pbr.2018.10.021

Knutson B, Huettel SA (2015) The risk matrix. Curr Opin Behav Sci 5:141–146. https://doi.org/10.1016/j.cobeha.2015.10.012

Kuhnen CM, Knutson B (2005) The neural basis of financial risk taking. Neuron 47(5):763–770. https://doi.org/10.1016/j.neuron.2005.08.008

Kurth F, Zilles K, Fox PT, Laird AR, Eickhoff SB (2010) A link between the systems: functional differentiation and integration within the human insula revealed by meta-analysis. Brain Struct Funct 214(5):519–534. https://doi.org/10.1007/s00429-010-0255-z

Lacuey N, Zonjy B, Kahriman ES, Marashly A, Miller J, Lhatoo SD, Lüders HO (2016) Homotopic reciprocal functional connectivity between anterior human insulae. Brain Struct Funct 221(5): 2695–2701. https://doi.org/10.1007/s00429-015-1065-0

Landtblom AM, Lindehammar H, Karlsson H, Craig AD (2011) Insular cortex activation in a patient with "sensed presence"/ecstatic seizures. Epilepsy Behav 20(4):714–718

Laukkonen RE, Slagter HA (2021) From many to (n)one: meditation and the plasticity of the predictive mind. Neurosci Biobehav Rev 128:199–217. https://doi.org/10.1016/j.neubiorev.2021.06.021

Laukkonen RE, Webb M, Salvi C, Tangen JM, Slagter HA, Schooler JW (2023) Insight and the selection of ideas. Neurosci Biobehav Rev 153:105363. https://doi.org/10.1016/j.neubiorev.2023.105363

Liakoni V, Lehmann MP, Modirshanechi A, Brea J, Lutti A, Gerstner W, Preuschoff K (2022) Brain signals of a surprise-actor-critic model: evidence for multiple learning modules in human decision making. Neuroimage 246:118780. https://doi.org/10.1016/j.neuroimage.2021.118780

Livneh Y, Sugden AU, Madara JC, Essner RA, Flores VI, Sugden LA, Resch JM, Lowell BB, Andermann ML (2020) Estimation of current and future physiological states in insular cortex. Neuron 105(6):1094–1111.e10. https://doi.org/10.1016/j.neuron.2019.12.027

Loued-Khenissi L, Pfeuffer A, Einhäuser W, Preuschoff K (2020) Anterior insula reflects surprise in value-based decision-making and perception. Neuroimage 210:116549. https://doi.org/10.1016/j.neuroimage.2020.116549

Luo L, Rodriguez E, Jerbi K, Lachaux J-P, Martinerie J, Corbetta M, Shulman GL, Piomelli D, Turrigiano GG, Nelson SB, Joëls M, De Kloet ER, Holsboer F, Amodio DM, Frith CD, Block ML, Zecca L, Hong J-S, Dantzer R et al (2010) Ten years of nature reviews neuroscience: insights from the highly cited. Nat Rev Neurosci 11(10):718–726. https://doi.org/10.1038/nrn2912

Martial C, Cassol H, Laureys S, Gosseries O (2020) Near-death experience as a probe to explore (disconnected) consciousness. Trends Cogn Sci 24(3):173–183. https://doi.org/10.1016/j.tics.2019.12.010

McGovern HT, Grimmer HJ, Doss MK, Hutchinson BT, Timmermann C, Lyon A, Corlett PR, Laukkonen RE (2024) An integrated theory of false insights and beliefs under psychedelics. Commun Psychol 2(69). https://doi.org/10.1038/s44271-024-00120-6

Mella N, Bourgeois A, Perren F, Viaccoz A, Kliegel M, Picard F (2019) Does the insula contribute to emotion-related distortion of time? A neuropsychological approach. Hum Brain Mapp 40(5): 1470–1479. https://doi.org/10.1002/hbm.24460

Menon V, Uddin LQ (2010) Saliency, switching, attention and control: a network model of insula function. Brain Struct Funct 214(5):655–667. https://doi.org/10.1007/s00429-010-0262-0

Nagel R, Brovelli A, Heinemann F, Coricelli G (2018) Neural mechanisms mediating degrees of strategic uncertainty. Soc Cogn Affect Neurosci 13(1):52–62. https://doi.org/10.1093/scan/nsx131

Nencha U, Spinelli L, Vulliemoz S, Seeck M, Picard F (2022) Insular stimulation produces mental clarity and bliss. Ann Neurol 91(2):289–292. https://doi.org/10.1002/ana.26282

Nguyen NT (2018) Model-reference adaptive control. In: Nguyen NT (ed) Model-reference adaptive control: a primer. Springer, pp 83–123. https://doi.org/10.1007/978-3-319-56393-0_5

Parr T, Friston KJ (2019) Attention or salience? Curr Opin Psychol 29:1–5. https://doi.org/10.1016/j.copsyc.2018.10.006

Picard F (2013) State of belief, subjective certainty and bliss as a product of cortical dysfunction. Cortex 49(9):2494–2500. https://doi.org/10.1016/j.cortex.2013.01.006

Picard F (2023) Ecstatic or mystical experience through epilepsy. J Cogn Neurosci 35(9): 1372–1381. https://doi.org/10.1162/jocn_a_02031

Picard F, Craig AD (2009) Ecstatic epileptic seizures: a potential window on the neural basis for human self-awareness. Epilepsy Behav 16(3):539–546. https://doi.org/10.1016/j.yebeh.2009.09.013

Picard F, Friston K (2014) Predictions, perception, and a sense of self. Neurology 83(12): 1112–1118. https://doi.org/10.1212/WNL.0000000000000798

Picard F, Kurth F (2014) Ictal alterations of consciousness during ecstatic seizures. Epilepsy Behav 30:58–61. https://doi.org/10.1016/j.yebeh.2013.09.036

Picard F, Bossaerts P, Bartolomei F (2021) Epilepsy and ecstatic experiences: the role of the insula. Brain Sci 11(11):1384. https://doi.org/10.3390/brainsci11111384

Picard F, Scavarda D, Bartolomei F (2013) Induction of a sense of bliss by electrical stimulation of the anterior insula. Cortex 49(10):2935–2937. https://doi.org/10.1016/j.cortex.2013.08.013

Preuschoff K, Quartz SR, Bossaerts P (2008) Human insula activation reflects risk prediction errors as well as risk. J Neurosci 28(11):2745–2752. https://doi.org/10.1523/JNEUROSCI.4286-07.2008

Prilutski Y, Livneh Y (2023) Physiological needs: sensations and predictions in the insular cortex. Physiology 38(2):73–81. https://doi.org/10.1152/physiol.00019.2022

Proust J (2019) From comparative studies to interdisciplinary research on metacognition. Anim Behav Cogn 6(4):309–328. https://doi.org/10.26451/abc.06.04.10.2019

Rao RPN, Ballard DH (1999) Predictive coding in the visual cortex: a functional interpretation of some extra-classical receptive-field effects. Nat Neurosci 2(1):79–87. https://doi.org/10.1038/4580

Rudorf S, Preuschoff K, Weber B (2012) Neural correlates of anticipation risk reflect risk preferences. J Neurosci 32(47):16683–16692. https://doi.org/10.1523/JNEUROSCI.4235-11.2012

Salomon R, Ronchi R, Dönz J, Bello-Ruiz J, Herbelin B, Martet R, Faivre N, Schaller K, Blanke O (2016) The insula mediates access to awareness of visual stimuli presented synchronously to the heartbeat. J Neurosci 36(18):5115–5127. https://doi.org/10.1523/JNEUROSCI.4262-15.2016

Schwartz BL, Metcalfe J (2011) Tip-of-the-tongue (TOT) states: retrieval, behavior, and experience. Mem Cognit 39(5):737–749. https://doi.org/10.3758/s13421-010-0066-8

Seeley WW, Menon V, Schatzberg AF, Keller J, Glover GH, Kenna H, Reiss AL, Greicius MD (2007) Dissociable intrinsic connectivity networks for salience processing and executive control. J Neurosci 27(9):2349–2356. https://doi.org/10.1523/JNEUROSCI.5587-06.2007

Seth AK, Friston KJ (2016) Active interoceptive inference and the emotional brain. Philosophical Transactions of the Royal Society B: Biological Sciences 371(1708):20160007. https://doi.org/10.1098/rstb.2016.0007

Sheldon KM, Lee W, Reeve J (2024) How people find better lifegoals: the goal breakthrough model and its neuroscientific underpinnings. Soc Personal Psychol Compass 18(6):e12974. https://doi.org/10.1111/spc3.12974

Simmons WK, Avery JA, Barcalow JC, Bodurka J, Drevets WC, Bellgowan P (2013) Keeping the body in mind: insula functional organization and functional connectivity integrate interoceptive, exteroceptive, and emotional awareness. Hum Brain Mapp 34(11):2944–2958. https://doi.org/10.1002/hbm.22113

Singer T, Critchley HD, Preuschoff K (2009) A common role of insula in feelings, empathy and uncertainty. Trends Cogn Sci 13(8):334–340. https://doi.org/10.1016/j.tics.2009.05.001

Stuyck H, Aben B, Cleeremans A, Van den Bussche E (2021) The Aha! Moment: is insight a different form of problem solving? Conscious Cogn 90:103055. https://doi.org/10.1016/j.concog.2020.103055

Tang Y-Y, Rothbart MK, Posner MI (2012) Neural correlates of establishing, maintaining, and switching brain states. Trends Cogn Sci 16(6):330–337. https://doi.org/10.1016/j.tics.2012.05.001

Thompson E, Varela FJ (2001) Radical embodiment: neural dynamics and consciousness. Trends Cogn Sci 5(10):418–425. https://doi.org/10.1016/s1364-6613(00)01750-2

Timmermann C, Bauer PR, Gosseries O, Vanhaudenhuyse A, Vollenweider F, Laureys S, Singer T, Antonova E, Lutz A (2023) A neurophenomenological approach to non-ordinary states of consciousness: hypnosis, meditation, and psychedelics. Trends Cogn Sci 27(2):139–159. https://doi.org/10.1016/j.tics.2022.11.006

Tse PU, Intriligator J, Rivest J, Cavanagh P (2004) Attention and the subjective expansion of time. Percept Psychophys 66(7):1171–1189. https://doi.org/10.3758/BF03196844

Tyrand R, Iannotti GR, Bartolomei F, Bratu F, Maliia MD, Nica A, Vulliemoz S, Spinelli L, Schaller KL, Picard F (2024) Brain-heart interaction during ecstatic seizures. Brain Spine. *Abstract in press.*

Vaccaro AG, Fleming SM (2018) Thinking about thinking: a coordinate-based meta-analysis of neuroimaging studies of metacognitive judgements. Brain Neurosci Adv 2:2398212818810591. https://doi.org/10.1177/2398212818810591

Varela FJ (1996) Neurophenomenology: a methodological remedy for the hard problem. J Conscious Stud 3(4):330–349

Walpola IC, Nest T, Roseman L, Erritzoe D, Feilding A, Nutt DJ, Carhart-Harris RL (2017) Altered insula connectivity under MDMA. Neuropsychopharmacology 42(11):2152–2162. https://doi.org/10.1038/npp.2017.35

Wang X, Wu Q, Egan L, Gu X, Liu P, Gu H, Yang Y, Luo J, Wu Y, Gao Z, Fan J (2019) Anterior insular cortex plays a critical role in interoceptive attention. Elife 8:e42265. https://doi.org/10.7554/eLife.42265

Xavier CH, Mendonça MM, Marins FR, Da Silva ES, Ianzer D, Colugnati DB, Pedrino GR, Fontes MAP (2018) Stating asymmetry in neural pathways: methodological trends in autonomic neuroscience. Int J Neurosci 128(11):1078–1085. https://doi.org/10.1080/00207454.2018.1473396

Zaki J, Davis JI, Ochsner KN (2012) Overlapping activity in anterior insula during interoception and emotional experience. Neuroimage 62(1):493–499. https://doi.org/10.1016/j.neuroimage.2012.05.012

Zhang Y, Zhou W, Wang S, Zhou Q, Wang H, Zhang B, Huang J, Hong B, Wang X (2019) The roles of subdivisions of human insula in emotion perception and auditory processing. Cereb Cortex 29(2):517–528. https://doi.org/10.1093/cercor/bhx334

Music in the Air and in the Body: An Interoceptive System's Perspective on Musical Emotions, Awareness, and Time

Clemens Wöllner

Contents

Abstract Performing and perceiving music requires the integration of multimodal information, including sensations of one's own body. Research on musical engagement investigated emotional responses and the peripheral physiological activations involved, as well as bodily action tendencies, effects on time perception, and the role of awareness and focus of attention. The concept of interoception dedicates a central role to the insular cortex, as suggested by Craig and others, and offers a unifying framework for studying music as an activity that has always had a key role for individuals, groups, societies, with a strong bodily component. Chances and challenges of an interoceptive perspective on music are discussed.

Keywords Dance · Focus of attention · Groove · Musical activities · Physiological responses · Time perception

C. Wöllner (✉)
University of Music Freiburg, Freiburg im Breisgau, Germany
e-mail: c.woellner@mh-freiburg.de

Curr Topics Behav Neurosci (2025) 70: 269–284
https://doi.org/10.1007/7854_2024_517
Published Online: 23 October 2024

1 Introduction

Engaging with music is genuinely a bodily activity. Manipulating physical objects to produce sounds, stretching one's vocal folds in singing, or interacting with others in joint clapping involve an awareness not only of the sounds reaching the ear, but also of multiple afferent sensations of one's body. Listening to music, moreover, evokes an array of feelings that have been empirically tested extensively over the past decades in terms of peripheral physiological sensations, neural activations in the brain, and subjective experiences. This holistic view of music performance and perception may have been uncontested in recent years. Yet, for long, cognitive psychology had considered music primarily as structured sounds with a grammar and a syntax, and Western historical musicology had often focused on musical scores rather than the living and interactive experience of musical performances (for a critique, see Cook 2014). Empirical and practice-informed approaches, on the other hand, realized the interdependencies between musical engagement and the performer's or listener's body, their movements and feelings. To name but a few early approaches, Seashore's (1938) seminal "Psychology of Music", based on extensive empirical research, begins by descriptions of "sensory capacities" and later, for instance, relates details of a singer's vibrato to feelings. Kurth (1931) described energy patterns of movement tensions in the music that are actively brought into the process by the listeners, who experience the music with their body including muscular sensations. Furthermore, Truslit (1938/2015) and Becking (1928) suggested that there are a number of gestural qualities in the music, varying for different compositions and composers, which are perceivable in an embodied way by performing conducting-like gestures along with the music.

More recent approaches to musical emotions and embodied music cognition have benefited from the evidence provided by Craig and others regarding the essential role of interoception. Described as "sensory representation of the condition of the body" including all bodily tissues (A. D. Craig 2015, p. xiii, 4f), this holistic concept of interoceptive integration of multiple types of sensory information offers a framework for studying an activity that is fundamentally shaped by sensory and motor integration in real time. Indeed, the array of information processed in music, from the ears, eyes, arms, and various parts of the body, requires a central integration mechanism. The mid-insula has been identified as a cortical area facilitating this mechanism, including "exteroceptive and vestibular inputs, ... preautonomic motor activity, visual and auditory sensory inputs, and hedonic inputs", all forming an "interoceptive image of the physiological condition of the body" (A. D. Craig 2015, p. 188; see A. D. Craig 2003). The multimodal integration in the insular cortex does not only facilitate efficiency and emotional processing, but has also been linked to awareness and sense of time (Fermin et al. 2023; Wittmann 2009). All of them are vital for engaging with music.

In the following, research on musical emotions, movement, music and dance training, attention and awareness, and sense of time is discussed. As I will argue, and as based on considerably growing empirical evidence, a bodily grounding is central for these processes, both for music perception and music performance.

2 Feeling the Music: Interoception and Emotions

Music listening engages networks in the brain that are similarly active in the processing of everyday emotions. The ability to evoke strong emotions is a key explanation for the important role of music in most people's lives, across historical epochs, ethnic groups, and cultural traditions. Other functions of music, such as rhythmic coordination, shaping rituals and social bonding, soothing newborns, enhancing the atmosphere, often involve an emotional component. Particularly when listening attentively, neuroimaging research showed activations in limbic areas such as the amygdala, hippocampus, or hypothalamus, and areas in the paralimbic cortex including the insula and cingulate cortex (for reviews, see Koelsch 2014; Vuust et al. 2022). Apart from central emotion processing, an increasing number of studies have investigated peripheral autonomic changes and physiological responses when listening to music (see Hodges 2016). The responses are manifold, ranging from tears, muscular tension, changes in pulse, respiration rate, or skin conductance, to the widely studied "chills" or "thrills" that are characterized as a shiver down the spine when listening to emotionally arousing music. In short, there is reason to assume that feeling the music involves an interoceptive mechanism.

While Craig did not develop a model of interoceptive music perception himself, he proposed a close connection between emotions, awareness, and music (A. D. Craig 2008). For him, music was a rhythmic succession of emotional moments which required awareness, could enhance social bonding, and also change our subjective sense of time. He further suggested that the anterior insula integrates emotions over a given time span, and that lesions in this region would reduce a patient's enjoyment of the emotional qualities in music. Indeed, acquired amusia, a rare condition affecting music perception and cognition, has been associated with lesions in the right hemisphere involving the insula, superior temporal gyrus, and the striatum (Sihvonen et al. 2016). Furthermore, the idea that emotions are experienced in a dynamic succession of their components rather than a fixed state was also proposed by Scherer (2009). The fluctuating emotion components of physiological responses, appraisal, motor expressions, as well as action tendencies require an integration mechanism that permits conscious emotion representation, communication, and regulation. This mechanism, according to Scherer, functions over time and fluctuates in the dynamic integration, such that the emotion components and particularly cognitive appraisal processes may vary within and between individuals.

Research in music psychology employed continuous measures of peripheral responses, self-reported experiences of chills, and the intensity of felt emotions to study these processes (Grewe et al. 2009). Listeners are reliable in telling when they feel a strong bodily sensation such a chill. In the Grewe et al. study, they listened to selected "chill-inducing" excerpts from the requiems by Mozart and Puccini and a further piece, and reported experiencing an average of nine chills over all musical pieces; yet it should be noted that the median was two, and one third of participants had no chill experiences at all. Chill responses were associated with felt emotional intensity as well as increases in skin conductance and heart rate. In fact, both physiological measures showed increases about 2 s before participants experienced chills, suggesting that interoception preceded cognitive appraisal. Hence, while this study did not directly assess interoceptive awareness, physiological responses evoked by the music were related to conscious responses of emotion effects fluctuating over time. Comparable results had been reported by D. G. Craig (2005; note the initials!), who found that skin conductance increased in moments of chill experiences (shiver or tingling sensation), which most often occurred at the arms, hands, face and head, neck, and spine. In chill moments, pilo-erections were additionally perceived on the legs, and participants often felt a sense of higher mental awareness, tension, as well as coldness. These findings indicate an array of bodily sensations during music listening that can be veridically described and measured, and represent an important component of perceived emotions (Hodges 2016; Juslin and Västfjäll 2008). A further study linked the appearance and intensity of chills during music listening to interoceptive sensibility (Schoeller et al. preprint 2023). In an online study using three emotional pieces of music (again, as in the Grewe et al. and D.G. Craig studies, all including choir and a Christian religious connotation), about 80% of participants reported that they had experienced chills. These participants, compared to those without chills, also had higher scores on the "Psychological Insight Scale" as well as higher emotional awareness in the "Multidimensional Assessment of Interoceptive Awareness" (MAIA, consisting of a total of eight components). The experienced intensity of chills correlated positively with psychological insight and emotional awareness, suggesting that psychological and interoceptive awareness are positively associated with the experience of physiological responses in music listening. At the same time, it should be noted that chills as sudden and very strong emotional moments, also occasionally referred to as "skin orgasms" (Panksepp 1995, p. 203), are not perceived by all individuals. Hence, more subtle emotional responses may also include an awareness of bodily changes and cognitive appraisal mechanisms.

A number of studies over the past years reveal a renewed interest in the topic of interoceptive effects of music listening. Tactile sensations of bass frequencies in music with high-groove ratings were investigated by Hove et al. (2020). Participants listening to the music were in some conditions additionally stimulated in their body by a subwoofer close to them, or a vibrotactile feedback system, which both led to higher enjoyment, more body movement, and also more pronounced finger tapping compared to listening only. The authors suggest that the tactile component in the multisensory condition enhanced the listening experience. In this experiment, listeners were undoubtedly aware of the different conditions by feeling the bodily sensations stemming from the bass frequencies in the music. Employing an even more explicit paradigm, a recent study used body maps that required listeners to highlight on a blank body silhouette the regions of their body that subjectively changed during listening (Putkinen et al. 2024). Western and East Asian songs representing six basic emotions were presented to listeners of similar origins. Self-indicated bodily sensations varied systematically between emotions, such as strong sensations in the chest/heart region for tender and sad songs, and higher activations in the feet for happy and danceable songs. There were high correlations (at least $r = 0.80$ across the six emotions) between Western and East Asian listeners in the body regions. This result suggests cross-cultural similarities in bodily sensations of emotions in music listening, which could be partially traced back to acoustical and musical features. Body maps were also used in an experiment with musical chord progressions of different levels of predictability (Daikoku et al. 2024). Contrary to previous research that posits stronger bodily responses for unexpected musical events, that is surprise moments including sudden changes in dynamics or harmony (Huron 2008), in the Daikoku et al. study predictable chord progressions resulted in stronger self-reported sensations in the abdomen. Valence ratings and aesthetic appreciation were also higher in predictable, low surprise progressions. These findings could be interpreted in a way suggesting that the listeners had more awareness of interoceptive responses with musical harmonies familiar to them.

There has been some controversy as to whether physiological sensations of the body are indeed needed for deep emotional feelings (Wiens 2005). Observations show that patients with disrupted autonomic sensation transmission or spinal cord injuries still experience a wide range of emotions. Wiens suggests that interoception as a central integration and representation mechanisms for bodily sensations still plays a dominant role even in these patients. According to this view, first-level emotions are based on actual feedback from the body, while second-level emotions are characterized by a degree of awareness and could be based on both actual and "illusory physiological changes" (p. 4), which may occur even in the absence of first-level emotion experiences. For instance, individuals may attribute self-perceived heart rate increases as anger without necessarily feeling the core affect in terms of a first-level basic emotion, that is, no feedback from the heart needs to be processed. It should be noted that the concept of second-level emotions without bodily feedback bears some resemblance with Damasio's (2004, p. 55) "as-if-body-loop", in which actual bodily processes are not involved yet subjective feelings still arise. Hence individuals' mental representations of emotions and their experience of emotions draw on simulated bodily activities that resemble those involved when processing emotional physiological signals from the body. In a similar vein, according to Wiens, interoception would still be at the core of all emotional experiences, whether illusory or actual in the bodily representation of physiological changes. Though further systematic research is needed in this area, applied approaches with spinal cord injury patients report benefits for mood improvement in music therapy interventions (Mercier et al. 2023), providing some support for the idea that emotional qualities can be experienced in the absence of peripheral sensations. Thus, paradoxically, interoceptive sensibility and awareness might still shape emotional responses in music listening and beyond, even if no afferent information is directly communicated from sensory organs.

3 Interoceptive Awareness and the Role of Groove, Motor Training, and Focus of Attention

Effects of music such as emotional experiences in listening are primarily shaped by musical features, differences between individuals, and the situational context facilitating background or attentive listening. Each of these dimensions has been studied extensively, yet research focussing on bodily components was typically carried out in attentive listening situations in laboratories. In the following, selected studies are discussed that address interoceptive sensations in the fields of music and dance, with a particular focus on feelings of musical groove, motor expertise, and attention. It should be noted that other factors may come into play in musical interoception, for instance in therapeutic or educational contexts, which are not reviewed here and would require more research.

The holistic concept of the interoceptive system, involving the integration of various sensory inputs, hedonic qualities, and motor or action tendencies (A. D. Craig 2015; Scherer 2009), presents striking parallels with the concept of groove. While "groove" is sometimes also used for referring to rock or funk genres in popular music, it has widely been researched as a specific experience in music listening. Definitions stress the central components of an urge to move one's body or parts of it, and a pleasurable response (Janata et al. 2012; Madison 2006; Senn et al. 2018, 2020). In other words, groove music is hedonically experienced via bodily sensations involving action tendencies that typically manifest themselves in overt movements along with the music. Different musical features have been investigated that give rise to groove feelings, including musical micro-timing, event density, or rhythmic and harmonic complexity (Senn et al. 2018; Witek et al. 2014). While apparently no study so far has stated a direct involvement of the insula in groove experiences, activities were shown in reward and motor networks, including the basal ganglia regions (Matthews et al. 2020), as well as motor-evoked potentials recorded in arms and hand muscles (Stupacher et al. 2013). Further musical features giving rise to bodily feelings apart from groove, such as in the sensation of strong physiological responses in chills, have been reviewed above. Yet the motor component in both groove experiences and a dynamic model of interoception and emotion leads to the question as to whether prolonged training in movement skills in relation to music enhances interoceptive accuracy and awareness. In the Stupacher et al. study, there was an effect of training such that musicians had higher motor-evoked potentials for high-groove and on-beat music compared to musically untrained participants.

Regarding different levels of interoceptive awareness, there is further evidence that musical training, or the prior decision to engage extensively in music making, is associated with higher interoceptive accuracy. In a study with professional musicians, Schirmer-Mokwa et al. (2015) employed a heartbeat discrimination paradigm (see Brener and Kluvitse 1988) that asked participants to indicate if their own heartbeats were synchronous to external auditory beats or not. The auditory signals were triggered by participants' own R peak in an ECG; synchronous conditions in this study were defined as a small delay of 300 ms, whereas asynchronous conditions were defined by delays of 600 ms between R peak and auditory signal. Musicians (singers and string instrument players) were more accurate in detecting synchronous tone-heartbeat presentations compared to a control group without any musical training. Schirmer-Mokwa et al. suggest that the high number of multisensory experiences accumulated through musical training may increase interoceptive accuracy and awareness. This conclusion is substantiated by a strong correlation between estimated duration of musical training and number of correct responses in the heartbeat discrimination task. A further study similarly showed higher sensibility in musicians compared to non-musicians in heartbeat discrimination as well as attention regulation – a subscale of an interoceptive awareness inventory (Hina et al. preprint 2020). Nevertheless, these behavioural results were recently challenged by findings from Herman et al. (preprint 2023) who found no differences between musicians playing keyboard instruments and a control, neither in heartbeat discrimination, heartbeat tracking (counting), self-report interoceptive awareness, nor an improvement with a musical training intervention for the non-musician control. Results from fMRI analyses involving multiple networks only showed higher functional connectivity between the insula and the left inferior temporal gyrus in musicians compared to the control, suggesting a stronger link between the activity in the insula and multimodal integration in musicians. The discrepancy between these studies may be partially due to differences between the types of professional musical activities, such that singers may have greater interoceptive awareness compared to keyboard players. Furthermore, they might also be due to issues in the definition of what constitutes a professional musician. Research should compare music training with other body-focused activities including acting or dancing, in order to estimate effects of the respective activity on interoception. On the other hand, musicians may benefit from training in interoceptive awareness in high-pressure performance situations through mindfulness practice and other mind-body training methods (Diaz 2022). Attending to one's body over years in musical training, which for most instruments involves high levels of motor control, undoubtedly requires awareness of the functioning of the hands, arms, lungs, and so on, simply to prevent oneself from injury by overexertion. Bodily awareness might also help in re-learning certain movement details in overcoming malfunctioning habits.

In a related field, there is reason to assume that the pronounced movement experience of dancers should be associated with high levels of interoceptive awareness. Dancers integrate sensations arising from various body parts, often synchronize with other dancers by means of visual and auditory information, and use auditory and emotional cues in many dance genres shaped by music (Bläsing 2023). In a qualitative study involving 13 professional dancers in the Western artistic tradition, Legrand and Ravn (2009) found out that dancers distinguished between perceiving their own body in terms of gaze checks, proprioception, and interoception, and their body in relation to others' bodies in terms of multimodal integration. Music as an external sensation is strongly related to proprioception, and is experienced and incorporated in the dynamics of the dancers' movements. Interoception plays an important role for movements and for preparing to move while being aware of one's bodily states, and dancers vary their attention from focussing on internal organs, muscles, or the skeleton—the latter allowing to focus less on feelings of pain compared to focussing on the muscles. A further study measured interoceptive accuracy in 20 dancers and a control group of similar size (Christensen et al. 2018). Using a heart-beat perception paradigm in which participants were required to count their heart beat in intervals from 25 to 100 s, professional dancers were more accurate in the ratio between counted and actual number of heart beats. More experienced dancers received even higher accuracy scores, compared to less experienced dancers, who on their own were still more accurate than individuals with no formal dance training. There were no significant correlations between interoceptive accuracy and self-reported emotional intensity, yet those scoring higher in the heart-beat task also had higher scores in an art-experience questionnaire that measures general interest and time spent with the arts. These results may point to training effects, but could equally be attributed to personality characteristics such that individuals with higher interoceptive awareness, and/or openness to experience (Chamorro-Premuzic et al. 2011), may be attracted more by the arts, including the performing arts. Longitudinal studies would be a potential avenue for pursuing this question further. Finally, the bodily movements of dancers, perceived in a multimodal way, often evoke interoceptive experiences in the audience; in the case of an extremely slow contemporary choreography by Myriam Gourfink, the audience felt bodily sensations in relation to the stomach, to breathing, and weight that lasted beyond the dance performance (Joufflineau 2023). The dancers in Gourfink's performances frequently engage in yoga training that should enhance their interoceptive attention and contemplation (Bläsing 2023). This undoubtedly affects the audience in turn.

Attention to bodily sensations in comparison to external goals was also investigated in recent music studies. It should be noted that at first, this approach appears to take an opposite stance regarding awareness of the body. The research was inspired by the constrained action hypothesis that posits diametral effects of an attentional focus on one's movements during performance. In high-pressure situations, performance could be impaired by a self-conscious focus on the production of actions due to limited resources of working memory (Wulf et al. 2001). Previous studies with athletes found beneficial effects of concentrating on action outcomes, hence advantages for an external rather than an internal focus of attention. A study with experienced and absolutely novice violinists tested this hypothesis by using EMG and error analysis in a specific, relatively novel bowing task (Allingham and Wöllner 2022). Since the action outcome in music might be less obvious and measurable compared to sports, in which a successful shot of a basketball in the basket may function as an external goal, a third focus condition was developed and implemented. This new "somatic focus", defined to participants as "focus your attention on the resistance of the bow against the string" (p. 176) seemed an adequate proxy for the external focus studied before in sports. The two other conditions involved focussing on the movements in the arm leading the bow (internal), and the sound produced (external). Results showed that experienced violinists benefited from the somatic compared to the internal focus in terms of number of errors made in the task, while novices did not show differences. The somatic focus also resulted in lower EMG triceps activity in the right arm across all participants, and no self-reported physical discomfort compared to the other two conditions. While a recent review recommends more research to be carried out before conclusive recommendations can be drawn (Hohagen and Immerz 2024), these findings show that during performance, focusing one's attention on bodily processes might limit oneself, since controlling automatized movements could constrain one's actions. On the other hand, as shown in the studies reviewed above, a general level of interoceptive awareness and cognitive control in situations beyond performance, for instance immediately before a high-pressure situation, could undoubtedly be beneficial. So, while high degrees of interoceptive sensibility and awareness might be reached by training in a multimodal integration activity such as music, and there is no reason to assume that general bodily awareness is limiting oneself in practice and improvement of skills, actual performance situations require different cognitive strategies with a rather non-interoceptive focus of attention.

4 Sense of Time in Music Listening Activities

Interoception, and in particular the experiences and activities reviewed above, clearly involve a temporal component. A recent meta-analysis has shown that the insula, in addition to the supplementary motor area, has a key role in explicit timing tasks (Mondok and Wiener 2023). Key psychological concepts of time perception have been investigated in the temporal art of music. Upper and lower limits were studied for perceiving synchrony of two or more sound events, for perceiving musical Gestalts, or regarding the sense of overall coherence in musical forms. For instance, the "psychological present", defined by Fraisse (1984, p. 9f) to last between 2 and 3 s, with an upper limit of 5 s due to working memory constraints, typically corresponds with the duration of motivic material in music that is still perceived as a unit or Gestalt. For an overview of inner-clock models and subjective time experiences in relation to music, see Wang and Wöllner (2020).

A bodily, emotional model of time perception was proposed by A. D. Craig (2009). Based on homeostatic afferent signals that are processed in the insular cortex and relate to awareness of oneself, Craig suggested that arousal, such as present in the heartbeat or respiration rates, shapes emotions and ultimately subjective time estimations. Hence, experiences of time are associated with emotions, can change dynamically in an adaptive way, and are based on endogenous bodily information. Both experiencing a "sentient self" (p. 1936) and experiencing presence would require an integration of moments of salience in a homeostatic progression of feelings across time. Neuroimaging research provided evidence for the role of interoception, and specifically the insula, in time perception and awareness (Richter and Ibáñez 2021; Wittmann 2013; Wittmann and Droit-Volet 2024).

The model offers two explanatory paths for the often-reported impact of music on time perception: attentional awareness and dynamic fluctuations over time that are shaped by emotions. In this regard, Craig suggested that "[s]tudies of interactions between subjective time, emotion and music could directly address this model and reveal new interactions that define the substrate that underlies our awareness" (A. D. Craig 2009, p. 1939). The dynamic and adaptive nature of the model may account for time dilations reported in a number of studies in the millisecond to second range, particularly when participants are asked to indicate the passage of time or the perceived duration of events. Moreover, it provides a link to the dynamic emotion model suggested by Scherer (2009) mentioned above, which emphasizes the integration of synchronized emotion components, including individually different levels of physiological arousal, motor cues, action tendencies, and appraisal. Yet Scherer assumed that individuals integrate these components widely without being aware of it. A study that tested directly interoceptive awareness found higher retrospective time distortions for individuals focussing on their body and bodily reactions, compared to other individuals who focussed on exteroceptive information (Pollatos et al. 2014). Since both groups watched film clips varying in emotion, an arousal effect could have mediated results.

Feeling arousal, constituting one of the components in the two-dimensional circumplex model of emotions, may be considered as a cognitive self-judgment of heightened bodily activity, perceived in an interoceptive way. Arousal in music has often been associated with increased tempo. The idea is that bodily processes including the heart rate would synchronize with musical tempo. A number of studies employed prospective duration estimations and found that participants judged time intervals in fast music to be longer compared to slower music (Droit-Volet et al. 2013; Hammerschmidt and Wöllner 2020; Oakes 2003). These findings were interpreted in relation to inner-clock models (Block and Zakay 1997; Grondin 2010), suggesting that more events per unit of clock time are perceived in faster tempo, and that in a putative pacemaker, more pulses were emitted due to higher arousal. Comparing the number of pulses accumulated with a reference number stored in memory, the higher number is then attributed to longer time spans. Again, the insular cortex has been shown to play a central role in the accumulation process (Wittmann 2013). Higher physiological arousal may have consequences for music listeners. A study asked one group of participants to engage in jogging, while the other group stayed still (Jakubowski et al. 2015). The first group's heart rate was faster after the exercise, yet they also preferred faster song tempi. Thus, their higher arousal had an impact on their musical tempo choice, which might reflect the higher number of pulses emitted in the inner-clock model.

To add complexity, increases in arousal did not always change sense of time, and higher tempo could simply have increased the number of events perceived in a certain span of time, without a strong bodily component. In the Droit-Volet et al. (2013) study, arousal was further manipulated by changing the musical sounds. However, listeners did not perceive differences in duration when being presented with orchestral versions of the music that should have increased arousal compared to piano versions played at the same tempi. A further study (Wöllner 2023) presented three different excerpts of the second movements of Beethoven's 7th symphony that are characterized by repetitions of the march-like theme, each time with a higher number of orchestra instruments involved, and a crescendo until a full tutti passage is reached. The number of bars and the duration of the excerpts remained the same – as did listeners' duration and passage of time judgments. Yet listeners clearly distinguished between the different arousal levels. Similarly, hip-hop songs that had been judged to be high in arousal in a pilot study did not lead to longer duration judgments compared to low-arousal songs at the same tempo (Wöllner & Hammerschmidt 2021). One could argue that cognitive appraisal (Scherer 2009) may have dominated judgments, and listeners may have possessed certain levels of awareness of different stimulus manipulations. A further explanation lies in the distinction between bodily arousal and the arousal level that is merely cognitively perceived in the music, but not necessarily experienced by the participants themselves with their bodies (cf. Song et al. 2016). While the former, in an interoceptive way, has an impact on sense of time, the latter does not automatically influence time perception.

Apart from higher tempo, awareness of features in the music has been shown to reliably influence time. In a series of experiments, we asked participants to focus on different metrical levels in musical drum patterns (Hammerschmidt and Wöllner 2020), and excerpts without lyrics taken from commercial hip-hop music (Wöllner and Hammerschmidt 2021). Participants were required to listen only, to tap with half notes (slow tapping rate), as well as with quarter notes (medium tapping rate; first study only) and eights notes (fast tapping). Time was perceived to be longest when listening only without engaging in the tapping task, while it was perceived to be shortest when focussing the attention on the highest metrical level, hence half notes with the slowest tapping rate. In other words, focussing on a slower event rate in the music gave rise to the illusion of shorter duration. These results indicate that it is not only the number of events perceived in a stimulus that changes time perception, but also the active and combined cognitive and motor engagement with it. Furthermore, as the second study revealed, the pure motor activity of tapping per se had no impact, since tapping with a simple metronome at the same beat rates, instead of the music, did not yield comparable results for time estimations. These results suggest that the number of musical events perceived and attentively processed in synchronous tapping had the strongest impact on time estimations. Other research has shown that synchronizing with the music by means of nodding the head or tapping the feet clearly helps participants in finding the pulse, and in being more consistent in attuning to the beat (Su and Pöppel 2012). This work provides further evidence for the role of the body in temporal tasks. There are promising approaches to musical traditions outside the Western isochronous beat schemes, with more complex metrical relationships (Polak et al. 2016), underlining the role of the body in finding the beat. It can be assumed that a central integration mechanism, encompassing both perceptions of external events and bodily sensations, and shaping them by cognitive appraisal, may have the explanatory power to model effects of music on time perception and joint behaviour more widely.

5 Conclusions

Considering musical engagement through an interoceptive system's perspective offers valuable explanations for the deep-rooted bodily components and action tendencies involved in music, as well as the integration of multimodal information. Music listening may evoke strong physiological responses, often experienced as chills or increases in the heartbeat, and individuals feel a pleasurable urge to move their body when listening to groove. Given that the insular cortex has been identified as an area for the integration of interoceptive sensations and emotional interpretations of feelings, including hedonic experiences, A. D. Craig's (2009) proposal for studying music in this context is well-justified and may deserve even more attention. There are some open questions in this context that call for more research.

Regarding the dynamics involved in emotion experiences, the role of awareness needs further consideration. A. D. Craig (2009) stated that "[t]he emergent capacity for music in this model [of the sentient self] can also explain the primal emotional effects of communal music making, because it indicates that music inherently involves the core of awareness" (p. 1937). Interoceptive sensibility and awareness, both reflecting attention to felt changes in one's body (Garfinkel et al. 2015), are presumably related to cognitive appraisal mechanisms. Individuals that are used to attend to changes in body sensations and to interpret them should then find it easier to indicate the valence of experienced emotions. According to Scherer (2009), nevertheless, the synchronized emotion components are likely to be integrated outside awareness. It could be investigated if individuals can actively facilitate the synchronization of the components or delay it, for instance by sensing bodily changes deliberately without cognitively appraising them, or experiencing them without emerging action tendencies. Music listening offers a rich arena for studying these processes, for instance by investigating the temporal dynamics in emotional responses, or by controlling the temporal-metrical levels individuals actively attend to via body movements.

Furthermore, differences in interoceptive accuracy and awareness between individuals point to the question if extensive training enhances interoception, or if individuals scoring high in interoceptive accuracy are prone to engaging in activities that partially depend on it. Musicians and dancers spend years to master their body, integrate and synchronize exteroceptive and interoceptive information, and deliver emotionally meaningful performances. Longitudinal research may help elucidating this question for expert performance, while intervention studies could yield some effects in therapeutical contexts. Regarding emotion processing, it has been suggested that sensing physiological arousal, developing action tendencies, and cognitively appraising emotions are shaped by individual differences in relation to the emphasis a person places on each of these components (Scherer 2009). Individual differences in interoception have also been linked to empathy (Fukushima et al. 2011; Grynberg and Pollatos 2015), an area that has received increasing attention in music research (e.g. Wöllner 2017). Thus, not only interoceptive accuracy could explain the findings for individuals engaging in music, dance and related activities, but also the higher-order integration of emotion components, and the processing of signals one receives from others.

Finally, it should be noted that interoception, seen as focussing on internal bodily states and often on muscular activation, may have detrimental effects on performance in high-pressure situations (Wulf et al. 2001; Allingham and Wöllner 2022), and that the desired experience of flow states in the performing arts may involve blending out the "here and now" as well as bodily constraints (see Wrigley and Emmerson 2011). Attentional focus and awareness should be further investigated both in preparing to perform and in actual performances outside the lab. Further open topics are time dilations in relation to arousal and other interoceptive and exteroceptive information that are integrated in the insula, since focussing on the body may enlarge the time dilations (Pollatos et al. 2014). Activities requiring high timing control would thus benefit more strongly from external sensory inputs rather than afferent body signals. In other words, it may not always be beneficial to follow one's heart, even in music.

References

Allingham E, Wöllner C (2022) Effects of attentional focus on motor performance and physiology in a slow-motion violin bow-control task: evidence for the constrained action hypothesis in bowed string technique. J Res Music Educ 70(2):168–189. https://doi.org/10.1177/00224294211034735

Becking G (1928) Der musikalische Rhythmus als Erkenntnisquelle. Filser. http://digitale-objekte.hbz-nrw.de/webclient/DeliveryManager?pid=2434319&custom_att_2=simple_viewer

Bläsing B (2023) Time experiences in dance. In: Wöllner C, London J (Hrsg.) Performing time: synchrony and temporal flow in music and dance. Oxford University Press, pp 13–31

Block RA, Zakay D (1997) Prospective and retrospective duration judgments: a meta-analytic review. Psychon Bull Rev 4(2):184–197. https://doi.org/10.3758/BF03209393

Brener J, Kluvitse C (1988) Heartbeat detection: judgments of the simultaneity of external stimuli and heartbeats. Psychophysiology 25(5):554–561. https://doi.org/10.1111/j.1469-8986.1988.tb01891.x

Chamorro-Premuzic T, Reimers S, Hsu A, Ahmetogluet G (2011) Who art thou? Personality predictors of artistic preferences in a large UK sample: the importance of openness. Br J Psychol 100(3):501–516. https://doi.org/10.1348/000712608X366867

Christensen JF, Gaigg SB, Calvo-Merino B (2018) I can feel my heartbeat: dancers have increased interoceptive accuracy. Psychophysiology 55(4):e13008. https://doi.org/10.1111/psyp.13008

Cook N (2014) Beyond the score: music as performance. Oxford University Press

Craig AD (2003) Interoception: the sense of the physiological condition of the body. Curr Opin Neurobiol 13(4):500–505. https://doi.org/10.1016/S0959-4388(03)00090-4

Craig AD (2008) Interoception and emotion: a neuroanatomical perspective. In: Handbook of emotions, 3rd edn. The Guilford Press, pp 272–292

Craig AD (2009) Emotional moments across time: a possible neural basis for time perception in the anterior insula. Philos Trans R Soc B Biol Sci 364(1525):1933–1942. https://doi.org/10.1098/rstb.2009.0008

Craig AD (2015) How do you feel? An interoceptive moment with your neurobiological self. Princeton University Press

Craig DG (2005) An exploratory study of physiological changes during "Chills" induced by music. Music Sci 9(2):273–287. https://doi.org/10.1177/102986490500900207

Daikoku T, Tanaka M, Yamawaki S (2024) Bodily maps of uncertainty and surprise in musical chord progression and the underlying emotional response. iScience 27(4):109498. https://doi.org/10.1016/j.isci.2024.109498

Damasio AR (2004) Emotions and feelings. A neurobiological perspective. In: Manstead ASR, Frijda N, Fischer A (eds) Feelings and emotions. Cambridge University Press, p 4957

Diaz FM (2022) Mindfulness. In: McPherson G (Hrsg.) The Oxford handbook of music performance, vol 2. Oxford University Press, pp 65–83

Droit-Volet S, Ramos D, Bueno JLO, Bigand E (2013) Music, emotion, and time perception: the influence of subjective emotional valence and arousal? Front Psychol 4:417. https://doi.org/10.3389/fpsyg.2013.00417

Fermin ASR, Sasaoka T, Maekawa T, Chan H-L, Machizawa MG, Okada G, Okamoto Y, Yamawaki S (2023) Insula neuroanatomical networks predict interoceptive awareness. Heliyon 9(8):e18307. https://doi.org/10.1016/j.heliyon.2023.e18307

Fraisse P (1984) Perception and estimation of time. Annu Rev Psychol 35:1–37. https://doi.org/10.1146/annurev.ps.35.020184.000245

Fukushima H, Terasawa Y, Umeda S (2011) Association between interoception and empathy: evidence from heartbeat-evoked brain potential. Int J Psychophysiol 79(2):259–265. https://doi.org/10.1016/j.ijpsycho.2010.10.015

Garfinkel SN, Seth AK, Barrett AB, Suzuki K, Critchley HD (2015) Knowing your own heart: distinguishing interoceptive accuracy from interoceptive awareness. Biol Psychol 104:65–74. https://doi.org/10.1016/j.biopsycho.2014.11.004

Grewe O, Kopiez R, Altenmüller E (2009) The chill parameter: goose bumps and shivers as promising measures in emotion research. Music Percept 27(1):61–74. https://doi.org/10.1525/mp.2009.27.1.61

Grondin S (2010) Timing and time perception: a review of recent behavioral and neuroscience findings and theoretical directions. Atten Percept Psychophys 72(3):561–582. https://doi.org/10.3758/APP.72.3.561

Grynberg D, Pollatos O (2015) Perceiving one's body shapes empathy. Physiol Behav 140:54–60. https://doi.org/10.1016/j.physbeh.2014.12.026

Hammerschmidt D, Wöllner C (2020) Sensorimotor synchronisation with higher metrical levels in music shortens perceived time. Music Percept 37(4):263–277. https://doi.org/10.1525/mp.2020.37.4.263

Herman AM, Olszewska A, Gaca M, Droździel D, Marchewka A (2023) Interoception and the musical brain: evidence from cross-sectional and longitudinal behavioral and resting-state fMRI study. Psychophysiology 60(12):e14402. https://doi.org/10.1111/psyp.14402

Hina F, Aspell J, Cardini F (2020) Enhanced behavioural and brain responses to interoceptive signals in musicians. OSF. https://doi.org/10.31234/osf.io/smdwc

Hodges DA (2016) Bodily responses to music. In: The Oxford handbook of music psychology, 2nd edn. Oxford University Press, pp 183–196

Hohagen J, Immerz A (2024) Focus of attention in musical learning and music performance: a systematic review and discussion of focus instructions and outcome measures. Front Psychol 15. https://doi.org/10.3389/fpsyg.2024.1290596

Hove MJ, Martinez SA, Stupacher J (2020) Feel the bass: music presented to tactile and auditory modalities increases aesthetic appreciation and body movement. J Exp Psychol Gen 149(6):1137–1147. https://doi.org/10.1037/xge0000708

Huron D (2008) Sweet anticipation: music and the psychology of expectation, Illustrated edn. MIT Press

Jakubowski K, Halpern AR, Grierson M, Stewart L (2015) The effect of exercise-induced arousal on chosen tempi for familiar melodies. Psychon Bull Rev 22(2):559–565. https://doi.org/10.3758/s13423-014-0687-1

Janata P, Tomic ST, Haberman JM (2012) Sensorimotor coupling in music and the psychology of the groove. J Exp Psychol Gen 141(1):54–75. https://doi.org/10.1037/a0024208

Joufflineau C (2023) Time through the magnifying glass of slowness: a case study in Myriam Gourfink's choreography. In: Wöllner C, London J (Hrsg.) Performing time: synchrony and temporal flow in music and dance. Oxford University Press, pp 127–133

Juslin PN, Västfjäll D (2008) Emotional responses to music: the need to consider underlying mechanisms. Behav Brain Sci 31(5):559–575. https://doi.org/10.1017/S0140525X08005293

Koelsch S (2014) Brain correlates of music-evoked emotions. Nat Rev Neurosci 15(3):170–180. https://doi.org/10.1038/nrn3666

Kurth E (1931) Musikpsychologie. Hesse

Legrand D, Ravn S (2009) Perceiving subjectivity in bodily movement: the case of dancers. Phenomenol Cogn Sci 8(3):389–408. https://doi.org/10.1007/s11097-009-9135-5

Madison G (2006) Experiencing groove induced by music: consistency and phenomenology. Music Percept 24(2):201–208. https://doi.org/10.1525/mp.2006.24.2.201

Matthews TE, Witek MAG, Lund T, Vuust P, Penhune VB (2020) The sensation of groove engages motor and reward networks. Neuroimage 214:116768. https://doi.org/10.1016/j.neuroimage.2020.116768

Mercier LJ, Grant C, Langelier DM, Plamondon S (2023) Scoping review of music therapy and music interventions in spinal cord injury. Disabil Rehabil 45(10):1736–1749. https://doi.org/10.1080/09638288.2022.2073391

Mondok C, Wiener M (2023) Selectivity of timing: a meta-analysis of temporal processing in neuroimaging studies using activation likelihood estimation and reverse inference. Front Hum Neurosci 16:1000995. https://doi.org/10.3389/fnhum.2022.1000995

Oakes S (2003) Musical tempo and waiting perceptions. Psychol Mark 20(8):685–705. https://doi.org/10.1002/mar.10092

Panksepp J (1995) The emotional sources of "chills" induced by music. Music Percept 13(2): 171–207. https://doi.org/10.2307/40285693

Polak R, London J, Jacoby N (2016) Both isochronous and non-isochronous metrical subdivision afford precise and stable ensemble entrainment: a corpus study of Malian Jembe drumming. Front Neurosci 10. https://doi.org/10.3389/fnins.2016.00285

Pollatos O, Laubrock J, Wittmann M (2014) Interoceptive focus shapes the experience of time. PLoS One 9(1):e86934. https://doi.org/10.1371/journal.pone.0086934

Putkinen V, Zhou X, Gan X, Yang L, Becker B, Sams M, Nummenmaa L (2024) Bodily maps of musical sensations across cultures. Proc Natl Acad Sci 121(5):e2308859121. https://doi.org/10.1073/pnas.2308859121

Richter F, Ibáñez A (2021) Time is body: multimodal evidence of crosstalk between interoception and time estimation. Biol Psychol 159:108017. https://doi.org/10.1016/j.biopsycho.2021.108017

Scherer KR (2009) The dynamic architecture of emotion: evidence for the component process model. Cognit Emot 23(7):1307–1351. https://doi.org/10.1080/02699930902928969

Schirmer-Mokwa KL, Fard PR, Zamorano AM, Finkel S, Birbaumer N, Kleber BA (2015) Evidence for enhanced interoceptive accuracy in professional musicians. Front Behav Neurosci 9. https://doi.org/10.3389/fnbeh.2015.00349

Schoeller F, Jain A, Christov-Moore L, Reggente N (2023) Musical chills induce psychological insight. OSF. https://doi.org/10.31234/osf.io/m9ea5

Seashore CE (1938) Psychology of music. McGraw-Hill

Senn O, Bechtold T, Rose D, Câmara GS, Düvel N, Jerjen R, Kilchenmann L, Hoesl F, Baldassarre A, Alessandri E (2020) Experience of groove questionnaire instrument development and initial validation. Music Percept 38(1):46–65. https://doi.org/10.1525/mp.2020.38.1.46

Senn O, Kilchenmann L, Bechtold T, Hoesl F (2018) Groove in drum patterns as a function of both rhythmic properties and listeners' attitudes. PLoS One 13(6):e0199604. https://doi.org/10.1371/journal.pone.0199604

Sihvonen AJ, Ripollés P, Leo V, Rodríguez-Fornells A, Soinila S, Särkämö T (2016) Neural basis of acquired Amusia and its recovery after stroke. J Neurosci 36(34):8872–8881. https://doi.org/10.1523/JNEUROSCI.0709-16.2016

Song Y, Dixon S, Pearce MT, Halpern AR (2016) Perceived and induced emotion responses to popular MusicCategorical and dimensional models. Music Percept 33(4):472–492. https://doi.org/10.1525/mp.2016.33.4.472

Stupacher J, Hove MJ, Novembre G, Schütz-Bosbach S, Keller PE (2013) Musical groove modulates motor cortex excitability: a TMS investigation. Brain Cogn 82(2):127–136. https://doi.org/10.1016/j.bandc.2013.03.003

Su Y-H, Pöppel E (2012) Body movement enhances the extraction of temporal structures in auditory sequences. Psychol Res 76(3):373–382. https://doi.org/10.1007/s00426-011-0346-3

Truslit A (2015) Gestaltung und Bewegung in der Musik: Ein tönendes Buch vom musikalischen Vortrag und seinem bewegungserlebten Gestalten und Hören (Reprint der Ausgabe Berlin 1938). Wißner

Vuust P, Heggli OA, Friston KJ, Kringelbach ML (2022) Music in the brain. Nat Rev Neurosci 23(5):287–305. https://doi.org/10.1038/s41583-022-00578-5

Wang X, Wöllner C (2020) Time as the ink that music is written with: a review of internal clock models and their explanatory power in audiovisual perception. Yearbook of music psychology, pp 1–22. https://doi.org/10.5964/jbdgm.2019v29.67

Wiens S (2005) Interoception in emotional experience. Curr Opin Neurol 18(4):442–447. https://doi.org/10.1097/01.wco.0000168079.92106.99

Witek MAG, Clarke EF, Wallentin M, Kringelbach ML, Vuust P (2014) Syncopation, body-movement and pleasure in groove music. PLoS One 9(4):e94446. https://doi.org/10.1371/journal.pone.0094446

Wittmann M (2009) The inner experience of time. Philos Trans R Soc B Biol Sci. https://doi.org/10.1098/rstb.2009.0003

Wittmann M (2013) The inner sense of time: how the brain creates a representation of duration. Nat Rev Neurosci 14(3):217–223. https://doi.org/10.1038/nrn3452

Wittmann M, Droit-Volet S (2024) Subjective time in ordinary and non-ordinary states of consciousness: how interoceptive feelings inform us about the passage of time. In: Wittmann M, Strigo I, Simmons A (eds) The bodily self, emotion, and subjective time: exploring interoception through the contributions of A.D. (Bud) Craig (current topics in behavioral neurosciences). Springer

Wöllner C (2017) Audience responses in the light of perception–action theories of empathy. In: King E, Waddington C (Hrsg.) Music and empathy. Routledge

Wöllner C (2023) Changes in psychological time when attending to different temporal structures in music. In: Wöllner C, London J (Hrsg.) Performing time: synchrony and temporal flow in music and dance. Oxford University Press, pp 251–268

Wöllner C, Hammerschmidt D (2021) Tapping to hip-hop: effects of cognitive load, arousal, and musical meter on time experiences. Atten Percept Psychophys 83(4):1552–1561. https://doi.org/10.3758/s13414-020-02227-4

Wrigley WJ, Emmerson SB (2011) The experience of the flow state in live music performance. Psychol Music 41(3):292–305. https://doi.org/10.1177/2F0305735611425903

Wulf G, McNevin N, Shea CH (2001) The automaticity of complex motor skill learning as a function of attentional focus. Q J Exp Psychol A 54(4):1143–1154. https://doi.org/10.1080/713756012

Body, Self, and Time: Bud Craig's Global Emotional Moments Theory

Julian Kiverstein

Contents

Abstract The topic of my chapter will be Bud Craig's theory of "global emotional moments" (henceforth the GEMs theory) and the relationship of GEMs to the experience of time. I connect three ideas prominent in Craig's writings: interoception, emotion, and time. Craig held that each GEM has as its neural substrate a large-scale network with the anterior insula cortex (AIC) serving as its central processing hub. This network integrates interoceptive signals that keep track of changes arising in the autonomic nervous system with hedonic and motivational signals based on the organism's sensory perception of its environment. Craig argued that GEMs function as moving windows of time within which "a phenomenal self" is

J. Kiverstein (✉)
Lemon Tree Center for Psychiatry, Psychotherapy, and Philosophy, Amsterdam, The Netherlands

Department of Psychiatry, Amsterdam University Medical Research, Amsterdam, The Netherlands
e-mail: j.d.kiverstein@amsterdamumc.nl

Curr Topics Behav Neurosci (2025) 70: 285–306
https://doi.org/10.1007/7854_2024_575
Published Online: 12 March 2025

experienced. By the "phenomenal self," I mean a material, embodied self that forms an organism's subjective point of view on the world. Craig proposed what he called a "cinemascopic" theory of GEMs. GEMs are combined over time to form a stream of consciousness, which Craig compared to a movie, with each GEM corresponding to a single snapshot of this movie. I will argue that Craig's cinemascopic theory has implications for our understanding of what I will call the "phenomenal now." There are three main theories of the phenomenal now in the philosophical literature. One point of contention between these theories is whether the phenomenal now has duration or temporal depth. I will argue that GEMs have duration and therefore count against so-called "cinematic" theories of the phenomenal now that take the contents of experience to be of discrete points or instances in time. However, there are different views within philosophy of how the phenomenal now can have duration. I end my chapter by considering how Craig's GEMs theory might bear on this debate.

Keywords Global emotional moments (GEMs) · Insular cortex · Interoception · Phenomenal now · Phenomenal selfhood · Specious present

1 Introduction

The topic of my chapter will be Bud Craig's theory of "global emotional moments" (henceforth GEMs) and the relationship of GEMs to the subjective experiences of time (Craig 2009a, b, 2015). At every moment of our lives, there is an emotional experience of some kind taking place. I may find myself absorbed in some activity or bored as I sit waiting for an appointment. I may feel overwhelmed by the busyness of my life, fearful about the reaction of a loved one to a difficult conversation, or angry at a perceived slight I have experienced. Even when the stormy seas of our emotional lives are temporarily calm, still this feels like something – it may feel good, for instance, to experience a brief respite from the turmoil.

Within each emotional experience, we feel time passing. When we are forced to sit through a boring talk at a conference, subjective time seems to pass painfully slowly. When we are enjoying a deep conversation with a friend, subjective time can seem to fly by in an instant. Moments of intense emotion can make subjective time apparently dilate. During a highly stressful event such as a traffic accident in which one's life is threatened, people report seeing events unfold in slow motion. They describe how their attention is sharpened, and how their thinking and acting speed up, enabling them to react fast to escape the threat to their life.

My chapter connects three ideas prominent in Craig's writings on the relationship between interoception, emotion, and time (see, e.g., Craig 2009a, b, 2015). The first concerns the neural basis of GEMs. Craig held that each GEM has as its neural substrate a large-scale network with the anterior insula cortex (henceforth abbreviated as "AIC") serving as its central processing hub. This network integrates interoceptive signals that keep track of changes arising in the autonomic nervous system with hedonic and motivational signals that are based on the organism's sensory perception of its environment. Interoceptive processes sense the changing internal physiological conditions of the body via signals carried by the cranial nerves, most notably the vagus nerve. Examples of interoceptive processes include cardiovascular, respiratory, gastrointestinal, and urogenital systems. These processes serve a homeostatic function of maintaining the health of our bodies, but they are also involved in the generation and regulation of emotions. The integration of bodily signals with hedonic and motivational signals in the AIC is the neurobiological basis for a GEM.

Craig argued that GEMs function as moving windows of time within which "a phenomenal self" is experienced. By the "phenomenal self," I mean to refer to a subject's experience of being a material, embodied self. Craig proposed what he called a "cinemascopic" theory of the phenomenal self. He compared the stream of consciousness to a movie, with each GEM corresponding to a single snapshot of this movie. I will argue that Craig's cinemascopic theory can be interpreted as having implications for a theory of consciousness more generally. This may seem like something of an argumentative leap. Craig was concerned with explaining the neuroanatomy of interoceptive pathways and how these pathways are involved in mapping out feelings of the body from the inside. However, I will argue that his GEMs theory can be interpreted as challenging the divide between interoception (as an inner-directed mode of perception) and exteroception (as describing outer-directed modes of sensory perception). This much is implied by Craig's theory of GEMs as constructed through the integration of interoceptive bodily feelings with exteroceptive sensory experiences with hedonic and motivational value. All exteroceptive sensory experiences are, I will argue, imbued with affect and emotion that is rooted in interoceptive awareness of the body.

I then switch gears from discussing phenomenal consciousness and selfhood in general and turn, more specifically, to the implications of Craig's GEMs theory for our understanding of what I will call the "phenomenal now" – our subjective experience of the present moment in time. Craig called his theory of GEMs, the "cinemascopic" theory. This might suggest that he thought of each GEM as a discrete snapshot of an instant in time – the now that serves as an interface between the past and the future. I will argue, however, that caution is warranted here. There are three main theories of the phenomenal now in the philosophical literature. The main point of contention between these theories is whether the phenomenal now has duration or temporal depth, or whether the now as we experience it is confined to a discrete instant or point in time, what William James described as a "knife-edge" that separates the past from the future. I will argue that GEMs have duration and therefore count against so-called "cinematic" theories of the phenomenal now that take the contents of experience to be of discrete points or instances in time.

I will argue that Craig's GEMs theory provides support for a view of the phenomenal now as the specious present. However, there are different views of the nature of the specious present in the philosophical literature. The debate concerns the relationship between what are called acts of consciousness and their contents.

The "extensionalist" theory holds that both acts of consciousness and their contents are extended through time. The "retentionalist" theory denies that temporally extended acts of consciousness and their contents coincide in time. Instead, they make a distinction between the temporality that is intrinsic to the contents of consciousness and a form of temporality that is intrinsic to acts of consciousness. I end my chapter by considering how Craig's GEMs theory might bear on this debate concerning the nature of the phenomenal now.

2 Interoception as the Basis for GEMs

Craig arrived at his GEMs theory based on decades spent studying the ascending sensory pathway in lamina I cells of the spinal cord that ascend toward the brain along a lateral spinothalamic pathway. This pathway connects the spinal cord to the thalamus, an area in the center of the forebrain, that sends signals to the cerebral cortex. Craig argued that this sensory pathway, made up of different types of modality-specific lamina I spinothalamic neurons, is the homeostatic, afferent side of the autonomic nervous system, a visceromotor system that regulates the physiological conditions of the body.[1] Homeostasis refers to a set of neural, endocrine, and behavioral functions that utilize metabolic energy efficiently with the aim of maintaining the body within a relatively stable, healthy state of balance across time. Examples of homeostatic processes include the regulation of body temperature, salt, glucose, insulin and water levels, blood pressure, metabolic rate, and so on. The effectors of homeostasis include neural, endocrine, and behavioral functions involving the coordination of brain, heart, lungs, kidneys, and other bodily organs. Changes in the mechanical, thermal, and chemical status of the tissues and organs of the body are signaled by lamina I neurons. These signals provide feedback on key variables that the homeostatic process regulates to keep the body's needs in balance, while at the same optimizing energy utilization.

Interoception provides the brain with sensory information about the conditions of the body that form the basis for bodily feelings such as warmth or cold, hunger and thirst, and the acute sensations a person experiences when in pain. Awareness of heart rate is often used as a measure of interoceptive awareness of the body. Heart rate awareness is measured by asking people to count the number of heartbeats they can sense in intervals of 30, 45, and 60 s. Interestingly, this measure is also predictive of how intensely a person experiences their emotions (Herbert and Pollatos 2012; Pollatos et al. 2007; discussed by Craig 2015, Chap. 1). Subjects

[1] Thus, he extended the classical concept of interoception, introduced by Sherrington, to include sensory input from nociceptors and thermoreceptors. Craig argued for such an extension on the grounds that nociceptors and thermoreceptors provide information about the physiological conditions of the body that can be used by homeostatic processes.

who score high in the measure of their heart rate awareness also tended to be good at reading their own and other's emotional states.

2.1 The Global Emotional Moments Theory

The GEMs theory argues that interoceptive processes involved in sensing the physiological conditions of the body also provide the basis for emotional awareness. According to this embodied theory of emotion awareness,[2] the neurobiological source of emotions is visceral sensations and activity in the autonomic nervous system. Emotions are generated by activity in the autonomic nervous system that readies the organism for adaptive action. Craig argues that human emotions have their evolutionary origin in brain networks that were originally built to optimally utilize metabolic resources in ways that take care of the body. Humans feel ongoing, homeostatic processes from the inside as they happen. At the core of emotions is a feeling of being alive. Feelings are an "interoceptive construct that the brain uses to represent the overall energy costs and benefits of any actual or potential emotional behavior … in order to facilitate choices that guide behaviour in the most energy-efficient manner" (Craig 2015, p. 12).

Based on his own functional magnetic resonance imaging (fMRI) studies and those of others, Craig proposed that the signals carried to the cortex by the afferent spinothalamic pathway are used by interoceptive cortex to form a map of bodily sensations (Craig 2002, 2015, Chap. 6). These sensations include the sensory – burning, pricking – component of pain experience, temperature, itch, muscle ache, sensual touch, thirst, hunger, oxygen starvation, and so on. Craig hypothesized that the dorsal posterior insula produces a somatotopic map of bodily sensations that provide almost continuous sensory feedback on autonomic and homeostatic motor activity. These sensory signals are mapped to form a moment-to-moment interoceptive image of the effects of homeostatic actions. In Craig's model, the interoceptive image generated by the dorsal posterior insula is hypothesized to be "re-represented" in the mid-insula that receives convergent sensory input about salient stimuli from high-order sensory areas, the frontal cortex, and the amygdala. Processing in the mid-insula is also modulated by the ventral striatum, which provides information about the hedonic value of sensory information (Menon and Levitin 2005). Craig proposed that bodily feelings are generated in the mid-insula by the integration of hedonic, exteroceptive sensory, and motor information with the current interoceptive image that is grounded in the states of the living body.

How do these integrative processes carried out in the mid-insula generate specific bodily feelings such as hunger, thirst, temperature, burning pain, muscle pain,

[2] Craig's embodied theory of emotion continues the tradition of the so-called James–Lange theory (James 1884) that conceptualizes emotions in terms of the behavioral and autonomic patterns of activity set in motion in response to the perception of events in the environment.

pleasurable touch, and so on? Craig offers the following answer to this question. He claims first that qualitatively distinct feelings of the body are caused by sensory pathways that are "labeled lines" by which he means that these sensory pathways carry signals that are specific to physical events in the sensory periphery such as heart rate, breathing, hunger, or pain. He proposed that specific sensory signals are transduced from a variety of sensory receptors distributed across the body including chemoreceptors, osmoreceptors, glucoreceptors, mechanoreceptors, and humoral receptors. This afferent activity indicates an actual or potential deviation from a homeostatic setpoint, which triggers a pattern of visceromotor activity that produces changes in the metabolic and vascular states of the body. Thus, Craig writes, "the labelled line sensory activity becomes a vivid feeling from the body by causing active changes in the living body that are 'felt' in the active feeling state of sentience" (2015, p. 196).

Thus, Craig takes bodily feelings that arise from interoception to represent "changes in the high-fidelity sensorium caused by autonomic activity" (Op cit). His account of bodily feelings relies on the finding of an ascending thalamocortical pathway, phylogenetically unique to primates, that serves homeostasis and is the basis for discriminating between changes in the body that are the consequences of autonomic activity. In Craig's account, bodily feelings are the outcome of constructing a map of bodily changes bottom-up from ascending sensory activity.[3]

3 From Bodily Feelings to Phenomenal Selfhood

Craig was careful to distinguish the bodily feelings that arise out of the integrative activity in the mid-insula from what he calls the "awareness" of emotions. Craig (2009a) defines emotion awareness as the "organism's experience of its own

[3]In recent years, an alternative picture has begun to take shape in which interoceptive feelings are largely a consequence of top-down prior beliefs or predictions (Seth 2013; Barrett and Simmons 2015; Seth and Tsakiris 2018; Feldman et al. 2024). In these models, processing in the mid- and posterior insula is hypothesized to compute prediction errors, which are then sent back to higher visceromotor regions that include the anterior insula as a central processing hub. The predictive processing models depart from Craig's account of how bodily feelings are based on interoception in a key respect. While Craig takes bodily feelings to be the outcome of mapping changes in the body bottom-up, based on sensory inflow, the predictive processing model claims that bodily feelings are largely the outcome of top-down predictions. Bodily feelings derive from what Barrett and Simmons (2015) describe as "simulated interoceptive sensations." They are modeled as the brain's best guess based on past experiences (its "posterior probability estimate") of the probable changes in the body. Homeostasis and allostatic motor actions have the goal of generating the bodily changes that have been predicted in a kind of self-fulfilling prophecy. What you feel is largely a consequence of what your brain predicts to be happening inside of your body based on past experience. The role of sensory afference from the body is to keep the brain's predictions in check. Predictive processing models propose that the brain relies more on top-down predictions than on continuously changing sensory afferents arising from the body to drive the body towards its anticipated state of homeostasis (Barrett and Simmons 2015, p. 6; see also Feldman et al. 2024).

existence as a sentient being" (p. 65). We have seen in the previous section how the homeostatic sensorimotor cycle in the living body forms the basis for a sentient self. Emotional actions do not only affect the body but also other people. What emotional awareness adds to this core layer of sentience is that the sensory feedback increases in quantity to include the feelings of pleasure or displeasure that are the consequences of acting in the social environment. Emotional awareness is based on processes of multimodal integration that combine feelings from the body with hedonic and motivational information relating to current and expected environmental conditions and social contexts.

Craig argues that emotion awareness is "engendered" by activity in the AIC, and the anterior cingulate cortex (henceforth "ACC").[4] Thus, Craig (2009a) reviews evidence of fMRI studies that have observed the co-activation of the AIC and ACC in studies of maternal and romantic love, anger, fear, sadness, happiness, sexual arousal, disgust, aversion, unfairness, inequity, indignation, uncertainty, social exclusion, trust, empathy, and beauty and in mystical-type experiences of the feeling of unity with God (Craig 2009a, pp. 60–62). Craig (2002) hypothesized that these areas form, respectively, the motor and sensory components of emotions that work together to provide the neural substrate of motivation and feeling, respectively. Craig proposes to understand subjective emotional experience in terms of the recognition of changes in the peripheral autonomic nervous system, and in facial musculature, that he claims are uniquely specifying signatures of each emotion (or at least of the six basic emotions of fear, sadness, disgust, surprise, and anger) (Craig 2015, Chap. 6). Craig further hypothesized that von Economo neurons (VENs) located in the AIC and ACC may serve as the anatomical correlates of emotional awareness. VENs allow for fast connections to form between physically separated limbic cortices (e.g., AIC) and motor cortex (e.g., ACC). Craig proposed that VENs generate a global emotional moment, which, he suggests, can be conceived of as an "image of a 'material me' at one moment in time – now" (Craig 2009a, b, p. 67) in which the person's feelings about things, events, and other people at that moment are represented. Each global emotional moment integrates information relating to the homeostatic condition of the body with information relating to sensory, environmental, hedonic, and social conditions.

[4] Craig (2015, p. 209) claims that activity in the AIC is necessary for the subjective experience of all emotions. He questions evidence from Damasio for instance of two encephalitic patients whose insular cortex was completely missing across left and right hemispheres, yet who nevertheless retained self-awareness (Damasio et al. 2013). However, the claim that cortex is necessary for consciousness remains controversial. Mark Solms (2021), for instance, has pointed to Bjorn Merker's work with hydraencephalic children who were born without a cortex (Merker 2007). Merker describes how these children display emotional, orienting reactions to environmental events, and express pleasure by laughing and smiling, and distress by fussing, arching their backs, and crying. One could reply that these children are behaving as if they are subjects of experience when, in fact, their behavior is the outcome of unconscious processes. Yet hydraencephalic children certainly appear to be awake and responsive. It is unclear that a child can be awake and capable of voluntary responsiveness but not be subjectively experiencing anything.

3.1 Emotions, Salience, and Phenomenal Selfhood

Craig presents his GEMs theory as a theory of emotional awareness, though sometimes he also suggests it could be thought of as a theory of awareness more generally which he defined as "an organism's experience of its own existence as a sentient being." I will interpret Craig as using the term "awareness" to refer to the experience of being a self that has its own subjective point of view on its world. This experience of being a self is sometimes referred to by talking of "phenomenal selfhood," an experience of mineness that is normally built into the structure of phenomenal experiences, though this structure can be disturbed to different degrees in some psychopathologies such as schizophrenia, depersonalization, and derealization (for discussion, see, e.g., Gallagher 2000; Zahavi 2005; Gallagher and Zahavi 2008; Blanke and Metzinger 2009; Gerrans 2015). The phenomenal self is typically not among the objects or entities a subject is aware of, or notices, in their experiences. For this reason, Metzinger (2003, 2009) has referred to the phenomenal self as being transparent: we look through the phenomenal self to the world.

Craig (2009a) proposed that awareness rooted in the AIC is a precondition for experiencing the existence and salience of anything else in the environment (see, e.g., p. 66). I will interpret him as claiming that it is only creatures that have a phenomenal self that can experience the environment as salient. This amounts to the claim that the term "salience" is a two-place relation. Situations are not intrinsically salient – they are salient for someone, a phenomenal self, or a subject of experience who is affected in some way by their situation in the world. "Salience" can be interpreted here in terms of an appraisal of motivational and hedonic value along different dimensions such as danger, interest, loss, and so on. The term "awareness" as Craig uses it can therefore be analyzed as referring to a phenomenal self that appraises the motivational and hedonic value of its current situation. This appraisal carries information about the world and how the creature is faring with respect to its goals and purposes. In other words, the AIC is keeping track of the salience of sensory information for the organism. It thereby provides the organism with a subjective point of view on the world that I have described in terms of phenomenal selfhood relative to which situations are sensed as salient.

A GEM represents the current conditions of the person – social, hedonic, and motivational – in terms of their salience for the person's goals both in the immediate present and going into the future. This present includes anticipated risks and rewards that allow motivation to influence emotions and vice versa (Craig 2009a, p. 62). The AIC is thus busily weaving together information about what is out there in the world with information about the subject's own internal physiological states and their readiness to act in the world.

Craig's GEMs theory has important implications for our understanding of phenomenal consciousness. I will use this term as is standard in philosophy and the science of consciousness to refer to what it feels like for a creature to be conscious (Chalmers 1996; Nagel 1974). The GEMs theory entails that every phenomenal conscious experience of the world will also be imbued with emotion. Some recent evidence for this comes from Feldman Barrett and colleagues' work on what they call "affective realism" effects: the tendency of individuals to perceive the world in the light of their affective states. When undergoing an elevated heart rate, for instance, individuals with poor interoceptive sensitivity tended to judge that people they were interacting with were aggressive, or lacking warmth as compared with people with more interoceptive acuity (Feldman et al. 2023). Similarly, police officers have been shown to display a tendency to perceive innocent objects such as smartphones as weapons because of their heightened state of bodily arousal, particularly when these objects are held by Black men (Fridman et al. 2019).

In the remainder of my chapter, I will turn to a debate within the philosophy of perception concerning the nature of the phenomenal now – our subjective experience of the present moment in time. Craig called his theory of emotion awareness a "cinemascopic theory." By "cinemascopic," he seems to have meant that each GEM can be compared to a snapshot that captures an "image" of what I have called the phenomenal self at a moment of time in much the same way as a single frame in a film does. Craig argued that this processing occurs in discrete epochs or temporal windows with a lower threshold of 100–150 ms corresponding to 1 cycle of 10–8 Hz oscillation and an upper threshold of seconds (Craig 2015, pp. 232–234). An integration period so understood is the period within which information from distinct sensory modalities – interoceptive and exteroceptive – is combined to form a GEM. The generation by the brain of a series of GEMs forms the basis for a "continuity of subjective emotional awareness" – that is to say, for an awareness of the phenomenal self across time from the past and into the anticipated future (Craig 2009a, p. 67). The hypothesis I will take up in the next section is that neural information processing mechanisms that integrate a series of successive GEMs are the neurobiological basis for the experience of the phenomenal now. Wittmann (2014) has pointed to evidence from research on motor control, perception, language, and music (see, e.g., Pöppel 1997; Wittmann and Pöppel 1999; Pöppel 2009; Wittmann 2011) that provides evidence for a temporal mechanism that integrates moments of experience of a sub-second duration to form perceptual gestalts of approximately 2–3 s in duration. The hypothesis I will explore, based on the GEMs theory, is that the experience of the phenomenal now depends on neurobiological processes that unify successive GEMs on the temporal scale of seconds. I will consider the implications of this hypothesis for philosophical debates concerning the nature of the phenomenal now.

4 Does the GEMs Theory Support a Cinematic Theory of the Phenomenal Now?

Consider an auditory experience of a melody, or a visual experience of a shooting star in the night sky. What these two experiences have in common is that they both involve an experience that is unfolding over a whole extended interval of time. In hearing a melody, we hear an organized series of musical notes that are extended through time. In seeing the shooting star, we experience the star moving across the sky. To experience something moving or changing seems at first glance to require experiencing a whole interval of time. (For arguments to the contrary, see Prosser 2016, Chap. 5; Arstila 2018.) This experience of a whole interval of time seems to be required for instance to see the shooting star at different positions in the sky at different times, or to hear the whole unfolding melody. The debate I will take up in the remainder of my chapter is about how it is possible to experience whole intervals of time within which events like the hearing of melodies can take place when our experience is confined to the present moment.

Philosophers have returned different answers to this question based in part on different theories of what it is to experience the present moment. In what follows, I will refer to the experienced present moment as the "phenomenal now," so-called because it relates to our subjective lived experience of the present moment. My aim in what remains of this chapter is to consider whether Craig's GEMs theory can be said to provide support for any of these philosophical theories.

The first of the theories, which I focus on in this section, are referred to as cinematic theories (Chuard 2011; Prosser 2016; Arstila 2018; for critiques, see Dainton 2000; Dorato and Wittmann 2020). Cinematic theories argue that conscious perceptual experiences can be compared to snapshot-like frames in a film that represent events, objects, and their properties at a single instant or point in time. Cinematic theories of time come in two forms. Static snapshot views hold that like a single frame in a film, the contents of experience at each instant in time are static or unchanging. Dynamic snapshot views allow the contents of each snapshot to change. I will not expand further on the debate between these two versions of the cinematic theory, since my concern will be with what they both hold in common: the claim that the contents of experience are always confined to a single point-like instant in time within which a subject can experience change, motion, and causation.[5]

4.1 Interoceptive Awareness and the Experience of the Passage of Time

Does Craig's GEMs theory provide support for a cinematic theory of the phenomenal now? The GEMs theory claims that emotions and bodily feelings are processed in discrete epochs that persist only for a finite interval of time. We saw at the end of the previous section how Craig seemed to have thought of each GEM as

[5]Cinematic theories face the challenge of explaining how experience seems to present us with continuous, unfolding processes such as melodies and moving objects if they are in fact instantaneous. To explain how we can experience a melody, for instance, a cinematic theory must posit a mechanism that somehow stitches together fast changing, instantaneous perceptual experience with short-term episodic memory that represents a recent experience. According to the cinematic theory then, we do not really experience change and motion. We only seem to do so because of memory-based inference (Le Poidevin 2007; Prosser 2016).

representing a relatively short-lived moment in time with a duration on the sub-second temporal scale. The phenomenal now might be argued to be the outcome of experiencing a succession of GEMs in line with the claim of cinematic theories. Craig also chose to refer to his theory as a "cinemascopic model of awareness," explicitly comparing each GEM to a single frame in a movie. These reasons for interpreting the GEMs theory as a cinematic theory notwithstanding, I suggest a more cautious approach to this question is needed. Craig is concerned with neural information processing, not with the phenomenal temporal structure of experience. Thus, to use his theory to make an intervention in the philosophical discussion will require first taking a stand on the question of how neural information processing relates to what the philosophers are discussing when they debate the nature of the phenomenal now. The GEMs theory clearly takes emotions and feelings to be processed in short-lived discrete epochs or snapshots. However, I will argue that Craig probably did not think of each GEM as representing a distinct, single isolated instant in time.

My argument is based on the role that Craig assigned to the AIC in accounting for our perception of duration. Craig wrote, for instance: that "feelings of the ongoing present moment accumulate in a new quantal storage unit," and that a sequence of such units "produces a natural flow of moments across time" (Craig 2015, p. 230). I will take his concept of "quantal storage units" to refer to the temporal processing windows within which interoceptive, hedonic, and motivational signals are integrated. Craig proposed that the number and rate of bodily signals accumulating in the posterior insula over an interval of time create the perception of duration. Thus, during an intense emotional experience such as falling down the stairs (Craig 2015, p. 234), signals from the body indicating danger rapidly accumulate, filling up what Craig referred to as a "quantal storage unit." Craig hypothesized that the brain would need to create more storage units to capture the emotional salience of you falling down the stairs. Your experience of the rate of the passage of time speeds up because of the increased rate in the production of quantal storage units.

Craig proposed to explain the experience of the passage of time as speeding up or slowing down in terms of the accumulation of interoceptive information signaling the occurrence of emotionally salient events. When interoceptive information rapidly accumulates, this results in an experience of time dilation in which time seems to come to a standstill, as we saw in Craig's example of your heart racing as you fall down the stairs. Additional support for the hypothesis that the insula is involved in the perception of duration comes from an fMRI study in which subjects were given the task of reproducing tone durations of 3, 9, and 18 s. The researchers measured "ramp-like" increases in activity in the insular cortex that corresponded to stimulus duration, and terminated with stimulus (Wittmann 2013). In a study with almost the same experimental design (Meissner and Wittmann 2011), subjects were asked to reproduce tones of different durations while their heart rate, breathing frequency, and skin conductance were measured. These three physiological measures track the changes in bodily arousal experienced along a dimension spanning agitation to calmness. Meissner and Wittmann showed a correlation between a calming pattern in heart rate and the persistence of a tone. They found similar effects for skin conductance and breathing frequency measures. They took this finding to indicate that the integration of bodily signals over time, especially relating to heart rate, plays a key role in the duration reproduction task.

There are some puzzle cases for this proposal I will return to in the next part of this section. When nothing much changes in the body because one is bored, and there is very little interoceptive information accumulating, the result is also an experience of time dilation in which the passage of time seems to pass painfully slowly. Conversely, when we enjoy ourselves in the company of friends or loved ones, a situation that is also highly emotionally salient, time does not slow down but speeds up. These seem like the opposite effects of what the GEMs theory should predict. Suppose for now that we accept Craig's hypothesis that the insular cortex functions as an internal clock that allows subjects to make duration estimations and to experience the passage of time. What would this imply for an interpretation of the GEMs theory as supporting a cinematic theory of the phenomenal now? I have made the following two claims in this section:

1. The cinematic theory claims that the experience of the phenomenal now consists in a succession of GEMs each representing a point in time that is without duration.
2. The accumulation of interoceptive bodily signals within and across processing epochs explains subjective variation in the experience of the passage of time.

It is difficult to see how these two claims can be consistently combined. The cinematic theory claims that the experience of the phenomenal now lacks duration. However, the GEMs theory seems to lead to the opposite claim that the experience of a succession of GEMs is what explains variability in the experience of duration. In other words, the GEMs theory seems to entail that the experience of the phenomenal now is an experience of a whole interval of time with duration.

4.2 A Connection Between the Phenomenal Self and the Experience of Duration

Before I move on to consider philosophical theories of the phenomenal now that take experiences to have duration, I will briefly return to the puzzle cases I described above. Recall how both when bored and when faced with a life-threatening situation, time can seem to pass slowly. In a life-threatening situation, it makes sense that the interoceptive signals originating from changes in bodily states rapidly accumulate. It is harder to see how this can be true in situations of boredom. By contrast in moments of joy and happiness, the passage of time can accelerate. In such contexts, we can suppose the brain is accumulating emotionally salient signals from the body, but instead of experiencing the passage of time as slowing, the subject has an experience of time as speeding by. Craig's solution to these puzzles was to point to the valence of the emotional experience. He wrote: "if we are emotionally engaged in a negative situation subjective time dilates, whereas in a positive situation subjective time contracts" (Craig 2009b, p. 1940). In a sympathetically arousing fight or flight situation, GEMs accumulate rapidly, with the consequence that the subject experiences the dilation of time. The relaxing situation with friends is a soothing situation in which the parasympathetic nervous system is more activated. In this situation, change in GEMs is slow with the consequence that time contracts and seems to pass by quickly. This is a promising resolution of the puzzle cases. However, Wittmann (2018, Chap. 3) points to an additional (but as we shall see related) factor that will allow us to return to the connection between GEMs and the phenomenal self I highlighted at the end of Sect. 2.

Consider, for instance, how in flow states when one is absorbed in what one is doing, time will pass relatively quickly (Csikszentmihaly and Csikszentmihalyi 1988). Flow experiences are triggered by activities such as sports, music, or play in which a person is able to act fluently and their attention is focused without effort. People report a loss of self-awareness and that the passing of time is hardly noticed. A similar kind of diminished self-awareness can also occur in states of deep meditation. Initially, one's attention may be focused on the activity of scanning the sensations one experiences in one's body. However, after a while, a shift may occur as one enters a deeper state of focused concentration in which one's sense of self and of time begin to recede from awareness (Berkovich-Ohana et al. 2013; Wittmann 2015).

The experience of the slowing of time occurs when attention is overly occupied with the phenomenal self. Thus, when one is bored for an instant, one has an experience of time passing frustratingly slowly. Wittmann (2018) has argued, drawing on the existential phenomenological writings of Martin Heidegger, that boredom is a mood one experiences when we are left completely to ourselves without anything to distract us. When self-awareness is high, and our focus is internal, as is the case during periods of boredom, time is experienced as passing painfully slowly. People with depression report a similar co-occurrence of the experience of increased self-awareness and a slowing of time. Depression is characterized by increased self-occupation characterized by feelings of negative self-worth, self-blame, and rumination. People with depression also report experiences of time passing excruciatingly slowly and an experience of being stuck in time (Vogel et al. 2018).

This tight relationship between attention to the phenomenal self and our experience of the passage of time passing makes perfect sense from the perspective of the GEMs theory. Recall, from Sect. 2, that each GEM is an experience of the phenomenal self at a moment in time. In the cases of negative emotion – intense fear in response to life-threatening danger, boredom, or low mood in the case of depression – attention is excessively focused on the phenomenal self. Our attention to the self is also an attention to the passing of time since in each GEM I am aware of myself at a moment in time – the phenomenal now. Thus, when I attend to myself, I am also attending to the passing of time. Perhaps it is this focusing of attention on the passing of time that makes it seem to us as if time is passing painfully slowly.

Conversely, in the case of positive emotions such as those experienced in flow states, in deep states of concentration in meditation, and in the company of friends, the passage of time seems to accelerate. Continuing the line of reasoning just advanced (and cf. Wittmann 2015, 2018), in such cases, one's attention is externally focused, rather than self-focused. In some positive emotional experiences, as in the examples of flow experience and in periods of deep concentration during meditation, one may also experience a diminished sense of phenomenal selfhood in and through time. I speculate that the passage of time may seem to accelerate in such contexts because attention is absorbed by an other-oriented or external activity that is not the self in time. It is the turning of attention away from the self in time that may have the effect of making time seem to pass much faster.

5 Philosophical Theories of the Specious Present

In the previous Sect. I outlined the hypothesis that the experience of the phenomenal now is the outcome of a succession of GEMs. However, as William James (1890) noted: "A succession of feelings, in and of itself, is not a feeling of their succession" (pp. 628–629). Succession takes place over time. To have a feeling of succession then seems to require that distinct moments of experience be somehow diachronically unified and experienced together over time. Recall that, at the end of Sect. 2, I outlined Wittmann's (2014) hypothesis that discrete GEMs come to be unified over time by a temporal integration mechanism with an upper temporal threshold of between 2 and 3 s. The hypothesis I take up in this section is that such a temporal window of processing functions as the neurobiological basis for an experience of the specious present. The term "specious present" was made famous by James (1890) who argued that the phenomenal now is a short duration of time that is "immediately and incessantly sensible" to us. The present conceived of as a point in time that interfaces between the past and future is "specious" then because such a present does not exist in experience. James held that the specious present we experience has a temporal structure. We can experience relations between phases of experiences such as the temporal order of these phases when the whoosh of a firework is followed by a bang. We can experience the notes in a melody as continuously changing. Proponents of the specious present argue that for our experiences to have this kind of organized temporal structure, our experiences must have as their contents structured temporally extended moments of time.

Philosophical theories of the specious present are commonly classified into two camps of extensionalist and retentionalist theories. The debate between these views concerns the relationship between what are called acts of consciousness and their contents. The extensionalist theory holds that both acts of consciousness and their contents are extended through time. The retentionalist theory agrees that the phenomenal present is a temporally extended field in which the perception of duration and of change is possible. However, the retentionalist theory denies that such temporally extended acts of consciousness and their contents coincide in time.

Instead, they make a distinction between the temporality that is intrinsic to the contents of consciousness and a form of temporality that is intrinsic to acts of consciousness. My aim in this final section is to consider how, if at all, Craig's GEMs might bear on this debate concerning the nature of the phenomenal now. Before we can address this question, it will be necessary to say more about each of these theories.

5.1 *The Extensionalist Theory of the Specious Present*

The extensionalist theory holds that acts of consciousness persist for the same duration as their contents (Dainton 2000, 2008; Hoerl 2009, 2013; Phillips 2008, 2014). If an experience represents an event such as melody that persists for half a second, then so also does the act of experience persist for this amount of time. The act of awareness of the unfolding melody marches in step with and runs concurrently with the content of the experience which represents the unfolding melody. In other words, both acts of awareness and the contents that represent events with temporal structure are extended in time. Barry Dainton, a philosopher who has perhaps done the most to develop the extensional theory, goes as far as to endorse what he calls a "simple conception" of experience that rejects a distinction between acts of consciousness and their contents. Dainton argues that "when a given phenomenal item comes into being, it comes into being as a conscious experience; to be an experience it does not need to fall under a separate awareness" (2000, p. 57). Acts of awareness and their contents are temporally extended and always coincide in time since, on the simple conception of experience, there is no distinction to be drawn between acts and contents.

The extensional theory owes an account of the difference between a succession of phases of experience and an experience of succession, which requires that one experience the phases of experience to be diachronically unified and to continuously flow into one another. Dainton (2000, 2008) has answered this challenge of accounting for the diachronic unity of experience by positing an overlap in the parts of temporally extended experiences. On his account, there is no need to introduce a separate act of awareness to unify different phases of experience. Distinct but successive events can be experienced together within a single specious present. Within a single specious present, different experiences can occur that are co-conscious. Normally co-consciousness is taken to be a synchronic relation. Co-consciousness consists in the simultaneous awareness of different experiences at a single time instance. However, Dainton argues that co-consciousness can also be understood diachronically as a relation that holds between experiences occurring at different times. The stream of consciousness seems to be diachronically unified because experiences occurring within a specious present S_1 are co-conscious, and two temporally distinct specious presents can share overlapping parts. This is to say that one can experience events A and B as part of specious present S_1 and one can experience the very same event B as part of specious present S_2.

5.2 The Retentional Theory of the Specious Present

The retentional theory draws on Husserl's theory of internal time consciousness (Husserl 1964, 1991, 2001), a term Husserl used to refer to a form of time that is internal or intrinsic to consciousness, lending to consciousness its dynamic, flowing character. The retentional theory argues that the experience of temporal objects like melodies and shooting stars would be impossible if the experience were composed of a series of unconnected, discrete snapshots. A succession of isolated, discrete snapshots would not allow us to perceive any objects that have temporal extension or duration. If perception of temporal objects is possible, which it manifestly is, the retentional theory argues that this must be because consciousness has a complex field-like structure. This temporal field comprises a primal impression that is directed toward the present, a retentional structure that reaches a short way back into the past to retain what has just been, and a protentional component that anticipates what is coming next. At any given moment, this threefold structure is simultaneously present as a whole field in which a now phase, a no longer now phase, and a not yet now phase occur together in a gestalt – a unified whole. While the contents of each structural element change over time, the internal threefold structure of retention–primal impression–protention is present in an act of consciousness, making it possible for a subject to be conscious of temporal objects.

In the retentional theory, consciousness does not occur in time as it is measured by clocks. I will distinguish lived time (e.g., the phenomenal now) from time as a unit of measurement – what I will call "clock time."[6] Nor would it be correct to say that clock time is the objective real time in which physical processes unfold, while lived time is merely subjective, existing only in the mind. Many would argue, based on the physical sciences, that our naïve experience of the flow and passage of time – of the present falling away into the past, and the future becoming the present and flowing into the past – exists only as a subjective projection. Many of the equations that describe physical reality can be written without making any reference to time at all (Rovelli 2018). To the extent that time belongs to the universe as it is described in physics, it is best conceived of as a spatial dimension in which events can be located as standing in relations of before, after, and simultaneous. The retentional theory would argue that clock time is, strictly speaking, an abstraction arrived at for the purposes of measurement. We use, for instance, points on a line or numbers on a clock face as our way of measuring, for a variety of purposes, the time of day. Each of these points in clock time is referred to as the present and can be considered distinct and separate from what has gone before and what will come after. There is therefore no duration in clock time or in the universe conceived of in terms of physics. Duration and the phenomenal now belong to what I am calling "lived time." The retentional theory would argue that it is lived time that has a concrete real existence. Time as it is conceived from within physics and the natural sciences, and

[6] See Frank et al. 2024, Chap. 3 for further discussion of this distinction between clock and lived time.

time as it is measured by clocks is an abstract idealization that derives its meaning from the more fundamental concrete reality of lived time.

The retentional theory claims that the phenomenal now is not an object of consciousness. Proponents of this theory would make a distinction between the temporality and dynamism of the contents of consciousness and the temporality of acts of consciousness. The relations between retention, primal impression, and protention are not relations between features of consciousness located in time. As phenomenological structures of consciousness, they are constitutive of temporal objects (Zahavi 2007; Gallagher and Zahavi 2014). Whereas the extensional theory claims that there is a strict coincidence between temporally extended acts of consciousness and their contents, the retentional theory makes a three-way distinction between temporal objects like melodies, the contents of consciousness that dynamically flow through time, and internal time consciousness that is the source of this flow and dynamism due to its threefold temporal form or structure. The retentional theory therefore distinguishes the temporal objects of consciousness that are extended in time from internal time consciousness that has an intrinsic temporal form that makes it possible for a subject to be conscious of temporal objects. While temporal objects are in time and have phases that can be described as past, present, and future, these phases of temporal objects should be distinguished from retention, primal impression, and protention. The latter are structures of internal time consciousness that are the source of the significance of lived time we make sense of using concepts of past, present, and future. Internal time consciousness should not be equated with the phenomenal now if we take this to be a moment in time that is distinct from past and future. It is a field of experience that because of its threefold temporal structure has duration and temporal thickness.

5.3 The GEMs Theory and the Nature of the Specious Present

I will close my chapter by considering if there is anything in the GEMs theory that might favor one of these accounts of the specious present over its rival. Dorato and Wittmann (2020) have proposed that when successive GEMs are processed within a temporal integration window they can be experienced as happening together but at different times within a specious present. The events may be experienced as having a temporal order, that is, as standing in relations of before, after, or as happening simultaneously. Events that occur outside of this window are experienced as having happened earlier (i.e., in the past) – a previous moment in time, or as about to happen later (i.e., in the future). Building on Wittmann's earlier work, Dorato and Wittmann (2020) have argued that integration processes form the neural basis for what they call the "experienced moment," and what I have been calling the phenomenal now. They propose that the phenomenal now is best conceived in extensional terms in which acts of consciousness and their contents coincide in time (p. 19). The phenomenal now is extended in time in a way that exactly coincides with the temporal window of processing. Thus, Dorato and Wittmann argue for what I will call a "psychophysical isomorphism" between neural information processing in the form of temporal integration mechanisms, temporally extended acts of consciousness, and their experienced contents.

Notice that such a claim of psychophysical isomorphism is premised on applying clock time to measure both the neural integration processes and the phenomenal now. Dorato and Wittmann base their claim of psychophysical isomorphism on the premise that both the phenomenal now and temporal integration windows in the brain have a duration in the range of seconds. Their argument for the extensionalist theory thus has its major premise that time as it is measured by clocks is what is objectively real.

The retentional theory, however, does not concede this premise. Proponents of this theory deny that clock time is what is most concretely real. They argue that it is lived time that has its origins in the temporal form of internal time consciousness that is concretely real. Clock time has, by contrast, the ontological status of an abstract conceptualization of time. The assumption that time as measured by clocks is what is objectively real is the result of what Husserl (1970) called a "surreptitious substitution" of an abstract measure of time for the concrete reality of lived time – the time that each of us lives through and subjectively experiences (Frank et al. 2024, Chap. 3). Such a substitution is, however, unjustified, the result of confusing an abstract measure that we employ for a variety of different human purposes for reality itself. From the perspective of the retentional theory, the units of measurement we use when we refer to clock time (or to the physical time of the block universe) are the outcome of abstracting away from lived time.

I suggest for these reasons that we should resist interpreting the GEMs theory along extensionalist lines. To do so would require accepting a relation of psychophysical isomorphism between the experience of the phenomenal now and the underlying neurobiological temporal processing windows that the phenomenal now depends on for its existence. However, I have argued that the existence of a psychophysical isomorphism of this kind would require assigning an ontological priority to clock time – time as it is measured by clocks, over lived time. In line with the retentional theory, I have argued for the ontological priority of lived time that has its origins in the dynamic internal structure of consciousness.

6 Conclusion

Where does Craig's research leave us today with regard to our understanding of the nature of the phenomenal now? The phenomenal now belongs to lived time given to us in experiences of duration. Craig's research illuminates the neural information processing mechanisms that can explain the subjective variations in the experience of duration based on our bodily feelings and our sense of self. Craig's research furthers our understanding of why this subjective variability should occur. His theory of the neurobiology of GEMs also allows us to understand why our experience of lived time, the sense of self, and bodily feelings turn out to be interrelated experiences. Part of the rich legacy of Craig's research is a deepened understanding of how our experience of lived time, self, and body all depend on the maps of the body that are constructed in the insular cortex for the purposes of homeostatic regulation.

I have argued that Craig's theory of global emotion moments also provides support for a view of phenomenal now as having duration and temporal depth, as contrasted with cinematic theories that take the phenomenal now to be confined to snapshot instants in time. Still, there are live debates within philosophy concerning how to conceive of the phenomenal now. I have finished my chapter by asking if Craig's theory of global emotion moments can be of any use for settling these philosophical questions. I ended my chapter by sounding a note of caution. We have seen how research in neurobiology on time perception and consciousness operates with an understanding of time as it can be measured in discrete units of time using clocks. Duration is measured as a determinate period or interval of time in which a stimulus is presented or reproduced. Thus, clock time is applied by the scientist as a means of measuring the experimental participant's subjective experience. Clock time is, however, distinct from lived time because it is made up of discrete points, each of which is an instant that can be referred to as now taken in isolation from what has gone before, or what will come later. Moments in clock time can be arranged along a line, or along a spatial dimension, in which a neat separation can be made between past, present, and future. Such a separation is, however, not possible in the case of lived time in which experiences have duration.

Could it be that our inner experience of duration is the outcome of temporal integration processes in the brain that lead to the construction of representations of duration on the scale of seconds (Wittmann 2009, 2011, 2013, 2014)? Such a proposal would seem to be premised on the possibility of reducing lived time to clock time. This is because both the experience of duration and the temporal integration processes in the brain are measured in the scientist's lab using clock time. The retentional theory claims that there can be no reduction of lived time to clock time. This is because clock time is a technique for measuring time that is arrived at by means of abstracting away from lived time. The very idea of such a reduction therefore relies on what Husserl called a surreptitious substitution of concrete lived experience by an abstraction (Husserl 1970; Frank et al. 2024). Clock time is an abstraction, a tool and technique that can be used for various purposes including doing neuroscience. In line with the arguments of Husserl (1970), and more recently of Frank et al. (2024), I have warned against confusing this tool for measuring reality with the reality we are measuring. What is most concretely real is lived time.

Nevertheless, I hope to have succeeded in demonstrating how Craig's research contributed to furthering our understanding of lived time in neurobiological terms. Lived time is what we described in Sect. 3 when we noted that depending on what one is feeling in the body, and whether one's attention is focused on the self, the phenomenal now can be experienced differently either as dilating, slowing down almost to a stand-still, or as contracting, speeding by in such a way that one is surprised by how much time has passed. In moments of boredom, time ticks by all too slowly, whereas in moments of joy, time flies. We have seen how our experience of lived time is profoundly shaped by our bodily feeling of being a phenomenal self. Indeed, if the argument of Sect. 2 is correct, all our conscious subjective experience comes bound together with emotional experiences. I therefore suggest that the GEMs theory is best interpreted as a theory of the neurobiological basis of the lived time we subjectively experience in which our lives unfold. However, if my argument is accepted, this should provide grounds for resisting reducing the reality of lived time to processes that can be described in third-person neurobiological terms.

References

Arstila V (2018) Temporal experiences without the specious present. Australas J Philos 96(2): 287–302

Barrett LF, Simmons WK (2015) Interoceptive predictions in the brain. Nat Rev Neurosci 16(7): 419–429

Berkovich-Ohana A, Dor-Ziderman Y, Glicksohn J, Goldstein A (2013) Alterations in the sense of time, space, and body in the mindfulness-trained brain: a neurophenomenologically-guided MEG study. Front Psychol 4:912

Blanke O, Metzinger T (2009) Full-body illusions and minimal phenomenal selfhood. Trends Cogn Sci 13(1):7–13

Chalmers D (1996) The conscious mind: in search of a fundamental theory. Oxford University Press, Oxford

Chuard P (2011) Temporal experiences and their parts. Philosophers' Imprint 11:1–28

Craig AD (2015) How do you feel? An interoceptive moment with your neurobiological self. Princeton University Press, Princeton

Craig AD (2009a) How do you feel now? The anterior insula and human awareness. Nat Rev Neurosci 10(1):59–70

Craig AD (2009b) Emotional moments across time: a possible neural basis for time perception in the anterior insula. Philos Trans R Soc Lond B Biol Sci 364(1525):1933–1942

Craig AD (2002) How do you feel? Interoception: the sense of the physiological condition of the body. Nat Rev Neurosci 3(8):655–666

Csikszentmihaly M, Csikszentmihalyi I (1988) Optimal experience: psychological studies of flow in consciousness. Cambridge University Press, Cambridge

Dainton B (2008) The experience of time and change. Philos Compass 3(4):619–638

Dainton B (2000) Stream of consciousness: unity and continuity in conscious experience. Routledge, New York

Damasio A, Damasio H, Tranel D (2013) Persistence of feelings and sentience after bilateral damage of the insula. Cereb Cortex 23(4):833–846

Dorato M, Wittmann M (2020) The phenomenology and cognitive neuroscience of experienced temporality. Phenomenol Cogn Sci 19(4):747–771

Feldman MJ, Bliss-Moreau E, Lindquist KA (2024) The neurobiology of interoception and affect. Trends Cogn Sci 28(7):643–661

Feldman MJ, MacCormack JK, Bonar AS, Lindquist KA (2023) Interoceptive ability moderates the effect of physiological reactivity on social judgment. Emotion 23(8):2231–2242

Frank A, Gleiser M, Thompson E (2024) The blind spot: why science cannot ignore human experience. The MIT Press, Cambridge

Fridman J, Barrett LF, Wormwood JB, Quigley KS (2019) Applying the theory of constructed emotion to police decision making. Front Psychol 10:1946

Gallagher S, Zahavi D (2014) Primal impression and enactive perception. In: Lloyd D, Arstila V (eds) Subjective time: the philosophy, psychology, and neuroscience of temporality. MIT Press, Cambridge

Gallagher S, Zahavi D (2008) The phenomenological mind. Routledge, New York

Gallagher S (2000) Philosophical conceptions of the self: implications for cognitive science. Trends Cogn Sci 4(1):14–21

Gerrans P (2015) All the self we need. In: Metzinger T, Windt J (eds) Open mind: 15(T). Frankfurt am Main MIND Group

Herbert BM, Pollatos O (2012) The body in the mind: on the relationship between interoception and embodiment. Top Cogn Sci 4(4):692–704

Hoerl C (2013) A succession of feelings, in and of itself, is not a feeling of succession. Mind 122(486):373–417

Hoerl C (2009) Time and tense in perceptual experience. Philosophers' Imprint 9:1–18

Husserl E (2001) In: Bernet R, Lohmar D (eds) Die Bernauer Manuskripte über das Zeitbewusstsein (1917/18). Kluwer Academic Publishers, Dordrecht

Husserl E (1991) On the phenomenology of the consciousness of internal time (1893–1917). Trans. J.B. Brough. Kluwer Academic Publishers, Dordrecht

Husserl E (1970) The crisis of European sciences and transcendental phenomenology. Northwestern University Press, Evanston

Husserl E (1964) *The Phenomenology of Internal Time Consciousness*. (Edited by M. Heidegger, Translated by J. Churchill), Bloomington, IN: Indiana University Press

James W (1890) The perception of time, ch.5. In: The principles of psychology, vol 1. Henry Holt and Co., pp 605–642

James W (1884) What is an emotion? Mind 9:188–205

Le Poidevin R (2007) The images of time: an essay on temporal representation. Oxford University Press, New York

Meissner K, Wittmann M (2011) Body signals, cardiac awareness, and the perception of time. Biol Psychol 86(3):289–297

Menon V, Levitin DJ (2005) The rewards of music listening: response and physiological connectivity of the mesolimbic system. Neuroimage 28(1):175–184

Merker B (2007) Consciousness without a cerebral cortex: a challenge for neuroscience and medicine. Behav Brain Sci 30(1):63–81. discussion 81–134

Metzinger T (2009) The ego tunnel: the science of mind and the myth of the self. Basic Books, New York

Metzinger T (2003) Being no one: the self-model theory of subjectivity. MIT Press, Cambridge

Nagel T (1974) What is it like to be a bat? Philos Rev 83:435–450

Phillips I (2014) Experience of and in time. Philos Compass 9(2):131–144

Phillips I (2008) Perceiving temporal properties. Eur J Philos 18(2):176–202

Pöppel E (2009) Pre-semantically defined temporal windows for cognitive processing. Philos Trans R Soc Lond B Biol Sci 364(1525):1887–1896

Pöppel E (1997) A hierarchical model of temporal perception. Trends Cogn Sci 1(2):56–61

Pollatos O, Gramann K, Schandry R (2007) Neural systems connecting interoceptive awareness and feelings. Hum Brain Mapp 28(1):9–18

Prosser S (2016) Experiencing time. Oxford University Press, Oxford

Rovelli C (2018) The order of time. Allen Lane, New York

Seth AK, Tsakiris M (2018) Being a beast machine: the somatic basis of selfhood. Trends Cogn Sci 22(11):969–981

Seth AK (2013) Interoceptive inference, emotion, and the embodied self. Trends Cogn Sci 17(11): 565–573

Solms M (2021) The hidden spring: a journey to the source of consciousness. Profile Books, London

Vogel DHV, Krämer K, Schoofs T, Kupke C, Vogeley K (2018) Disturbed experience of time in depression – evidence from content analysis. Front Hum Neurosci 12

Wittmann M (2018) Altered states of consciousness: experiences out of time and self. MIT Press, Cambridge

Wittmann M (2015) Modulations of the experience of self and time. Conscious Cogn 38 (C):172–181

Wittmann M (2014) Embodied time: the experience of time, the body and the self. In: Arstilla V, Lloyd D (eds) Subjective time: the philosophy, psychology, and neuroscience of temporality. MIT Press, Cambridge, pp 507–523

Wittmann M (2013) The inner sense of time: how the brain creates a representation of duration. Nat Rev Neurosci 14(3):217–223

Wittmann M (2011) Moments in time. Front Integr Neurosci 5:66

Wittmann M (2009) The inner experience of time. Philos Trans R Soc Lond B Biol Sci 364(1525): 1955–1967

Wittmann M, Pöppel E (1999) Temporal mechanisms of the brain as fundamentals of communication – with special reference to music perception and performance. Musicae Scientiae 1999–2000:13–28

Zahavi D (2007) Perception of duration presupposes duration of perception – or does it? Husserl and Dainton on time. Int J Philos Stud 15(3):453–471

Zahavi D (2005) Subjectivity and selfhood: investigating the first-person perspective. MIT Press, Cambridge

Interoception: Synthesizing Insights and Charting New Frontiers

Alan N. Simmons, Marc Wittmann, and Irina A. Strigo

Contents

Abstract This chapter is dedicated to the intellectual legacy of A.D. Bud Craig and advancing the understanding of interoception. The chapter is organized into three critical domains of interoceptive research: Bodily Self, Emotion, and Subjective Time. Beginning with foundational neuroanatomical research on temperature and pain processing, the volume progresses to explore how interoceptive signals integrate to form our bodily self-representation and emotional experiences. This leads to discussion of the interoceptive self and its foundation role in emotional experience. Finally, this culminates in how interoception shapes our perception of time, including experiences in altered states of consciousness. This exploration into interoception points to promising future research focused on computational modeling of interoceptive processing, personalized treatment approaches based on interoceptive principles, and clinical applications for psychiatric, neurological, and pain

A. N. Simmons (✉)
Veterans Affairs & University of California, San Diego, San Diego, CA, USA
e-mail: ansimmons@health.ucsd.edu

M. Wittmann
Institute for Frontier Areas of Psychology and Mental Health, Freiburg, Germany
e-mail: wittmann@igpp.de

I. A. Strigo (✉)
Veterans Affairs & University of California, San Francisco, San Francisco, CA, USA
e-mail: irina.strigo@ucsf.edu

Curr Topics Behav Neurosci (2025) 70: 307–317
https://doi.org/10.1007/7854_2025_587
Published Online: 25 July 2025

conditions. In this chapter, we aim to outline the goals of this collection and the remaining gaps in the continuance of Bud Craig's seminal career.

Keywords Bodily self · Emotions · Insula · Interoception · Pain · Time perception

1 Introduction

Bringing together this edited volume entitled "The Bodily Self, Emotion, and Subjective Time: Exploring Interoception Through the Contributions of A.D. (Bud) Craig" was an ambitious project. When we first discussed our anticipated endeavor, combining the broad scientific, conceptual, and, at times, deeply personal set of topics, we were confident of the exceptional quality of the contributing authors and hopeful that the sum of these chapters would lead to a volume that was more than the sum of its parts. As we reached out to contributors that had worked with, or had been influenced by, Bud through the course of his career a narrative built. Bud always said that "brain is not a mystical place" and described himself as a "functional neuroanatomist," a term he used to encapsulate his belief that anatomy and physiology of the nervous system were inextricably linked. When asked why he used this term rather than "anatomist," "physiologist," or "neuroscientist," his reply was "when you are going somewhere new, the first thing you need is a map", as such, his career mirrored the neural projections that he skillfully and painstakingly mapped through the brain. Fittingly, this book starts with understanding the beginnings of the interoceptive experience grounded in precise and exacting nerve fiber paths from which all constructs of interoception are built. As his career progressed, much like the propagating signal through the body, through each terminus more information is integrated.

To follow the interoceptive pathway that Bud Craig mapped so caringly is to follow the path of his impactful career; it is to follow the chapters of the book; it is to follow the growth of the field of interoception; it is the temporal nature of experience from a base signal, to a contemporaneous moment, to prediction, and, eventually, awareness of the passage of time. It became evident in the editing of this volume, that while this book would, inevitably, have a final chapter, the propagation of the ideas that Bud was instrumental in bringing to a wide variety of fields would continue long after the completion of this project.

We will walk through chapters to show the linkage and narrative implicit within this edited volume and we will conjecture as to the areas where there is still important scientific territory that future scientists can explore and build their own map.

Fig. 1 *Bodily Self*: Word cloud of the Chapters 2–6 that focus on the foundation of interoception and the bodily self

2 Bodily Self

This edited volume begins where the story of interoception begins and at the beginning of Bud Craig's career. This first section is visually depicted in the word cloud in Fig. 1. This work initiates with Anders's discussion of the scientific and personal impact Bud Craig had. Intrinsic in this chapter is the striving for an understanding and unerring focus on building a scientific model for interoception that was founded in exacting research. This work came from building a construct concerning temperature and pain processing that could be traced neuron by neuron, thus building the foundation for the future work. In this chapter, the importance lies in the processing of the signals from the body (including skin) that pass through the dorsal horn of the spinal cord to the thalamus and the posterior insula (chapter "The Life and Legacy of Arthur Bud Craig: A Pioneer in Interoception"). This moves forward to a discussion by Henry Evrard of the foundational work focusing on the way these interoceptive signals are pathed up through complimentary pathways to the insula for better integration and, thus, facilitation of the bodily-self representation (chapter "The Isle of Craig: Neuroanatomical and Functional Evidence for a Role of the Insular Cortex in Subjective Feelings"). This further expands to the impact of this pathway that was largely predicated on pain research into a broad set of interoceptive processes that are foundational to our emotional experience. Specifically, this work shows the importance of the homeostatic afferent pathway and the linkage of Bud Craig's concepts of interoception and Antonio Damasio's concepts of somatic markers. The implications toward mental health and therapeutic interventions are based on these models (Critchley and Patchitt 2024) (chapter "Interoception, Insula, and Autonomic Integration: Relevance to the Expression and Treatment of Psychiatric Symptoms"). As the bodily self extends to interact with the outside world, the critical role of touch is emphasized. Touch plays a fundamental role in establishing and maintaining the bodily self. Through the course of our life, tactile experiences shape our concept of our physical boundaries, the separation of ourselves from others, and the experience of being embodied. Importantly, the C-fibers and A-delta fibers of the skin carry affectively pertinent aspect of touch that are also critical to the experience of pain (Rebecca Böhme's chapter "The Touched Body and the Experience of Self"). Disruption to the bodily self can substantially affect the resultant body image. This can be explored through the lenses of clinical conditions such as anorexia nervosa, body dysmorphic disorder, and phantom limb syndrome (chapter by Emily Choquette and Sahib Khalsa: "Sensing the Self: The Role of the Insula and Interoception in Body Image"). In each of these conditions, the varied pathway signals are not effectively integrated in the insula.

3 Emotion

The construction of a bodily sense as identified in the first third of the chapter provides a foundation from which emotions are based. It should be noted that "without the bodily states following on the perception, the latter would be purely cognitive in form" (James 1884). Thus, "emotions are constructed from the integration of interoceptive signals from the body with cognitive appraisal, creating a unified perception of one's physiological condition and its implications for personal well-being" (Craig 2015). In this context, we can view pain as a progenitor of

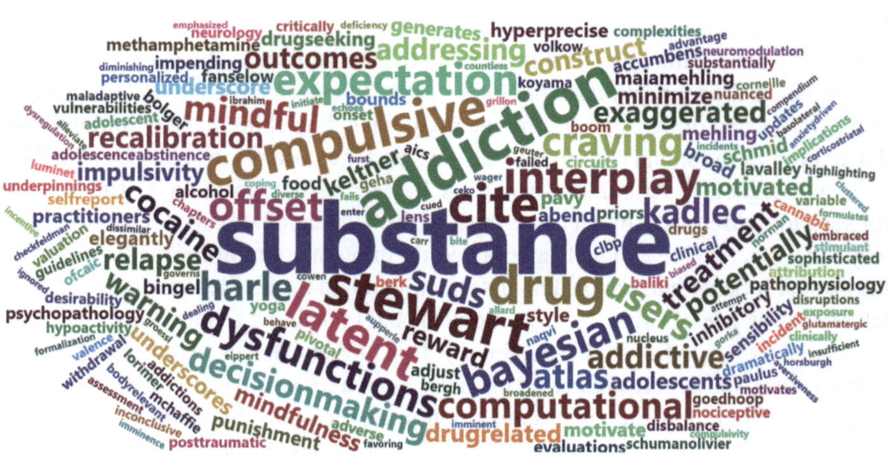

Fig. 2 *Emotion*: Word cloud of the Chapters 7–10 that focus on the emotional and psychiatric models that have been established from the interoceptive perspective

emotion, a contextualized interoceptive state integrated with perception that motivates action. Thus, we can better understand the affective experience of pain specifically through the predicted and integrated bodily self. As depicted in Fig. 2, the manifestation of the interoceptive signal is viewed as a contemporaneous signal, a computationally based experience, that affects bodily urges. In this way, we exist as a predicted self, experiencing anticipated states informed by prior beliefs, contemporaneous context, and the predicted future state (chapter by Irina Strigo and Alan Simmons: "The Intersection of Interoception and Anticipation Related to Pain"). Disruption to this concept of self is vital in understanding the experience of chronic pain. This model of pain allows for a host of therapeutic interventions that, through mindful assessment of the current moment and exposure to interoceptive experiences in a controlled and contextualized way, can reduce chronic pain (Mehling 2024) (chapter "How Bud Craig's Insights Reshape the Research on Pain and Mind–Body Therapies"). This interoceptive model of the insula has also been implicated in the context of substance abuse. This can be effectively understood in Bayesian terms and has been effectively modeled in animal research as well as functional brain imaging in humans. In substance abuse, this disrupted interoceptive self is exacerbated by disruptions to inhibition, reward sensitivity, and salience attribution such that this behavior which implicates long-term harm is habitually and compulsively selected (chapter by Martin Paulus: "Interoceptive Processing in Substance Use Disorders (SUD): From the Neuroanatomy to Insights from Computational Models and Predictive Coding Frameworks"). Disruption in this interoceptive process over the course of maturation can lead to specific vulnerabilities and to the acquisition of substance use disorders (chapter by April May and Susan Tapert: "Interoception in Adolescence: Impacts on Mental Health and Adaptive Functioning").

4 Subjective Time

In the third section of the volume (Fig. 3), we move from the contemporaneous moment to an appreciation that this contemporaneous moment exists only in the context of prior and future moments in a dynamic continuum, a bodily and emotional self that is as linked to what will be as well as to what has been (chapter by Marc Wittmann and Sylvie Droit-Volet: "Subjective Time in Ordinary and Non-Ordinary States of Consciousness: How Interoceptive Feelings Inform us bout the Passage of Time"), (Wittmann and Droit-Volet 2024). This prior self exists in the interplay between the memory from our explicit (hippocampal) memory and our interoceptive/bodily (insula) memory. That is, Alice Teghil in her chapter ("Interoceptive and Bodily Processing in Prospective and Retrospective Timing") shows that next to the interoceptive-insular system which builds a primary representation of time passage, as experienced in the here and now (prospective timing), her recent empirical studies show a tight link between episodic memory and interoceptive as well as emotional states which are linked to duration judgments in retrospect. These features of the

Fig. 3 *Subjective Time*: Word cloud of the Chapters 11–15 that focus on the constructs of subjective and interoceptive experience of time

insula that make it critical to predict body states and, in the case of pain, tissue damage, has allowed it to be the hub for the experience of time in a uniquely human way. In our modern world, we often appreciate time as an objective unerring existent truth. This view, however, can be inverted such that attempts to define an objective measure lose the critical essence of time as subjective experience (Julian Kiverstein's chapter "Body, Self and Time: Bud Craig's Global Emotional Moments (GEMs) Theory"). Subjective time is highly dependent on interoceptive coding and is influenced by the bodily self (see Sect. 1) and the contemporaneous emotional moment (see Sect. 2). As Julian Kiverstein argues, Craig's "global emotional moments" which are built on large-scale networks with the anterior insula cortex have implications for our understanding of what Kiverstein calls the "phenomenal now." The critical role of the insula in understanding this experience of time is highlighted in ordinary and altered state of consciousness (Wittmann and Droit-Volet 2024) (Chapter "Subjective Time in Ordinary and Non-Ordinary States of Consciousness: How Interoceptive Feelings Inform us about the Passage of Time"). Following Craig's theory, Wittmann & Droit-Volet argue that dynamic physiological signals are essential for human time perception. Neuroimaging and psychophysiological studies show that time judgment is rooted in the physical and emotional self. The sense of time is embodied. Altered states of consciousness are not just critically dependent on the modulation of the self in time, as associated with the insula, but in the case of ecstatic epilepsy, these altered states or "aha" moments can be repeatedly induced from a seizure that specifically targets the anterior insula (chapter by Nina Sooter, Federico Seragnoli, and Fabienne Picard: "Insights from Ecstatic Epilepsy: From Uncertainty to Metacognitive Feelings"). This experience often manifests as the perception of the connectedness of all things across time and space, where the insula has a pivotal role as an integrator of this interoceptive experience. The nexus of interoceptive meaning, the capacity to experience the dynamics of time, and predictive activation make the insula a critical location for the appreciation of the experience of music (Clemens Wöllner's chapter "Music in the Air and in the Body: An Interoceptive System's Perspective on Musical Emotions, Awareness, and Time"). The experience of music ties together this long chain of interoceptive appreciations of body, emotion, and time into a dynamic experience, where we can appreciate the elegant interplay of a complex story of interoception. Craig's ideas provide a unified framework for studying music as a deeply embodied activity essential to individuals, groups, and societies.

5 Projections Going Forward (Figs. 4 and 5)

This volume aims to not only highlight the importance of the work of Bud Craig but also to present the work that is continuing in the field of interoception in the work of the contributors. As the interoceptive pathways in building a representative interoceptive self that can experience, predict, and move through time, these processes have to be investigated in complex interdisciplinary studies. One of the key legacies of Bud Craig's work is the appreciation of this broad wealth of knowledge. He was as likely to cite from insights hundreds and even thousands of years ago in conceptualizing a construct. Humans have not changed markedly in these thousand years, and the implicit and explicit appreciation of the flow of interoception in the experience of person-hood has been present for our scientific history. However, this

Fig. 4 In 2011, together with Dorothe Poggel, Marc Wittmann organized the International Symposium on 'Time and the Conscious Brain' at the Hanse-Wissenschaftskolleg (HWK) Delmenhorst, Germany (31.10.–2.11.2011), where the invited Bud Craig (yellow circle) gave a talk on his insula-theory of time perception in front of the (at the time) biggest time perception crowd gathering at a conference

Fig. 5 Hugo Critchley, Marc Wittmann, Anders Blomquist, and Henry Evrard (Fabienne Picard via Zoom) in front of Vadstena castle in Sweden for the Insular cortex symposium in memory of A.D. (Bud) Craig at the Linköping University Neuroscience Retreat 2024

contemporaneous moment feels special as new techniques and approaches allow greater insight into the underpinnings of experience. Throughout this volume, the authors have highlighted interventions that employ our ever-developing priors of scientific investigations to predict ways in which we can use this knowledge to enhance the human experience.

Last, but not least, the unsolved question of phenomenal consciousness can find a solution in Bud Craig's embodiment idea. Visceral and somatosensory input from the peripheral nervous system are the foundation for subjective feelings and self-consciousness. Several conceptualizations in this book build a direct link between bodily signals, subjective time, and self-consciousness (chapter "Subjective Time in Ordinary and Non-Ordinary States of Consciousness: How Interoceptive Feelings Inform us about the Passage of Time"), (chapter "Interoceptive and Bodily Processing in Prospective and Retrospective Timing", chapter "Body, Self and Time: Bud Craig's Global Emotional Moments Theory"). The experiences of self and time would be constituted through the interoceptive system including the insular cortex. That is, successive moments of self-realization over extended time, consciously felt as passage of time, would be based on continuous input originating within the body. Moreover, the philosopher Julian Kiverstein in his chapter analysis Craig's "cinemascopic" idea of "global emotional moments" which combined over time form a stream of consciousness.

Three areas that should be highlighted for future growth of applied knowledge are (1) the role of computational modeling in interoceptive processing, (2) the role of interoception in personalized treatment approaches for psychiatric, neurological, and pain conditions, and (3) the role of treatments in clinical psychology and psychiatry based on the understanding of interoceptive principles. Regarding the new wave of interoception research which has accelerated through the integration of Bayesian statistical models (Paulus et al. 2019; Petzschner et al. 2021), these models provide a platform for understanding how interoceptive signaling can combine prior, contemporaneous, and future predictions in the context of psychiatric conditions (Strigo et al. 2024a). Of principle importance to this computational model is that it can provide a framework to combine multifactorial models of interoceptive experiences with an adaptive, predictive, and testable hypothesis. While the broad principles of this approach to understanding the brain is quite general (Friston 2010), it has specific application to interoception (Seth et al. 2011), not only because of interoception's central role in broad aspects of the human experience, but because interoception is the key linkage with pain which involves the prediction of future tissue damage (Raja et al. 2020).

The importance of personalized medicine in psychiatry has been more and more appreciated in recent years (Menke 2018). In the context of the understanding of the complex presentation of interoceptive processing, the construct of subsamples of individuals that differentially process interoceptive signals, thus require differing approaches, has gained more traction in recent years (Khalsa and Verdonk 2024; Strigo et al. 2023, 2024b). There are a number of highly effective treatments that leverage knowledge about interoception (Weng et al. 2021), but more work needs to be done to maximize these approaches (Lavretsky and Feldman 2021). As outlined in this collection, the interoceptive path is both complex and broadly effects so many aspects of how we experience life and interoception itself (Brewer et al. 2021). Interoception is repeatedly modeled as a Bayesian or state space model that is highly personalized. Thus it is no surprising that a highly individualized interoceptive system is best addressed through a personalized approach. Interoceptive treatments include some relatively well-established interventions that provide evidence-based care, as well as some potentially interesting intersectional approaches (Weng et al. 2021). Meditative and/or breathing approaches (Farb et al. 2013; Haase et al. 2015, 2016; Zeidan et al. 2011), and vagal nerve stimulation (Fang et al. 2016; Lerman et al. 2019, 2022) has been shown to activate and modulate interoceptive pathways and cortically activate the insula. The potential for utilizing a combination of neuromodulation with interoceptive therapies has been explored in various psychiatric conditions. This complimentary approach in which the interoceptive pathway can be accessed in top-down and bottom-up directions may be effective to break new ground in the clinical application of the current research. We eagerly anticipate that the trail blazing legacy of Bud Craig will continue to live on in this ever-evolving field.

Acknowledgments This work would not have been possible without the work of a large team of individuals. We would like to thank the contributing authors who submitted chapters for this volume. This work would also not be possible without the vital assistance staff at Springer from reaching out to us regarding this book, working with the editors and authors to see this work to fruition, and displays of astounding grace with our ever-expanding timeline. We would like to thank Lora. Finally, we would like to thank Bud Craig for his continued influence and inspiration and to his wonderful wife Lora Sanders for assisting with completing our collections of photos and memories in making this book.

References

Brewer R, Murphy J, Bird G (2021) Atypical interoception as a common risk factor for psychopathology: a review. Neurosci Biobehav Rev 130:470–508

Craig AD (2015) How do you feel?: an interoceptive moment with your neurobiological self. Princeton University Press

Critchley HD, Patchitt J (2024) Interoception, insula, and autonomic integration: relevance to the expression and treatment of psychiatric symptoms. Curr Top Behav Neurosci

Fang J, Rong P, Hong Y, Fan Y, Liu J, Wang H, Zhang G, Chen X, Shi S, Wang L, Liu R, Hwang J, Li Z, Tao J, Wang Y, Zhu B, Kong J (2016) Transcutaneous Vagus nerve stimulation modulates default mode network in major depressive disorder. Biol Psychiatry 79:266–273

Farb NA, Segal ZV, Anderson AK (2013) Mindfulness meditation training alters cortical representations of interoceptive attention. Soc Cogn Affect Neurosci 8:15–26

Friston K (2010) The free-energy principle: a unified brain theory? Nat Rev Neurosci 11:127–138

Haase L, May AC, Falahpour M, Isakovic S, Simmons AN, Hickman SD, Liu TT, Paulus MP (2015) A pilot study investigating changes in neural processing after mindfulness training in elite athletes. Front Behav Neurosci 9:229

Haase L, Thom NJ, Shukla A, Davenport PW, Simmons AN, Stanley EA, Paulus MP, Johnson DC (2016) Mindfulness-based training attenuates insula response to an aversive interoceptive challenge. Soc Cogn Affect Neurosci 11:182–190

James W (1884) On some omissions of introspective psychology. Mind 9:1–26

Khalsa SS, Verdonk C (2024) Interoception and mental health, interoception: a comprehensive guide. Springer, pp 265–316

Lavretsky H, Feldman JL (2021) Precision medicine for breath-focused mind-body therapies for stress and anxiety: are we ready yet? Glob Adv Health Med 10:2164956120986129

Lerman I, Davis B, Huang M, Huang C, Sorkin L, Proudfoot J, Zhong E, Kimball D, Rao R, Simon B, Spadoni A, Strigo I, Baker DG, Simmons AN (2019) Noninvasive vagus nerve stimulation alters neural response and physiological autonomic tone to noxious thermal challenge. PLoS One 14:e0201212

Lerman I, Klaming R, Spadoni A, Baker DG, Simmons AN (2022) Non-invasive cervical vagus nerve stimulation effects on reaction time and valence image anticipation response. Brain Stimul 15:946–956

Mehling WE (2024) How Bud Craig's insights reshape the research on pain and mind-body therapies. Curr Top Behav Neurosci

Menke A (2018) Precision pharmacotherapy: psychiatry's future direction in preventing, diagnosing, and treating mental disorders. Pharmgenomics Pers Med 11:211–222

Paulus MP, Feinstein JS, Khalsa SS (2019) An active inference approach to interoceptive psychopathology. Annu Rev Clin Psychol 15:97–122

Petzschner FH, Garfinkel SN, Paulus MP, Koch C, Khalsa SS (2021) Computational models of interoception and body regulation. Trends Neurosci 44:63–76

Raja SN, Carr DB, Cohen M, Finnerup NB, Flor H, Gibson S, Keefe FJ, Mogil JS, Ringkamp M, Sluka KA, Song XJ, Stevens B, Sullivan MD, Tutelman PR, Ushida T, Vader K (2020) The revised international association for the study of pain definition of pain: concepts, challenges, and compromises. Pain 161:1976–1982

Seth AK, Suzuki K, Critchley HD (2011) An interoceptive predictive coding model of conscious presence. Front Psychol 2:395

Strigo IA, Simmons AN, Giebler J, Schilling JM, Moeller-Bertram T (2023) Unsupervised learning for prognostic validity in patients with chronic pain in transdisciplinary pain care. Sci Rep 13: 7581

Strigo IA, Craig ADB, Simmons AN (2024a) Expectation of pain and relief: a dynamical model of the neural basis for pain-trauma co-morbidity. Neurosci Biobehav Rev 163:105750

Strigo IA, Kadlec M, Mitchell JM, Simmons AN (2024b) Identification of group differences in predictive anticipatory biasing of pain during uncertainty: preparing for the worst but hoping for the best. Pain 165:1735

Weng HY, Feldman JL, Leggio L, Napadow V, Park J, Price CJ (2021) Interventions and manipulations of interoception. Trends Neurosci 44:52–62

Wittmann M, Droit-Volet S (2024) Subjective time in ordinary and non-ordinary states of consciousness: how interoceptive feelings inform us about the passage of time. Curr Top Behav Neurosci

Zeidan F, Martucci KT, Kraft RA, Gordon NS, McHaffie JG, Coghill RC (2011) Brain mechanisms supporting the modulation of pain by mindfulness meditation. J Neurosci 31:5540–5548